Sustainable Development

The current focus on sustainable development opens up debates surrounding our relationship with the natural world, about what constitutes social progress and about the character of development, both in the Global North and the Global South. The promotion of sustainable futures is taking on a new urgency in the context of climate change and biodiversity loss.

This concise and accessible text explores how the international community is responding to the challenge of sustainable development. It also investigates the prospect for, and barriers to, the promotion of sustainable development in high-consumption societies of the industrialised world, from the USA and the EU to the economies of transition in Eastern Europe and Southern Asia. This global coverage is balanced by investigating how local action, ranging from the transition towns movement in the UK to the Green Belt movement in Kenya, can contribute to the pursuit of sustainable development.

The second edition has been extensively revised and updated and benefits from the addition of three new chapters: sustainable development in China; the governance of sustainable development; and sustainable production and consumption. Climate change and biodiversity management have also been expanded into full chapters.

Providing an up-to-date and comprehensive treatment of the issues surrounding the promotion of sustainable development, this unique, internationally focused book combines a strong conceptual analysis with wide-ranging empirical focus and a wealth of case material. Including summary points and suggestions for further reading, as well as web resources and an extensive bibliography, it is ideal for students, scholars and researchers in the fields of environmental sciences, politics, geography, sociology and development studies.

Susan Baker is a Professor in the Cardiff School of Social Sciences and Co-Director of the Sustainable Places Research Institute, Cardiff University, UK. Her current research focuses on the interrelationship between social and ecological processes, particularly in relation to climate change and biodiversity loss.

Routledge Introductions to Environment Series
Published titles

Environmental science texts

Atmospheric Processes and Systems
Natural Environmental Change
Environmental Biology
Using Statistics to Understand the Environment
Environmental Physics
Environmental Chemistry
Biodiversity and Conservation, 2nd Edition
Ecosystems, 2nd Edition
Coastal Systems, 2nd Edition

Series Editor:
Timothy Doyle

Environment and society texts

Environment and Philosophy
Energy, Society and Environment, 2nd Edition
Gender and Environment
Environment and Business
Environment and Law
Environment and Society
Representing the Environment
Environment and Social Theory, 2nd Edition
Environmental Values
Environment and Tourism, 2nd Edition
Environment and the City
Environment, Media and Communication
Environmental Policy, 2nd Edition
Environment and Economy
Environment and Food
Environmental Governance
Environment and Citizenship
Sustainable Development, 2nd Edition
Environment and Politics, 4th Edition

Sustainable Development

Second edition

Susan Baker

LONDON AND NEW YORK

First published 2016
by Routledge
2 Park Square, Milton Park, Abingdon, Oxon OX14 4RN

and by Routledge
711 Third Avenue, New York, NY 10017

Routledge is an imprint of the Taylor & Francis Group, an informa business

British Library Cataloguing in Publication Data
A catalogue record for this book is available from the British Library

Library of Congress Cataloging in Publication Data
Baker, Susan, 1955-
Sustainable development / Susan Baker. – Second edition.
(Routledge introductions to environment series)
Includes bibliographical references and index.
1. Sustainable development. 2. Globalization–Economic aspects.
3. Globalization–Environmental aspects. 4. Economic policy–Environmental aspects. 5. International organization. I. Title.
HC79.E5B347 2015
338.9'27–dc23
2015010679

ISBN: 978-0-415-52291-5 (hbk)
ISBN: 978-0-415-52292-2 (pbk)
ISBN: 978-0-203-12117-7 (ebk)

Typeset in Times New Roman
by Cenveo Publisher Services

MIX
Paper from
responsible sources
FSC
www.fsc.org FSC® C013604

Printed and bound by CPI Group (UK) Ltd, Croydon, CR0 4YY

In memory of my father Micheál Bácaoir, for showing me the signs of the ebbing tide, and how to watch in silence for the *spideog* to land. For my daughter Niamh, whose respect for animals and passion for politics continues to inspire me.

Contents

Illustrations

Figures

Tables

Preface to Routledge Introductions to Environment Series

Series Editor: Timothy Doyle

Professor of Politics and International Studies, University of Adelaide, Australia;
Professor of Politics and International Relations, Keele University, United Kingdom;
Distinguished Research Fellow, Curtin University, Australia;
Chair, Indian Ocean Rim Association Academic Group, Port Louis, Mauritius;
Chief Editor, *Journal of the Indian Ocean Region*, Routledge, Taylor & Francis, London, UK.

It is openly acknowledged that a critical understanding of socio-economic, political and cultural processes and structures is central in understanding environmental problems and establishing alternative modes of equitable development. As a consequence, the maturing of environmentalism has been marked by prolific scholarship in the social sciences and humanities, exploring the complexity of society–environment relationships.

This series builds on the work of the founding series editor, David Pepper, to continue to provide an understanding of the central socio-economic, political and cultural processes relating to environmental studies, providing an interdisciplinary perspective to core environmental issues. David initiated the series by celebrating the close connections between the academic traditions of environmental studies and the emergence of the Green movement itself. Central to the goals of the movement were social and environmental change. As the 'new science' of ecology was interdisciplinary, seeking to understand relationships within and between ecosystems, so too was the belief within the academy (informed by the movement), that real environmental change could only emerge if traditional borders and boundaries of knowledge and power were bypassed, transgressed and, where necessary, challenged.

This bid for engaged knowledge and interdisciplinarity also informs the structure and 'pitch' of these books. For it is no good communicating with just one particular group within society. It is equally important to construct forms of knowledge which can cross over demographic and market borders, bringing together communities of people who may never 'meet' in the usual course of events. So, the epistemological design of this series is oriented around three particular audiences, providing an unparalleled interdisciplinary perspective on key areas of environmental study: (1) students (at undergraduate and coursework post-graduate levels); (2) policy practitioners (in civil society, governments and corporations); and (3) researchers. It is important to note, therefore, that these books – though strongly used in diverse levels of tertiary teaching – are also built, in large part, on the primary and often ground-breaking research interests of the authors.

In his own ground-breaking work, David Pepper was particularly interested in exploring the relationships between capitalism, socialism and the environment. David argued that the modern environmentalist movement grew at a rapid pace in the last third of the twentieth century. It reflected popular and academic concerns about the local and global degradation of the physical environment which was increasingly being documented by scientists. It soon became clear, however, that reversing such degradation was not merely a technical and managerial matter: merely knowing about environmental problems did not of itself guarantee that governments, businesses or individuals would do anything about them. Since David wrote his last series preface, this focus has continued to be important, but with special permutations as time has worn on. One more recent, key feature of these society–environment relationships has been the clear differentiation between the environmentalisms of the majority worlds (the Global South) and environmentalisms of the minority worlds (the more affluent Global North). Whereas environmentalism came to the less affluent world later (in the 1980s), key environmental leadership is now being provided by activists in the South, oriented around a post-colonial environmentalism, with its key issues of human dispossession and survival: water, earth (food security and sustainability), fire (energy), and air (not climate). Much of the focus in environmentalism in the South relates to a critique of capitalism and its big business advocates as being the major perpetrators of severe environmental problems which confront the Earth. In the global North (where the modern movement began in the 1960s), there has been far more emphasis on post-materialism and post-industrialism and, more recently, building a *sustainable capitalism.*

Climate change is now the neoliberal cause célèbre of this approach, with its heavily biased focus on market mechanisms and green consumerism as answers to environmental crises. In fact, climate change, in the Global North, has now become so powerful and omnipresent that many more affluent world green activists and academics now comprehend *all* environmental problems within its rubric, its story. Of course, climate change issues will continue to be crucial to the planet's continued existence, but more importantly, it must be acknowledged that in living social movements – like the Green movement – issues will come and go; will be reordered and rearranged on the issue attention cycle; be rebadged under different symbols, signs and maps; and new green narratives, issues and stories will emerge. The environment movement, born in the North – and its associated academic studies – will continue to be the foremost global social movement for change for many years to come – if it can continue to truly engage with the Global South – utilising these new and revised banners, issues and colours to continually and creatively mark out its territories, constructing versions of environmentalism for all, not just for the few. And it is within these new sites of politics and knowledge that some of the most exciting advances in the relationships between societies and 'nature' will continue to emerge and be celebrated. Much still is to be learned from our universe, the planet Earth, its human and non-human communities.

Tim Doyle
October 2014

Acknowledgements

This first edition of this book was written while I held a royal appointment as King Carl XVI Gustaf Professor in Environmental Science. I remain indebted to His Majesty King Carl XVI Gustaf of Sweden for providing me with the opportunity to engage in this task. A special thanks also to Umeå University, Sweden for hosting me in situ, and with such generosity and kindness.

My daughter Niamh deserves special thanks. For the second edition of this book she work diligently, digging out much needed new material on the UN, LA21 and the USA, while also meeting a host of other demands.

Jackie Swift kindly took on the task of proof reading, while Richard Bloor managed to trawl the text to extract a list of abbreviations. I am very grateful to both.

I would also like to thank the series editor, Dr David Pepper, for his initial suggestion that I write this book. A special thanks to the production team at Routledge, especially Andrew Mould, for his ongoing patience and understanding as deadlines were set and missed.

Susan Baker
Cardiff, December 2014

Abbreviations

10YFP	Ten-Year Framework Programme of the UN
BRIC	Brazil, Russia, India and China
CAP	Common Agricultural Policy of the European Union
CBD	UN Convention on Biological Diversity
CCAD	Central American Commission for Environmental Development
CCS	Carbon Capture and Sequestration/Carbon Capture and Storage
CDM	Clean Development Mechanism of the UN FCCC
CDR	Common but Differentiated Responsibility
CITES	Convention of International Trade in Endangered Species Flora and Fauna
CO_2	Carbon dioxide
CoP	Conference of Parties
CSD	Commission of Sustainable Development
CSR	Corporate Social Responsibility
DG	Environment Director-General of the Environment, European Commission
DDT	Dichlorodiphenyltrichloroethane
EAP	Environmental Action Programme of the European Union
EEA	European Economic Area
EEA	European Environment Agency
EEB	European Environment Bureau
EFCA	Ecological Function Conservation Area
EIT	Parties to UN FCCC, countries with economies in transition
EMS	Environmental Management Systems
EPI	Environmental Policy Integration
EU	European Union
EU ETC	EU Emission Trading Scheme
FAO	Food and Agriculture Organisation
FDI	Foreign Direct Investment
FLO	Fairtrade International

FYP	Five-Year Plan
FYRM	Former Yugoslav Republic of Macedonia
GBO	Global Biodiversity Outlook
GEF	Global Environment Facility
GDP	Gross Domestic Product
GHG	Greenhouse Gas Emissions
GMO	Genetically Modified Organisms
GONGO	Government Organised Non-Governmental Organisation
GRI	Global Reporting Initiative
GTGP	Grain to Green Programme
HDR	Human Development Report
HFCs	Hydrofluorocarbons
ICLEI	International Council for Local Environmental Initiatives – Local Governments for Sustainability
IPA	Instrument for Pre-Accession Assistance of the European Union
IPBES	Inter-governmental Platform on Biodiversity and Ecosystem Services
IPCC	Intergovernmental Panel on Climate Change
ISO	International Organisation for Standardisation
IUCN	International Union for the Conservation of Nature and Natural Resources
IUCN	International Union for Conservation of Nature
JNNURM	Jawaharlal Nehru National Urban Renewal Mission
LA21	Local Agenda 21
LCA	Lifecycle Assessments
LDCs	Least Developed Countries
LMO	Living Modified Organisms
LULUCF	Land Use, Land-Use Change and Forestry
MDG	Millennium Development Goals of the United Nations
MEA	Multilateral Environmental Agreements
MOSAICC	Micro-Organisms Sustainable Use and Access Regulations International Code of Conduct
NFCP	Natural Forest Conservation Programme
NGO	Non-Governmental Organisation
NIMBY	Not In My Back Yard
NMC	National Model City
NSDS	National Sustainable Development Strategy
ODA	Official Development Assistance
OECD	Organisation for Economic Co-operation and Development
OSCE	Organisation for Security and Co-operation in Europe

PES	Payments for Ecosystem Services
RBMP	River Basin Management Plan
REDD & Redd+	Reducing Emissions from Deforestation and Forest Degradation
SAPARD	Special Accession Programme for Agriculture and Development of the European Union
SCP	Sustainable Consumption and Production
SIDS	Small Island Developing States
SIP	Sustainable Industrial Policy
SOER	The European Environment – State and Outlook 2010
TEEB	The Economics of Ecosystems and Biodiversity
ULB	Urban Local Body
UN	United Nations
UNCED	United Nations Conference on Environment and Development
UNCSD	United Nations Conference on Sustainable Development
UNEP	United Nations Environment Programme
UNESCO	United Nations Educational, Scientific and Cultural Organisation
UN FCCC	UN Framework Convention on Climate Change
UNGASS	United Nations General Assembly Special Session
US EAP	US Environment Protection Agency
USDA	US Department of Agriculture
WBCSD	World Business Council for Sustainable Development
WCED	World Commission on Environment and Development
WEDO	Women's Environment and Development Organisation
WEHAB	Water and Sanitation, Energy, Health, Agriculture, Biodiversity Protection and Ecosystem Management
WFD	European Union Water Framework Directive
WSSD	UN World Summit on Sustainable Development
WTO	World Trade Organisation

Introduction
The environment and sustainable development

Key issues

- Re-conceptualising development
- Ultimate limits to growth
- The common good
- Promoting sustainable development: sustainability
- Three pillars of sustainable development.

Promoting sustainable development opens up debates about our relationship with the natural world, about what constitutes social progress and about the character of development, both in the North and the South, in the present and into the future. These interrelated issues form the main themes of this book. The book explores the prospects for, and barriers to, the promotion of sustainable development as presented in the classic Brundtland formulation and promoted by the United Nations (UN). It looks at policy developments and actions in different socio-economic contexts: the high consumption societies of the industrialised world, including in Europe and North America; the Third World; the economies in transition in East and Central Europe; and the emerging economies, especially that of China. The exploration is international in its focus, because it recognises that promoting sustainable development is a quintessentially global task. At the same time, it takes account of the fact that policy implementation, and its related actions, take place at lower scales, including at the regional, national and sub-national levels.

Challenging the dominant model of development

Sustainable development represents a direct challenge to the conventional form of economic development. Conventional approaches see development as simply modernisation of the globe along Western lines.

Modernisation theory holds that the more structurally specialised and differentiated a society is, the more modern and progressive it is (Pepper, 1996). To be modernised, a society has to become more technically sophisticated and urbanised and to make increased use of markets for the distribution of economic goods and services. Modernisation also brings social changes, including the development of representative democracy, increased mobility and the weakening of traditional kinship groups and communities. Modernisation is closely tied to the promotion of individual self-advancement. The transformation of nature, such as taming wilderness into natural parks, harnessing wild rivers to make energy and clearing forests for agricultural production, is one of the hallmarks of modernisation.

In the conventional model, society is understood to go through different 'stages of economic growth' (Rostow, 1960). Traditional societies develop to a stage of economic 'take-off'. This sees new industries and entrepreneurial classes emerge, as they did in Britain in the nineteenth century. In 'maturity', steady economic growth outstrips population growth. A 'final stage' is reached when high mass consumption allows the emergence of social welfare (Pepper, 1996). This model of development assumes a linear progression, in which it becomes necessary for Third World societies to 'catch up' with the Western style of development. This means opening up their economies to Western values, influences, investment and trade, thereby becoming more integrated into the global market system.

Modern environmentalism has emerged as a critique of this Western-centric development model, although it takes different forms and has different expressions (Barry, 2006). Environmentalism points to the failure of a model of development that results in ecological destruction, and at a global scale. Social exclusion, economic alienation and rising levels of stress and ill health in OECD countries are accompanied by socially disruptive, politically unstable and at times violent transitions in the countries of the former Soviet Union, ecologically disastrous industrialisation in emerging economies, while the human tragedies of the failed development strategies continue to mar life for many in the Third World. Environmentalism challenges many of the basic assumptions that the Western model of development makes about the use of nature and natural resources, the meaning of progress and the ways in which society is governed, including both the traditional patterns of authority within society and how public policy is made and implemented.

Several other social and political movements, such as Marxism and the *dependencia* theories of Third World under-development have made similar critiques. However, while environmentalism may find common cause with these arguments, it can be distinguished by its focus on the economic, social *and* ecological dimensions and repercussions of development. Seven key arguments form the backbone of the environmentalist challenge. First, environmentalism takes issue with the understanding of progress found in the Western model. Progress is understood in a limited way, primarily in terms of increased domination over nature and the use of her resources solely for the benefits of humankind. In this model, the domination of nature has become a key indicator of human progress (Macnaghten and Urry, 1998). Progress is seen, for example, in the clearance of forested land for agricultural production or in the use of natural resources, such as coal, oil and gas, to produce energy in the form of electricity that, in turn, drives production and its related consumption. This, in turn, results in pollution and related climate change. This path of development is bolstered by public policies or practices that induce behaviour that is harmful to the environment. Known as 'perverse incentives', such policy failures include government grants, subsidies or tax incentives that fail to take into account the creation of environmental externalities.

Underlying this domination is a reduction of nature merely to a natural resource base, a reduction that values nature only in terms of the use that these resources have for human beings. This gives nature only 'instrumental value', ignoring the 'intrinsic value' of the natural world, that is, the value that nature has over and above its usefulness to humans. Viewing nature instrumentally also leads to neglect of the needs of other, non-human species and life forms.

Second, the Western development model prioritises economic growth, even though the heightened consumption patterns that it stimulates now threaten the very resource base upon which future development depends. This model assumes environmental deterioration to be an inevitable consequence of development. Although Western society has seen enhanced legal and technical efforts to address environmental pollution, its model of development is nonetheless premised on the acceptance of a 'trade-off' or unequal exchange between economic development and the environment, where the environment is relinquished for the sake of development. Third, the model assumes that consumption is the most important contributor to human welfare. Here, it is common practice to measure welfare by means of the 'standard of living', that is, the amount of disposable income that an individual has to purchase goods and services. A development model based on individualistic consumption, rather than fostering social cohesion, leads to increased inequality, especially in an economic system subject to cyclical recession (Ekins, 2000). It prioritises individual self-attainment at the expense of consideration of the common good (Baker, 2012). In contrast, environmentalism focuses not on the 'standard of living' but on the 'quality of life'. Quality of life refers to the collective, not the individual, level and to enhancing the quality of the public domain, such as through the provision of public education, health care and environmental protection.

Fourth, the model ignores that fact that social stability requires the maintance of natural resources. The deterioration of the natural environment causes social disruptions, insecurity, and damage to human health. For example, loss of wild biodiversity in agricultural systems increases the vulnerability of local communities, especially with respect to food supply, which in turn, leads to social unrest that can undermine social and political institutions (Gowdy, 1999).

Fifth, the traditional understanding of development ignores the fact that Western development was, and continues to be, based upon the exploitation not only of their own natural resource base but that of many Third World societies, including their plant and animal genetic resources, land, timber and mineral ore. The human resources of the Third World have also been exploited, including but not limited to the practice of slavery. Exploitation has caused under-development in the Third World, not least by creating resource poverty and a culture of dependency. In this view, poverty is caused by the penetration of Western, environmentally destructive development models into Third World societies, not alleviated by it. This condemns Third World societies to 'backwardness', while ignoring their long traditions of community resource management. These traditions have built a body of indigenous knowledge, which has enabled many traditional societies to live in harmony with their natural surroundings, although of course not all traditional societies have managed to live in this way.

Sixth, the model is blind to the fact that is not possible to achieve a *global* replication of the resource-intensive, affluent lifestyle of the high consumption economies of the North. The planet's ecosystem cannot absorb the resultant pollution, as witnessed by global environmental change, that is climate change and biodiversity loss. Furthermore, there are not enough natural resources, including water, to support such development. In other words, the model of development pursued by Western industrial society cannot be carried into the future, either in its present forms or at its present pace.

Finally, and closely related to the previous point, is that the environmental critique points to the failure of the Western development model to acknowledge that there are limits to economic growth. Limits to growth are imposed by the carrying capacity of the planet, especially the ability of the biosphere to absorb the effects of human activities, and the fact that the amount of resources that the planet contains, including water and minerals, is finite and that ecosystem services are reduced or eliminated through overuse. Technological advancement, while it may enable society to produce goods with more resource efficiency, will not overcome this limitation. There are thus *ultimate* limits to growth. This means that development has to be structured around the need to adopt lifestyles within the planet's ecological means. Several recent assessments, including by the United Nations, warn of serious consequences for human societies as ecosystems become incapable of providing the goods and services on which hundreds of millions of people depend (Rockström et al., 2009). Such *thresholds* have already been passed in certain coastal areas where 'dead zones' now exist, including a range of coral reefs and lakes that are no longer able to sustain aquatic species; and some dryland areas that have been effectively transformed into deserts. Similarly thresholds have been passed for some fish stocks. The expansion of the Western, consumerist model of development, coupled with population growth, sees human demands on ecosystems increase at the risk of further weakening the natural infrastructure on which all societies depend. The relationship between population growth and resource use is not, however, straightforward, as discussed in this book.

What is significant about this multiple environmental critique of the traditional model of development is that it has shown that the post-World War II experience of economic growth and prosperity was both exceptional and contingent (Redclift and Woodgate, 1997). It was exceptional in that it cannot be replicated across space (from the West to the global level) or across time (into the future). It was contingent upon a short-term perspective, the prioritisation of one region of the globe over another, and upon giving preference to one species (humans) over the system as a whole. Environmentalism has also undermined the assumption of a progressive view of society's evolution (Redclift and Woodgate, 1997). The environmental critique of development shows that there is no continuous linear development guaranteed for modern society, nor is this development necessarily harmonious (Barry, 2006).

> Protecting and improving our future well-being requires wiser and less destructive use of natural assets. This in turn involves major changes in the way we make and implement decisions. We must learn to recognize the true value of nature – both in an economic sense and in the richness it provides to our lives ... Above all, protection of these assets can no longer be seen as an optional extra, to be considered once more pressing concerns such as wealth creation or national security have been dealt with.
>
> (MEA, 2005)

It is in this critical context that the model of sustainable development has gained traction.

Emergence of a new model of development

Many environmental development models have emerged to replace the old development paradigm. These promote forms of social change that are aimed at fulfilling human material and non-material needs, advancing social equity, expanding organisational effectiveness and building human and technical capacity towards sustainability (Roseland, 2000). They share in common the objectives of protecting the natural resource base upon which future development depends. For many, valuing nature and non-human life forms in an intrinsic way is also essential. The environmental development model is not just aimed at protecting nature, but creating an ecological society that lives in harmony with nature. This means reconciling economic activity, social progress and environmental protection. Well-being is promoted, as opposed to a mere focus on measures such as GDP as indicators of progress. In this understanding, the promoting of human well-being and happiness does not have to depend upon the destruction of nature.

The 'sustainable development' model represents an important example of the new environmentalist approach. It seeks to reconcile the ecological, social and economic dimensions of development, now and into the future, and adopts a global perspective in this task. It aims at promoting a form of development that is contained within the ecological carrying capacity of the planet, which is socially just and economically inclusive. It focuses not upon individual advancement but upon protecting

the common future of humankind. Put this way, sustainable development would appear to be an aspiration that almost everyone thinks is
desirable: indeed, it is difficult *not to* agree with the idea.

> Sustainable development is part of new efforts, albeit tentative, to integrate
> environmental, economic and (more recently) social considerations into a
> new development paradigm. There are many versions of this new
> approach. They are united in their belief that there are ultimate, biophysical
> limits to growth. This challenges industrial societies not only to reduce the
> resource intensity of production (sustainable production) but to undertake
> new patterns of consumption that not only reduce the levels of consump-
> tion but change what is consumed and by whom (sustainable consumption).
> This creates the conditions necessary for ecologically legitimate develop-
> ment, particularly in the Third World.

However, there are many versions of the sustainable development
model and not all of them are mutually compatible. There is very little
agreement on what sustainable development means and even less
agreement on what is required to promote a sustainable future (Redclift
and Woodgate, 1997).

The Brundtland model of sustainable development

Despite ambiguity in meaning, the term 'sustainable development' has
been prominent in discussions about environmental policy since the
mid-1980s. Following the central role it played in the UN-appointed
Brundtland Commission (1984–87) and its report, *Our Common Future*
(WCED, 1987), it has appeared with increasing frequency in academic
studies and in reports of international agencies and of governments.
The Brundtland formulation of sustainable development has come to
represent mainstream thinking about the relationship between environment and development. It now commands authoritative status, acting as
a guiding principle of economic and social development (Lafferty and
Meadowcroft, 2000).

An increasing number of international organisations and agencies subscribe to at least some, and often most or all, of its objectives (Lafferty and
Meadowcroft, 2000). These include the European Union (EU), the United
Nations Environment Programme (UNEP) and the World Bank. National
governments, sub-national regional and local authorities, as well as groups
within civil society and economic actors, have all made declaratory and

practical commitments to the goal of promoting sustainable development. This is also recognised as a crosscutting policy task, that is, it cuts across many areas of public policy, including international development, trade, urban and land use planning, environmental protection, energy policy, and agricultural and industrial policy, to name but a few.

The UN has played a particularly prominent role in stimulating this engagement. It has organised several world summits, including the UN Conference on Environment and Development (UNCED), which took place in June 1992 in Rio de Janeiro, otherwise known as the Rio Earth Summit; the Johannesburg World Summit on Sustainable Development (WSSD), held in 2002; and more recently the Rio+20 United Nations Conference on Sustainable Development, held in Rio de Janeiro in 2012. The Rio Declaration, which arose from the original Rio Earth Summit, provides an authoritative set of normative principles, that is, principles that deal with moral issues, including gender equality, intra-generational equity (within a generation) and inter-generational equity (between generations) and matters relating to environmental justice. The Declaration also details the governance principles needed so as to best manage and organise the promotion of sustainable development within society, institutions, and at the political level. Both the Rio Declaration and subsequent activity organised under the UN have advanced the understanding of what sustainable development means. Summits have also led to several international legally binding environmental agreements, including the UN Framework Convention on Climate Change (UN FCCC) and its related Kyoto Protocol, as well as the Convention on Biological Diversity (CBD). UN engagement has also led to a proliferation of institutions and organisations, including within civil society and from the business community, with a remit to promote sustainable development, such as the Women's Environment and Development Organisation (WEDO) and the World Business Council for Sustainable Development (WBCSD). The 'UNCED process' is used as a shorthand way to indicate the range of activities that has taken place under the auspices of the UN since the publication of the Brundtland Report. Each of these activities, from the World Summits to the development of legally binding agreements, from the engagement of states to the role of civil society and business interests, is explored in this book. In addition, the normative and governance principles that have come to be associated with the term sustainable development are explained and critically discussed.

Clarifying the terms used

There is need for some conceptual clarity at this point. This book is about sustainable development. It is not about the concept of sustainability. The term sustainability originally belongs to ecology, and referred to the potential of an ecosystem to subsist over time (Reboratti, 1999). More recently, the term sustainability has come to be associated with the *goal* of policy.

With the addition of the notion of development to the notion of sustainability, the focus of analysis shifted from that of ecology to that of society. The chief focus of sustainable development is on society, and its aim is to include environmental considerations in the steering of societal change, especially through changes to the way in which the economy functions.

Sustainable development refers to the many processes and pathways to reconcile the ecological, economic and social dimensions of life. This can include, for example, the promotion of sustainable agriculture and forestry, sustainable production and consumption, good government, research and technology transfer, education and training, recognition of cultural values and different forms of knowledge.

Sustainability is the long-term goal, that is, a more sustainable world.

Promoting sustainable development is about steering societal change at the interface between:

1 **The social**: this relates to human mores and values, relationships and institutions.
2 **The economic**: this concerns the allocation and distribution of scarce resources.
3 **The ecological**: this involves the contribution of both the economic and the social and their effect on the environment and its resources.

These are known as the three dimensions, or pillars, of sustainable development (Ekins, 2000).

Sustainable development is a dynamic concept. It is not about society reaching an end state, nor is it about establishing static structures or about identifying fixed qualities of social, economic or political life. It is better to speak about *promoting*, not achieving, sustainable development. Promoting sustainable development is an ongoing process,

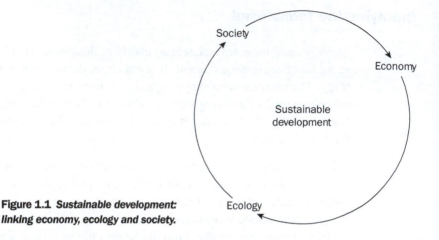

Figure 1.1 *Sustainable development:*
linking economy, ecology and society.

whose desirable characteristics change over time, across space and
location and within different social, political, cultural and historical
contexts. Therefore, in this book the expression 'promoting sustainable
development' is used to show sensitivity to evolving understandings of,
and judgements about, what constitutes sustainable development
(sensitive across time) and to the fact that different societies, cultures
and groups may aspire to different sustainable development pathways
(sensitive across place).

Adopting a dynamic understanding of sustainable development also
helps emphasise that alternative futures lie before society. The promotion of sustainable development is about envisioning these alternatives and, through attitudinal and value changes, policy innovations,
political transformations and economic restructuring, embracing a
future that is sustainable. While this will be different for different
societies, across space and time, there are certain global or common
'baseline' conditions that are required if humanity is to embrace sustainability. These include a healthy ecosphere and biosphere. They
also include adherence to certain normative principles and acceptance
of guidelines about what constitutes good governance practice, issues
that are given particular attention in this book. This is what is meant
by saying that promoting sustainable development requires recognition of the common good (Baker, 2012). Its challenge is to ensure
that society moves along a social trajectory – one that avoids both
pathways that lead to a direct deterioration of the social state
and those that lead to a situation from which further progress is
impossible (Meadowcroft, 1999).

The governance challenge

Governance can be understood as the steering of society towards collective goals. However, the environmental challenge to the traditional model of development has led to a questioning of the traditional modes of governance within society, including at the international level. Environmentalism has, for example, challenged the ability and legitimacy of traditional forms of government intervention and policy making to address the complex issues posed by the promotion of sustainable development. Furthermore, rather than being the task of national governments acting alone and using traditional policy means, promoting sustainable development is seen to require engagement across all levels of social organisation, from the international, the national, the sub-national and the societal to the level of the individual.

The rise of global environmental problems, such as climate change, biodiversity loss and deforestation, has led to a growing demand for international interventions to deal with both transboundary and global environmental matters. This has stimulated the rapid growth of international environmental laws, and management and administration (governance) regimes (Gupta, 2002). The rise of global environmental governance has been accompanied by pressure to try new and innovative procedures, including expanding the range and role of non-state actors involved in steering societal change. It embraces efforts to enhance the involvement of business interests as well as non-governmental organisations (NGOs), that is, organisations operating at the national and, increasingly, international level, which have administrative structures, budgets and formal members, and are non-profit making. Thus, while states continue to serve as the primary repositories of authority in relation to environmental governance (Young, 1997; 2003), these are increasingly combined with participatory practices aimed at enhancing both the legitimacy and democratic nature of policy making. At the same time, sub-national, regional and local engagement also act to press for the development of new forms of governance, not least so that regional and local variations, capacities and needs can be taken into account in sustainable development plans.

> The term 'new practices of environmental governance' refers to the participation of non-state actors, alongside state and international organisations, as well as the utilisation of a wide range of policy instruments (including legal, voluntary and market instruments) and normative and governance principles to promote sustainable development.

There is a clear relationship between the type and mode of governance and the success of efforts to promote sustainable development. With this in mind, this book explores whether, and to what extent, the commitment by an increasing number of international organisations and agencies to the objectives of the Brundtland model of sustainable development has resulted in any changes in the power relations between state institutions and societal actors.

The structure of the book

This book is divided into three parts. Part I presents a theoretical and conceptual exploration of sustainable development. Part II looks at multi-level engagement, including international efforts and the involvement of the sub-national, local level. Part III looks at the promotion of sustainable development in different social, political and economic contexts.

Part I

This introductory Chapter 1 sets out the themes of the book, placing the study of sustainable development in the wider context of understanding social development and change. Chapter 2 provides the conceptual framework that informs the discussions in the rest of the book. It explores the evolution in the meaning and use of the concept of sustainable development. It begins by briefly tracing the development of the concept from its early use in resource ecology to its eventual adoption as a norm of global environmental politics. This discussion is then expanded to take account of new ways of theorising the relationship between ecological and social systems that take a planetary focus. This pays attention to the coupled nature of socio-ecological relationships, as reflected in the notion of planetary boundaries. The concept of 'ecological footprint' is also included in the discussion on limits to growth. The chapter then proceeds to explore the variations in meaning and subsequent disputes over the value of the concept, but pays particular attention to the authoritative Brundtland formulation. It also examines the claims that sustainable development is premised upon a strong 'anthropocentric' approach that endorses a managerial relationship with nature. The elaboration of a unifying or precise definition of the concept is less important than understanding the political, economic and social challenges presented in practice. The ladder of sustainable development, as elaborated by Baker et al. in 1997, is updated and expanded to take account of current thinking in the literature, including in relation to eco-system functioning and ecosystem services and matters of governance. This ladder is used to explore the range of normative and governance principles, as well policy issues associated with the promotion of sustainable development at the global level.

Chapter 3 examines how society is and can be steering towards a more sustainable future. It begins by detailing the role of the three classical styles of governance, namely markets, networks and hierarchies. The discussion on markets includes critical analysis of the growing use of market-led environmental policy instruments, including voluntary agreements and tradable permits. This theme is taken up in several other chapters of the book, including Chapter 6, where carbon trading is explored. Chapter 3 also explores the use of policy networks, including public/private partnerships. The role of hierarchies, in particular top-down regulations and thus of the state and state-like actors, is also given attention. New ways of theorising the governance of sustainable development have emerged, including the notions of both 'meta governance' and 'reflexive governance'. These developments are covered in the chapter, as are normative arguments about the type of governance arrangements that *ought* to be put in place to promote sustained development.

Chapter 4 looks at sustainable development from the point of view of its two main component parts: sustainable production and sustainable consumption. It begins by explaining how the promotion of sustainable development requires changes in both production and consumption activity. It then deals with issues in relation to sustainable production, while the section dealing with sustainable consumption provides a new opportunity to shift from the focus on institutional engagement to investigate the role of individual behaviour, including consumer behaviour. The 'lifestyle' approach, including voluntary simplicity, is briefly examined. The discussion on sustainable consumption opens up investigation of the role of cultural values. This provides a forum for the discussion of issues in relation to environmental justice and fair trade, issues addressed throughout the book. The key distinctions between ecological modernisation and sustainable development are itemised and critically analysed.

Part II

Chapter 5 explores the rationale behind, significance of and theoretical explanations for the construction of a global regime for the promotion of sustainable development. It develops a historically informed, critical awareness of the role played by the UN. It looks at how recognition of the global nature of the challenge has been stimulated by and, in turn, has stimulated a new era of global environmental governance. Attention is also paid to the association of sustainable development with specific governance principles. The steps from the Brundtland Commission's Report, *Our Common Future*, to the Rio Earth Summit, Rio+5 and onwards to the 2002 WSSD and the 2012 Rio+20 Conference, alongside the ongoing reportage, monitoring and evaluation regimes established under the auspices of UNCED, are followed. This will familiarise the reader with historical developments at the global level and their institutional expressions.

The next two chapters provide the reader with an understanding of the links between the promotion of sustainable development and the resolution of certain critical global environmental problems. The Rio Earth Summit led to two binding conventions, on climate change and on biological diversity. These chapters show how climate change and biodiversity loss present key challenges that need to be addressed as part of the promotion of sustainable development. Both conventions are examined in some detail, as they raise a number of key issues that are of concern for this book. In relation to Chapter 6 and the discussion on climate change, these include the marked imbalance in resource use between the industrialised and the Third World and hence differences in the burden each is placing on the limited carrying capacity of the planet. It also presents an ideal opportunity to explore whether and in what ways the principles of sustainable development help shape the concrete responses taken to particular environmental problems. Chapter 6 includes discussion on energy policy, explaining the links between energy policy and climate change and the challenges involved in making the transition to a post-carbon future. The focus on institutional responses is combined with discussion on climate justice, including in relation to climate change adaptation. Extensive treatment of the USA is given, including a focus on the state and city levels, which are seeing vibrant local initiatives and responses.

Chapter 7 examines the loss of biodiversity, as this throws into sharp relief the tension between economic development and environmental protection, both within the developing world and also within the high consumption societies of the West. There are also growing disputes between the interests of the biotechnology industry of the industrialised world and Third World countries over who should have access to and use of plant and animal genetic resources. A shift in the conceptualisation of biodiversity as 'ecosystem services' has contributed to a major change in how biodiversity is perceived, and has helped to heighten its policy salience. The loss of ecosystem services is now receiving considerable attention at international, EU and member-state levels. The launch of The Economics of Ecosystems and Biodiversity (TEEB) Report 2010, completion of national ecosystem assessments and, in the UK, the publication of the Lawton Report 2011, to take some key examples, have contributed to this development. These developments have enhanced our understanding of the ecological pillar of sustainable development.

Chapter 8 explores the tensions involved in global regimes seeking to facilitate bottom-up engagement with sustainable development. The promotion of sustainable development is being encouraged by top-down, global environmental management regimes. At the same time, UNCED is also encouraging bottom-up engagement. Local Agenda 21 (LA21) is the most important action-orientated, bottom-up initiative to emerge from the UNCED process. This chapter begins with an outline of the aims and objectives of LA21. It then goes on to explore the experiences within several countries in organising LA21. This will include short case studies of industrialised, transition, Third World, and emerging economies. The extent to which LA21 contributes to new forms of participatory governance that help promote sustainable development, and the structural challenges involved in that undertaking, are critically assessed.

Part III

Part III looks at the promotion of sustainable development in different contexts. Chapter 9 examines high consumption societies, paying particular attention to the role and actions of the EU. The EU is an important exemplar of efforts to translate into practice the declaratory statements issued after the Rio Earth Summit. The extent to which EU practice is in keeping with the spirit and principles of Rio is examined. The discussion points to the need for new patterns of sustainable consumption and sustainable production. In the EU context, social actors play a key role in the shift to sustainable consumption; firms and industry, including business interest associations, play a vital role in shifting to more sustainable forms of production.

Chapter 10 looks at the economies in transition in Eastern Europe. Many of these countries are now EU member states, and the EU's influence on transition states continues to deepen. The chapter focuses attention on the challenges involved in the countries in transition in Eastern and Central Europe. It asks, in the context of marketisation and democratisation, what are the prospects for the promotion of sustainable development in transition countries?

Chapter 11 looks at the Third World. Here, the issues raised stand in contrast to the challenges facing high consumption societies. Protection of the environment and achieving necessary economic development are closely linked with the need to address issues of global justice, poverty and equity in resource use, and the terms of global trade. Both the trade agreements promoted by the World Trade Organisation (WTO), and the financial instruments controlled by the World Bank and the Global Environmental Facility are included in the analysis. An additional aim of the chapter is to infuse gender awareness into the study of sustainable development. Finally, Chapter 12 examines the rise of China and considers the global impact of China's industrial boom, as well as how Chinese investments are shaping the prospects for sustainable development beyond its own borders. The Chinese model of sustainable development is examined in detail.

The Conclusion to the book returns to the conceptual and theoretical issues raised in this Introduction. Having exposed the reader to detailed and critical discussions of the multifaceted challenges involved, it asks whether and to what extent the adoption of sustainable development as a norm of global, regional, national and sub-national politics is helping society to find ways to overcome the tension between economic development and environmental protection.

Summary points

- Environmentalism challenges the dominant, Western model of economic development. This model has a limited understanding of progress, prioritises growth and fails to recognise the interrelationship between economic, social and ecological systems.
- The sustainable development model represents a new approach to development and the steering of social change.
- The Brundtland formulation of sustainable development has attained authoritative status.

Further reading

Barry, J. (2006), *Environment and Social Theory* (London: Routledge, 2nd edn).

Blühdorn, I. and Welsh, I. (eds) (2008), *The Politics of Unsustainability: Eco-Politics in the Post-Ecologist Era* (London and New York: Routledge).

Dodds, F., Strauss, M. and Strong, M. F. (2012), *Only One Earth: The Long Road via Rio to Sustainable Development* (London: Earthscan).

Helm, D. and Hepburn, C. (eds) (2009) *The Economics and Politics of Climate Change* (Oxford: Oxford University Press).

McMichael, P. (2012), *Development and Social Change: A Global Perspective* (London: Sage, 5th edn).

Sachs, J. (2014), *Speaking on Sustainable Development Goals at Monash Sustainability Institute*, available online at http://jeffsachs.org/category/sustainable-development/

Stephens, P., Barry, J. and Dobson A. (2009), *Contemporary Environmental Politics: From Margins to Mainstream* (London: Taylor and Francis).

WCED (World Commission on Environment and Development) (1987), *Our Common Future* (Oxford: Oxford University Press).

References

Baker, S. (2012), 'Climate Change, the Common Good and the Promotion of Sustainable Development', in J. Meadowcroft, O. Langhelle and A. Ruud (eds), *Democracy, Governance and Sustainable Development: Moving Beyond the Impasse* (Cheltenham: Edward Elgar), 249–271.

Baker, S., Kousis, M., Richardson, R. and Young, S. (eds) (1997), *The Politics of Sustainable Development: Theory, Policy, and Practice within the European Union* (London: Routledge).

Barry, J. (2006), *Environment and Social Theory* (London: Routledge, 2nd edn).

Ekins, P. (2000), *Economic Growth and Environmental Sustainability: The Prospects for Green Growth* (London: Routledge).

Gowdy, J. (1999), 'Economic Concepts of Sustainability: Relocating Economic Activity within Society and Environment', in E. Becker, and T. Jahn (eds), *Sustainability and the Social Sciences: A Cross-Disciplinary Approach to Integrating Environmental Considerations into Theoretical Reorientation* (London: Zed Books), 162–181.

Gupta, J. (2002), 'Global Sustainable Development Governance: Institutional Challenges from a Theoretical Perspective', *International Environmental Agreements: Politics, Law and Economics*, 2: 361–388.

Lafferty, W. M. and Meadowcroft, J. (2000), 'Introduction', in W. M. Lafferty and J. Meadowcroft (eds), *Implementing Sustainable Development: Strategies and Initiatives in High Consumption Societies* (Oxford: Oxford University Press), 1–22.

Macnaghten, P. and Urry, J. (1998), *Contested Natures* (London: Sage).

Meadowcroft. J. (1999), 'Planning for Sustainable Development: What Can Be Learnt from the Critics?' in M. Kenny and J. Meadowcroft (eds), *Planning Sustainability* (London: Routledge), 12–38.

Millennium Ecosystem Assessment (MEA) (2005), *Living Beyond Our Means: Natural Assets and Human Well-Being, Statement from the Board* (Rome: MEA Secretariat), available online at www.maweb.org/documents/document.429.aspx. pdf, p. 3.

Pepper, D. (1996), *Modern Environmentalism: An Introduction* (London: Routledge).

Reboratti, C. E. (1999), 'Territory, Scale and Sustainable Development', in E. Becker, and T. Jahn (eds), *Sustainability and the Social Sciences: A Cross-Disciplinary Approach to Integrating Environmental Considerations into Theoretical Reorientation* (London: Zed Books), 207–222.

Redclift, M. and Woodgate, G. (1997), 'Sustainability and Social Construction', in M. Redclift and G. Woodgate (eds), *The International Handbook of Environmental Sociology* (Cheltenham: Edward Elgar), 55–70.

Rockström, J., Steffen, W., Noone, K., Persson, A., Chapin, F. S., Lambin, E., Lenton, T. M., Scheffer, M., Folke, C., Schellnhuber, H., Nykvist, B., De Wit, C. A., Hughes, T., van der Leeuw, S., Rodhe, H., Sörlin, S., Snyder, P. K., Costanza, R., Svedin, U., Falkenmark, M., Karlberg, L., Corell, R. W., Fabry, V. J., Hansen, J., Walker, B., Liverman, D., Richardson, K., Crutzen, P. and Foley, J. (2009), 'A Safe Operating Space for Humanity', *Nature* 461: 472–475, doi: 10.1038/461472a.

Roseland, M. (2000), 'Sustainable Community Development: Integrating Environmental, Economic, and Social Objectives', *Progress in Planning*, 54: 73–132.

Rostow, W. (1960), *The Stages of Economic Growth: A Non-communist Manifesto* (Cambridge: Cambridge University Press).

WCED (World Commission on Environment and Development) (1987), *Our Common Future* (Oxford: Oxford University Press).

Young, O. R. (1997), 'Rights, Rules and Resources in World Affairs', in O.R. Young (ed.), *Global Governance: Drawing Insights from the Environmental Experience* (Cambridge, MA: MIT Press), 1–25.

Young, O. R. (2003), 'Environmental Governance: The Role of Institutions in Causing and Confronting Environmental Problems', *International Environmental Agreements: Politics, Law and Economics*, 3: 337–393.

Part I

Theoretical and conceptual exploration of sustainable development

The concept of sustainable development

Key issues

- Sustainable development
- Brundtland formulation
- Limits to growth
- Planetary boundaries
- Ecological footprint
- Ladder of sustainable development
- Normative principles
- Ecological modernisation.

Introduction

This chapter explores the concept of sustainable development, paying particular attention to the Brundtland conceptualisation because it has achieved authoritative status. An increasing number of international organisations and agencies subscribe to at least some, and often most or all, of the objectives and principles embedded in the Brundtland approach (Lafferty and Meadowcroft, 2000). The chapter begins by briefly outlining the historical origins of the concept of sustainable development before examining the Brundtland formulation in detail. This is followed by discussion of how the concept has evolved since the original Brundtland engagement. A ladder of sustainable development is used to organise these different interpretations and to relate them to specific policy imperatives. The key normative principles that are associated with the concept are then discussed. Attention is also given to the rejection of the concept by certain green theorists and environmental activists. Finally, the relationship between ecological modernisation and sustainable development is discussed, not least because the literature often assumes that the terms are synonymous.

Early use of the term sustainable development

Concerns about sustainability can be traced back to Malthus (1766–1834) and William Stanley Jevons (1835–1882), both of whom were troubled about resource scarcity, especially in the face of population rise (Malthus) and energy (coal) shortages (Jevons). More recently, the decade of the 1950s saw both Fairfield Osborn (Osborn, 1948; 1953) and Samuel Ordway (Ordway, 1953) raise concerns about limits to growth, especially given rising pollution, species loss and decline in natural resources. It was not until the 1960s and the 1970s, however, that a significant segment of public opinion expressed unease about resource scarcity and unsustainable patterns of use. These decades were also marked by intensification of anxiety about the environment, particularly the health hazards caused by industrial pollution. This led, in turn, to the advancement of environmental critiques of the conventional, growth-orientated, economic development model promoted in the post-war period.

Initially, these concerns led some to call for zero-growth strategies, especially following the publication of the 1972 Club of Rome Report, *The Limits to Growth* (Meadows et al., 1972), which was updated in 2004 (Meadows et al., 2004). The Report, undertaken by a group from the Massachusetts Institute of Technology, concluded that if present trends in population growth, food production, resource use and pollution continued, then the carrying capacity of the planet would be exceeded within the next 100 years. The result would be ecosystem collapse, famine and war. The 'limits to growth' argument was also taken up by Herman Daly, building a model of 'steady state economics' on recognition of the absolute limits to economic growth (Daly, 1977). However, the 'limits to growth' argument was subject to considerable criticism. It concentrated only on the physical limits to growth, ignoring the possibility of technological innovations leading to new ways, for example, to address pollution or to use resources more efficiently in production. It was also seen to present an overly pessimistic view of the rate of resource depletion globally. In its place, a new belief developed that environmental protection and economic development could become mutually compatible, not conflicting, objectives of policy. This did not necessarily undermine the 'limits to growth' argument but modified its focus, pointing to the need to limit growth in some areas, to allow for necessary growth in others (Paehlke, 2001). It is in this context that the concept of sustainable development gained policy traction.

The term sustainable development came into the public arena in 1980 when the International Union for the Conservation of Nature and Natural Resources published the *World Conservation Strategy* (IUCN, 1980). This strategy aimed at achieving sustainable development through the conservation of living resources. However, its focus was rather limited, primarily addressing *ecological* sustainability, as opposed to linking sustainability to wider social and economic issues.

The Brundtland formulation

It was not until 1987, when the World Commission on Environment and Development (WCED) published its report, *Our Common Future*, that the links between the social, economic and ecological dimensions of development were explicitly addressed (WCED, 1987). The WCED was chaired by Gro Harlem Brundtland, then Norwegian prime minister, and *Our Common Future* is sometimes known as the Brundtland Report. The establishment of the WCED and its links to the emerging system of international environmental management, or governance, is discussed more fully in Chapter 3. In the present chapter, the task is to clarify and discuss the principles, values and norms that have come to be associated with the concept of sustainable development, and to explore their related policy imperatives.

The Brundtland Report makes four key links between the economy, society and the environment. An example of environmental linkages is when deforestation leads to soil erosion, causing silting of rivers and

---

Box 2.1

Causal links between the economy, society and environment

1. Environmental stresses are linked to one another;
2. Environmental stresses and patterns of economic development are linked to one another;
3. Environmental and economic problems are linked to social and political factors;
4. These influences operate not only within, but also between nations.

(WCED, 1987: 37–40)

lakes that can, in turn, reduce fish stocks, and thus threaten livelihoods. Similarly, the linkages between environmental stresses and patterns of economic development are seen when agricultural policy encourages over-use of chemical fertilisers, which, in turn, can lead to soil degradation and water pollution; or when energy policies rely on coal-fired electricity stations which results in greenhouse gas (GHG) emissions that, in turn, are linked to climate change. These linkages operate not only within, but also between nations and upwards to the global scale. The highly subsidised agriculture of the North, for example, can erode the viability of agriculture in the developing countries, as can the terms of international trade. These linkages raise a series of issues that are addressed throughout this book. While Chapter 6 deals with matters relating to climate change, both the gender dimension and the issues of trade and the environment are examined in Chapter 11.

In making the links between the economy, society and the environment, the Brundtland Report puts 'development', a traditional economic and social goal, and 'sustainability', an ecological goal, together to devise a new development model – that of sustainable development. Sustainable development is a model of societal change that, in addition to traditional developmental objectives, has the objective of maintaining ecological sustainability (Lélé, 1991). This differs from the previous IUCN approach, mentioned above, which linked the environment to conservation, not to development. In addition, the Brundtland Report made it explicit that social and economic factors, especially processes operating at the international level, such as trade, influence whether or not the interactions between society and nature are sustainable. Brundtland hoped that agreement on what type of growth is or is not acceptable, and under what circumstances, could be reached through the development of mutual understanding, through dialogue and through the negotiation of new, and strengthening of existing, international environmental conventions and agreements, as discussed in Chapter 5.

The now famous and much popularised Brundtland definition of sustainable development is 'development that meets the needs of the present without compromising the ability of future generations to meet their own needs' (WCED, 1987: 43). What is often forgotten is that Brundtland went on to argue that sustainable development:

> contains within it two key concepts: the concept of 'needs', in particular the essential needs of the world's poor, to which priority should be given; and the idea of limitations imposed by the state of

technology and social organisation on the environment's ability to
meet present and future needs.

(WCED, 1987: 43)

Several factors combined to help the Brundtland formulation become
the dominant concept in international discussions on environment and
development. First, the formulation offered a way of reconciling what
had hitherto appeared to be conflicting societal goals, namely economic
development and environmental protection. Second, it came at a time
when the problem of environmental deterioration, especially pollution,
was high on the political agenda. This followed both the discovery of
the ozone hole above Antarctica and the Chernobyl nuclear accident.
Third, the Brundtland approach supported developing countries in their
pursuit of their goals of economic and social improvements.

The Brundtland concept of sustainable development is global in its
focus and makes the link between the fulfilment of the needs of the
world's poor and the reduction in the wants of the world's rich. It is difficult to distinguish needs from wants, as they are socially and culturally
determined. However, in most cultures fundamental needs are similar,
and include subsistence, protection, affection, understanding, participation, creation, leisure, identity and freedom (Pepper, 1996). The industrialised world consumes in excess of these basic needs, understanding
development primarily in terms of ever increasing material consumption
that in turn drives and is driven by economic growth. This excess threatens the planet's ecological resource base and ecosystem health. There is,
however, an emerging engagement with consumption reduction among
certain social groups in high consumption societies, as discussed in
Chapter 4. The Brundtland Report challenges the industrialised world to
keep consumption patterns within the bounds of the ecologically possible and set at levels to which all can reasonably aspire. This requires
changes in the understanding of well-being and what is needed to live a
good life. This change allows for necessary development in the South.
Here economic growth can have, in some contexts, net positive environmental, as well as social and economic, benefits.

> Growth must be revived in developing countries because that is
> where the links between economic growth, the alleviation of pov-
> erty, and environmental conditions operate most directly. Yet devel-
> oping countries are part of an interdependent world economy: their
> prospects also depend on the levels and patterns of growth in indus-
> trialised nations.

(WCED, 1987: 51)

Limits to growth

The Brundtland focus on limitations, imposed by the state of technology and social organisation, presents an optimistic view of our common future. It presents a vision of the future that holds the promise of progress, opened up through technological development and societal change. *Our Common Future* nonetheless argues that, while technology and social organisation can be both managed and improved to make way for a new era of economic growth, limits are nonetheless imposed. These are imposed 'by the ability of the biosphere to absorb the effects of human activities' (WCED, 1987: 8) and by the need to 'adopt life-styles within the planet's ecological means' (WCED, 1987: 9). There are thus ultimate limits to growth. The Brundtland conception of sustainable development does not assume that growth is both possible and desirable in all circumstances and holds that there are ultimate limits to growth.

The idea of ultimate limits is linked to the notion of 'ecosystem health'. Here the health of the ecosystem is the 'bottom line' guiding development, because if the health of the environment is compromised then everything else is undermined. In this approach, the environment can be seen as a form of 'natural capital', that is, a resource that can be put to human use. Sustaining this natural capital is a precondition for human life, because

> ecological processes underpin the rest of human activity, and, if these are impaired, then a condition for the very possibility of human activity is impaired too.
>
> (Dobson, 1998: 44)

There is a danger in this ecosystem approach, however, in that it sustains what is of instrumental value for human beings, and does not protect nature for its own sake. This approach has strong anthropocentric underpinnings, a matter discussed later in this chapter.

Other approaches to conceptualising limits to growth have subsequently developed. The idea of 'environmental space' acknowledges that there are limits to the amount of pressure that the earth's ecosystem can handle without irreversible damage. This leads to a search for the 'threshold level', the level beyond which damage occurs, and to use this level to set operational boundaries, for example the level of permitted GHG emissions. The idea of limits imposed by environmental space was taken forward in 2009, when scientists identified and quantified a set of

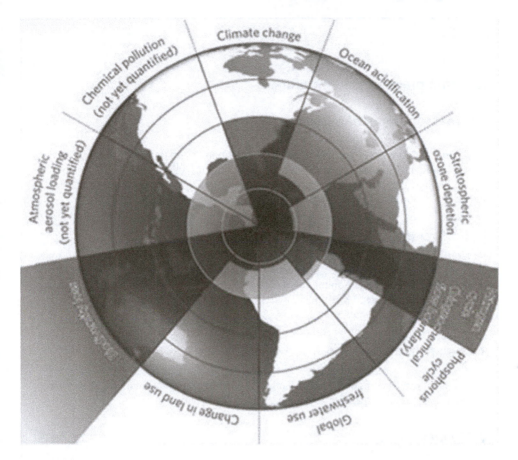

Figure 2.1 *Planetary boundaries.*

nine planetary boundaries (Rockström et al., 2009). Crossing these boundaries could generate abrupt or irreversible environmental changes. To avoid such catastrophic environmental change, humanity must stay within defined planetary boundaries for a range of essential Earth-system processes. If one boundary is transgressed, then safe levels for other processes could also be at serious risk. Staying within these boundaries will allow humanity to develop and thrive for generations to come.

The planetary boundaries research has been criticised for not being well adapted to policy. Despite this, however, the planetary boundaries concept played a prominent role in discussions at Rio+20, providing a scientifically based framework to support the elaboration of sustainable development goals, as discussed in Chapter 5. The concept is new and further research is needed to define a safe operating space for humanity.

Box 2.2

The nine planetary boundaries

1 Stratospheric ozone layer

If this layer deteriorates, increasing amounts of ultraviolet radiation will reach ground level, causing a higher incidence of skin cancer in humans and damage to terrestrial and marine biological systems. Because of the Montreal Protocol, we appear to be staying within this boundary.

2 Biodiversity

The drivers of change that cause biodiversity loss and lead to changes in ecosystem services are either steady, showing no evidence of declining over time, or are increasing in intensity. Further research is underway to determine whether a boundary based on extinction rates is sufficient.

3 Chemicals dispersion

Emissions of toxic compounds drive changes to the planetary environment. At present, scientists are unable to quantify the chemical pollution boundary.

4 Climate change

Evidence suggests that the Earth has already transgressed the planetary boundary and is approaching several Earth-system thresholds. A major question is how long we can remain beyond this boundary before large, irreversible changes become unavoidable.

5 Ocean acidification

Around a quarter of the CO_2 humanity emits into the atmosphere is ultimately dissolved in the oceans. Here it forms carbonic acid, altering ocean chemistry and decreasing the pH of the surface water. This increased acidity reduces the amount of available carbonate ions used by many marine species for shell and skeleton formation. Compared to pre-industrial times, surface ocean acidity has already increased by 30 per cent.

6 Freshwater consumption and the global hydrological cycle

Human pressure is now the dominant driving force determining the functioning and distribution of global freshwater systems. Water is becoming increasingly scarce.

A water boundary related to consumptive freshwater use has been proposed to maintain the overall resilience of the Earth system and reduce the risk of 'cascading' local and regional thresholds.

7 Land system change

Forests, wetlands and other vegetation types have been converted to agricultural land. This is driving reductions in biodiversity, and has impacts on water flows and on the bio-geochemical cycling of carbon, nitrogen and phosphorus and other important elements.

8 Nitrogen and phosphorus inputs to the biosphere and oceans

The biogeochemical cycles of nitrogen and phosphorus have been radically changed through industrial and agricultural processes. Human activities now convert more atmospheric nitrogen into reactive forms than all of the Earth's terrestrial processes combined. Much of this new reactive nitrogen is emitted to the atmosphere. When it is rained out, it pollutes waterways and coastal zones or accumulates in the terrestrial biosphere. Similarly, much of the phosphorus used by humans makes its way to the sea, and can push marine and aquatic systems across ecological thresholds of their own.

9 Atmospheric aerosol loading

Aerosols influence the Earth's climate system and have adverse effects on many living organisms. Inhaling highly polluted air causes roughly 800,000 people to die prematurely each year. However, many causal links are yet to be determined, so it is not possible at this stage to set a specific threshold value.

(*Sources*: Rockström et al., 2009; SRC, 2009)

There is also a need to discuss how to translate planetary boundaries downward to the relevant scale, including the regional, national and sub-national levels where both environmental and governance processes occur. Despite its limitations, the concept has proved highly relevant for returning the issue of limits to growth to the policy agenda, including through heightening the need to engage with the task of setting a safe operating space for humanity.

The second concept that has advanced discussions on limits is the concept of 'ecological footprint'. Ecological footprint refers to the impact of a community on natural resources and ecosystems, by taking account of the land area and the natural capital on which it draws to sustain its population and production structure (Wackernagel and Rees,

1996). The ecological footprint measures one key dimension that contributes to the sustainability or unsustainability of human activities: the extent to which the Earth's productive ecosystems have sufficient regenerative capacity to keep up with society's metabolic demands. It provides a method for measuring how people's competing demands for

---

Box 2.3

Calculating ecological footprints

Aim

The ecological footprint is a method to answer the following research question: 'How much of the regenerative capacity of the biosphere is occupied by human activities?'

It measures how much biologically productive land and sea an individual, population or activity requires to produce the renewable resources it consumes and to absorb the waste.

It expresses the consumption of renewable resources (crops, animal products, timber, and fish), the result of the consumption of energy and the use of built-up areas in standardised units of biologically productive area, so-called 'global hectares' (gha). The global hectare is a common unit that quantifies the biocapacity of the Earth. One global hectare measures the average productivity of all biologically productive areas on Earth in a given year.

Calculating the footprint

The global yield factor by type of consumption (e.g. crops, pasture, fisheries, timber) translates a product (or waste) into an area (local hectares) required to produce the product (or to absorb the waste). It describes the resource productivity for the selected period (e.g. one year), the selected product (e.g. crops, pasture-fed animal product, fish) and the connected land type (e.g. cropland, pasture, fisheries area). The equivalence factor (in gha/ha) translates a specific land type (such as cropland or forestland) into a global hectare. This equivalence factor represents the world's average potential productivity of a given bioproductive area relative to the world average potential productivity of all bioproductive areas.

Critique of calculation

The ecological footprint measures the demand on nature that results from specific human activities. The question arises whether the full impact of the resource use is included (the geographical principle) or only that part of the impact which is attributable to the population within the region (the responsibility principle). This issue has yet to be resolved.

(*Source*: Schaefer et al., 2006)

resources and the capacity of the planet cope with this, through, for example, assimilating carbon dioxide waste from fossil fuel use (Wackernagel, 2014).

The concept of ecological footprint has proved particularly useful for looking at the environmental impact of urban development (Mcmanus and Haughton, 2006). The more populous and rich a city is the larger is its ecological footprint, both in terms of its demands on resources and the size of the area from which these resources are drawn. Many cities not only appropriate resources and carrying capacity from their own regions but also from other locations, including globally (Roseland, 2000). The concept has also been applied in national environmental planning, including in the Netherlands (Dresner, 2002).

Criticisms of the ecological footprint approach have tended to focus on the calculations, such as the assumptions made in drawing up the figures, or raising concerns that the footprint accounts are incomplete, notably in relation to water and waste streams. In addition, there have also been arguments that the calculations need to find ways to take account of human capacity to use technology to change limits (Mcmanus and Haughton, 2006).

Irrespective of these criticisms, there is general agreement that the use of ecological footprints casts new light on how much of the planet's productive capacity is demanded to support human activities. This has helped to document what is referred to as 'overshoot', that is, the excess global demand over global supply of nature's resources.

> Global ecological overshoot refers to the footprint of consumption in excess of biocapacity.

This focus on overshoot draws attention sharply to the need to live within ecological limits. As Wackernagel, the founder of the ecological footprint approach, argues, 'the enormous social and economic inequities that exist within countries, and even more blatantly worldwide, are not only inhumane, but also threaten everybody's security' (Wackernagel, 2001: 4). It shows that the minimum requirement for global sustainable development is that humanity's footprint be smaller than the biosphere's biological capacity (ibid.: 11). With its focus on equity and measuring overshoot, ecological footprint calculations can support the application of the Brundtland understanding of sustainable development, in particular through sharpening focus on the need to apply limits to growth.

A new development paradigm

While placing emphasis on the limits to growth, the Brundtland formulation nevertheless presents an optimistic view, especially in relation to the capacity of humankind to engage collectively to support our sustainable futures. It also places strong emphasis on, and hope in, technological development. However, Brundtland envisages building a common future on more fundamental processes of change, which involve not just technological and institutional changes, but also social, economic, and cultural and lifestyle transformations.

> Sustainable development is a process of change in which the exploitation of resources, the direction of investment, the orientation of technological development, and institutional change are all in harmony and enhance both current and future potential to meet human needs and aspirations.
>
> (WCED, 1987: 46)

What is significant about the Brundtland Report is that it addresses not just the causes of unsustainable development but also puts forward solutions or pathways to support the construction of sustainable futures. It provides a framework for the integration of environmental policies and development strategies into a new development paradigm, one that breaks with the perception that environmental protection can only be achieved at the expense of economic development. The new development paradigm contains many features.

While the Brundtland model provides a set of guidelines, it is not detailed enough to determine actual policies. These have to be worked out in practice through, for example, international negotiations. However, as will be seen, a distinction needs to be made between what Brundtland argues ought to be the case and what actually is the case in practice, as actors, including governments at the international, national or sub-national levels, engage with the promotion of sustainable development – a gap revealed in several of the forthcoming chapters. As will become clear throughout this book, many actors, while adopting a commitment to sustainable development, have not embraced the full agenda of change that was envisioned by Brundtland.

Box 2.4

The Brundtland development paradigm

Reviving growth

- Changing the quality of growth: making it less material- and energy intensive and more equitable in its impact;
- Meeting essential needs for jobs, food, energy, water and sanitation;
- Merging environmental and economic considerations in decision making.

Population and human resources

- Reducing population growth to sustainable levels;
- Stabilising population size relative to available resources;
- Dealing with demographic problems in the context of poverty elimination and education.

Food security

- Addressing the environmental problems of intensive agriculture;
- Reducing agricultural subsidies and protection in the North;
- Supporting subsistence farmers;
- Linking agricultural production to conservation;
- Shifting the terms of trade in favour of small farmers;
- Addressing inequality in access to and distribution of food;
- Introducing land reform.

Loss of species and genetic resources

- Maintaining biodiversity for moral, ethical, cultural, aesthetic, scientific and medical reasons;
- Halting the destruction of tropical forests;
- Building up a network of protected areas;
- Establishing an international species convention;
- Funding biodiversity preservation;
- Conserving and enhancing the natural resource base.

Energy

- Establishing safe and sustainable energy pathways;
- Providing for substantially increased primary energy use by the Third World;
- Ensuring that economic growth is less energy intensive;
- Developing alternative energy systems;
- Increasing energy efficiency, including through technological developments and pricing policies.

Industry

- Producing more with less;
- Promoting the ecological modernisation of industry;
- Accepting environmental responsibility, especially by trans-national corporations;
- Achieving tighter control over the export of hazardous material and wastes;
- Ensuring a continuing flow of wealth from industry to meet essential human needs;
- Reorienting technology and the management of risk.

Human settlement and land use

- Confronting the challenge of urban growth;
- Addressing the problems caused by population shifts from the countryside;
- Developing settlement strategies to guide urbanisation;
- Ensuring that urban development is matched by adequate services provision.

Summary

Box 2.5

Summary of the Brundtland approach to sustainable development

- Links environmental degradation to economic, social and political factors;
- Presents sustainable development as a model of social change;
- Constructs a three-pillar approach: reconciliation of the social, economic and ecological dimensions of change;

- Takes a positive attitude towards development: environmental protection and economic development can be mutually compatible goals and may even support each other;

- Argues that the state of technology and social organisation limit development, but progress in these areas can open up new development possibilities;

- Adopts a global focus, including in relation to planetary limits;

- Recognises that there are ultimate biophysical limits to growth;

- Makes explicit the needs of the poor, especially in the developing world;

- Argues that the planetary ecosystem cannot sustain the extension of the high consumption rates currently enjoyed in industrialised countries upwards to the global level;

- Holds that the consumption patterns of the North are driven by wants not needs;

- Challenges the North to reduce its consumption levels to within the boundaries set by ecological limits and by considerations of equity and justice;

- Calls for new models of environmental governance, ranging across all levels, from the local to the global;

- Acknowledges the responsibility that present generations have towards future generations;

- Has achieved authoritative status within international environmental and development discourse and international environmental governance structures and legal frameworks.

The ladder of sustainable development

Proliferation of terms

Since the publication of *Our Common Future*, there have been numerous attempts to specify exactly what is meant by the term sustainable development. There is a ten-page listing of the most common definitions of sustainable development used in the decade of the 1980s alone (Prezzey, 1989; Lélé, 1991). In addition to the myriad attempts to define sustainable development, the word 'sustainable' has been combined with an array of terms, to create such concepts as 'sustainable growth', 'sustainable cities' and 'sustainable culture'. Not all of these applications can be explored, even though some of them, like the concept 'sustainable growth', are particularly problematic.

The broadening of the concept of sustainable development, coupled with its popularity, has given rise to ambiguity and lack of consistency in the use of the term. Some have argued that the concept's ambiguity severely diminishes its usefulness. There is concern among environmentalists that the lack of clarity in the definition allows anything to be claimed as 'sustainable' (Jacobs, 1991). It also makes it difficult to devise a set of measurable criteria with which to evaluate whether concrete development programmes are helping to promote sustainable development. Attempts to overcome this problem have led to the elaboration of sustainable development indicators.

Sustainable development as a political concept

The proliferation in meaning and in application of the term 'sustainable development' does not necessarily undermine the usefulness of the term. Rather, it reflects the complexity of issues that are invoked when development and environment are juxtaposed (Meadowcroft, 1999). As Donella Meadows, one of the authors of *Limits to Growth*, has argued when discussing some of the linguistic confusion surrounding the use of the word:

> We are struggling for the language now for a whole set of concepts that are urgent in our conversation ... It's a mess. But social transformations are messy.
>
> (Interview quoted in Dresner, 2002: 66)

The lack of clarity has also been politically advantageous, because it has allowed groups with different and often conflicting interests to reach some common ground upon which concrete policies can be developed. This is particularly the case within the UNCED process, as discussed in Chapter 5.

More importantly, the search for a unitary and precise meaning of sustainable development may well rest on a mistaken view of the nature and function of political concepts (Lafferty, 1995). As many commentators have argued, sustainable development is best seen as similar to concepts such as 'democracy', 'liberty' and 'social justice' (Lafferty, 1995; Jacobs, 1995; O'Riordan, 1985). For concepts such as these, there is both a readily understood 'first level meaning' and general political acceptance, but also a deeper contestation. Sustainable development, it is argued, is an essentially contested concept (Lafferty, 1995). In liberal democracies, the debates around such contested concepts form an essential component of the political struggle over the direction of social and economic development, that is, of change (Lafferty 1995). Substantive political arguments are part of the dynamics of democratic politics and the process of conscious steering of societal change.

Such arguments are important as they can stimulate creative thinking and practice, such as in relation to the idea of 'development', as discussed in this book's Introduction.

The ladder of sustainable development

The diversity of policy options associated with sustainable development can best be seen in terms of a ladder (Table 2.1), originally developed by Baker et al. in 1997. The ladder offers a useful heuristic device for understanding the variety of policy imperatives that are associated with different approaches to the promotion of sustainable development. These approaches can be adopted by governments, by organisations, or by individual green thinkers or activists. The original version of the ladder has been extensively modified, its organising principles have been amended, the rungs of the ladder altered, the components for each rung have been made more distinct, and overall the ladder has been given a more global focus.

Each column focuses on a different aspect of sustainable development. Reading across the ladder identifies the political scenarios and policy implications associated with each rung. The ladder also tracks the connection between these positions and particular philosophical beliefs about nature and about the relationship between human beings and the natural world. This helps put flesh on the environmental ethics that underpins practical sustainable development actions.

The philosophical underpinning

The variety of approaches to sustainable development is an indication of differing beliefs about the natural world held within different societies, cultures, and historical settings and at the individual level. The values that are attributed to nature range across a broad spectrum, from an 'anthropocentric' to an 'ecocentric' position. At the extreme end of the anthropocentric view, the wealth of nature is seen only in relation to what it can provide in the service of humankind (O'Riordan, 1981). The focus on nature as provider of ecosystem services provides an example, as discussed in detail in Chapter 7. In contrast, the ecocentric position holds that nature has intrinsic value. Sustainable development, in this view, aims at creating a partnership, based on reciprocity, between human beings and nature.

These two different perspectives have important implications for the design and implementation of policies. The ecocentric approach focuses

Table 2.1 *The ladder of sustainable development: the global focus*

Model of sustain-able development	Normative principles	Type of development	Nature	Spatial focus
Ideal model	Principles take precedence over pragmatic considerations	Right livelihood; simplicity meeting needs not wants; biophysical/ planetary limits guide development	Nature as intrinsic value; no substitution allowed; strict limits on resource use, aided by population reductions	Bioregionalism; extensive local self-sufficiency
Strong sustainable development	Principles enter into international law and governance	Changes in patterns and levels of consumption; shift from growth to non-material aspects of development; necessary development in Third World	Maintenance of critical natural capital and biodiversity	Heightened local economic self-sufficiency, promoted in the context of global markets; green and fair trade
Weak sustainable development	Declaratory commitment to SD principles stronger than practice	Decoupling; reuse, recycling and repair of goods; product life-cycle management	Substitution of natural capital with human capital; harvesting of biodiversity resources	Initial moves to local economy; minor initiatives to alleviate power of global markets
Pollution control	Pragmatic not principled approach	Exponential, market-led growth	Resource exploitation; marketisation and further closure of the commons; nature as use-value	Globalisation; shift of production to less regulated locations

on the community level and espouses small scale, locally based technology. The objective is to maintain social and communal well-being and not merely a harmonious use of natural resources (Baker et al., 1997). In contrast, the anthropocentric approach can be distinguished by its optimism over the successful manipulation of nature and her resources in the interest and to the benefit of humankind. An extreme example of the anthropocentric approach can be found in the USA's Wise Use Movement, a coalition of ultra-conservative politicians, interest groups, scientific institutions and consumers, that promotes economic growth and rejects the need to consider the environment in economic development.

Making too sharp a distinction between the anthropocentric and ecocentric positions on nature, however, is not wise. This is because the main motivation behind any conception or theory of sustainable development is human interest in human welfare (Dobson, 1998). This is certainly true of the Brundtland formulation, with its emphasis on human needs and finding a

Governance	Technology	Policy integration	Policy tools	Civil society/ state relationship	Philosophy
Decentralisation of political, legal, social and economic institutions	Labour-intensive appropriate, green technology; new approach to valuing work and well-being	Environmental policy integration; principled priority of environment	Internalisation of sustainable development norms through ongoing socialisation, reducing need for tools	Bottom-up community structures and control; equitable participation; deliberative democracy	**Ecocentric**
Partnership and shared responsibility across multi-levels of governance; use of good governance principles	Ecological modernisation of production; mixed labour- and capital intensive technology	Integration of environmental considerations at sector level; green planning and design	Sustainable development indicators; wide range of policy tools; green accounting	Democratic participation; open dialogue to envisage alternative futures	
Some institutional reform and innovation; move to global regulation	End-of-pipe technical solutions; mixed labour- and capital intensive technology	Addressing pollution at source; some policy co-ordination across sectors	Environmental and SD indicators; market-led policy tools; economic valuing/pricing	Top-down initiatives; limited state/ civil society dialogue; elite participation	
'Command and control' state-led regulation of pollution	Capital-intensive technology; progressive automation	'End of pipe' approach to pollution management	Conventional accounting	Dialogue between the state and economic interests	**Anthropocentric**

way to ensure that development (a human activity) is sustainable over time. While this may involve the protection of the natural resource base, the rationale for this protection is essentially a human-centric one: it is protected because it is necessary for our well-being. Nevertheless, ranging attitudes towards nature along a continuum from anthropocentric to ecocentric is useful because it exposes sustainable development perspectives that go beyond the Brundtland approach. At one extreme, nature is seen only in relation to its use to human beings. Moving along the continuum, sustainable development becomes a challenge to develop a more environmentally friendly approach to planning and resource management. Moving further along, nature is allowed to set the parameters of economic behaviour, so that sustainable development becomes an 'externally guided' development model, based on planetary limits. Reaching the other extreme, so deep is the green philosophy that sustainable development is viewed as managerial interference with nature and her natural cycles.

Grouping the different policy imperatives on the ladder

At the bottom of the ladder is the pollution control approach. It is not that the environment is given no consideration, but rather that there is an underlying assumption that, given the freedom to innovate, human ingenuity, especially expressed through technology, can solve any environmental problem (Simon and Kahn, 1984). A good example of this approach is found in the so-called Heidelberg Appeal, released by a group of business interests during the Rio Earth Summit in 1992. This accepts that environmental protection is an integral part of development, but argues that it should not put limits upon that development, nor should it form our main priority (Sourcewatch, 2009).

In evidence of the capacity of human being to manage their environment, supporters of the pollution control approach point to the empirical claim that pollution typically arises in the early stages of industrial development, followed by a stage where pollution is no longer regarded as an acceptable side effect of economic growth and when pollution control policies are introduced (Arrow et al., 1995). This idea has been expressed in the so-called environmental 'Kuznet curve'. This curve graphically represents how pollution starts out low, increases at the early stages of development, but then diminishes as the economy shifts into a less resource intensive, post-industrial stage.

However, despite its popularity, the Kuznet curve ignores the fact that high pollution activities can be displaced from the industrialised to the

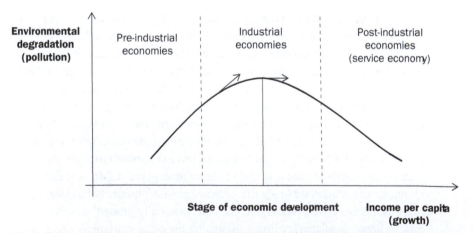

Figure 2.2 *Kuznet Curve: Pollution rises at earlier stages of industrial development then diminishes as societies shift into a post-industrial age.*

Source: Panayotou, 1993.

developing world, thereby reducing pollution in one place but not the overall, global levels. There are many such examples of pollution displacement activity, including the case of Japan, which uses the forest resources of other countries to provide the packing needed for the consumer goods it produces, while maintaining high levels of forest protection at home. Global environmental change is also not captured by the curve.

Immediately above the pollution control approach on the ladder is the concept of 'weak' sustainable development, whose aim is to integrate capitalist growth with environmental concerns. This position is closely associated with David Pearce and the highly influential *Blueprints* for a green economy reports (Pearce et al., 1989; Pearce, 1994; 1995; Pearce and Barbier, 2000). These argue that the best way to preserve critical 'natural capital', important natural resources or processes such as forests or the climate system, is to give it an economic value or price. The price is based on what people would be willing to pay to protect that natural capital. This figure can then be used to undertake a 'cost–benefit analysis' to determine the gains and losses involved in using that natural capital. If the gains outweigh the losses, then the natural capital should be used, or 'drawn down'. However, this idea of 'putting a price on the planet' has been severely criticised (Dresner, 2002), an issue that is taken up again in Chapter 7, where the economic valuation of ecosystem services is discussed. The anthropocentric basis of this position is clear, and some argue that much of nature is 'beyond price'. It also breaches many of the normative principles associated with sustainable development:

> Pearce's work shows quite well the way in which the application of cost–benefit analysis to global environmental issues works against the principles of inter-generational equity and intragenerational equity that lay at the core of the Brundtland Commission's definition of sustainable development. Because decisions are based on ability to pay, less weight is given to the interests of the poor and the future. ... It is hard to see how [money] ... is a good measuring rod when comparing the preferences of Americans and Bangladeshis, or people today and people a hundred years from now.
>
> (Dresner, 2002: 116)

The objective of policies to promote weak sustainable development remains economic growth, but environmental costs are taken into consideration through, for example, accounting procedures. This is possible because the environment is considered as a measurable resource. Weak sustainable development has a growing influence on international agencies, including the World Bank. It has led to the development of many new environmental policy tools, including tradable permits, as discussed in Chapter 3.

On the third rung of the ladder is the 'strong' sustainable development position. Whereas Pearce asserts that economic development is a precondition for environmental protection, strong sustainable development asserts that environmental protection is a precondition for economic development (Baker et al., 1997). One of the major differences between strong and weak forms of sustainable development is in relation to whether 'natural capital' (oil, for example) can be drawn down and technology used as a substitute (replacing oil by solar technology), or whether there is such a thing as 'critical' natural capital, which cannot be substituted for by technological and should be preserved absolutely. This disagreement is also about how development is structured: should the proceeds from running down natural resources, such as the UK's North Sea gas, be invested in the development of new technologies that can substitute for these resources, or be invested in other forms of capital, such as human capital, through investment in education? (Dresner, 2002). Weak sustainability assumes almost total substitutability by technology, whereas strong sustainability assumes some substitutability, but imposes strict limits on how much human capital can compensate for running down natural capital.

The precautionary principle supports strong forms of sustainable development. The precautionary principle holds that, in the face of risk but uncertain scientific knowledge, policy makers should err on the side of caution. Promoting this form of sustainable development requires that strong state intervention (government) be combined with new forms of participation (governance), as discussed in Chapter 3. In relation to climate change, for example, governments need to ensure adequate market regulation and develop new energy and transport policies to deal with climate change. The involvement of consumers, economic interests and local communities is also needed to bring about changes in consumption patterns and to ensure that society makes more use of environmentally friendly transport modes. Thus, the stronger form of sustainable development does not give free rein to market forces to determine behaviour.

Strong sustainable development also seeks a shift from quantitative growth, where growth is seen as an end in itself and measured only in material terms, to qualitative development, where quality of life is prioritised. This has led several green economists to develop alternative indicators to the traditional GNP measure of human welfare. Herman Daly's 'Index of Sustainable Economic Welfare' is among the best known of these. It includes calculations for depletion of natural capital, the cost of pollution and social issues such as unemployment and inequality (Daly and Cobb, 1990). The weak form of sustainable development cannot perpetuate itself indefinitely, as it permits the draw-down of natural resources in order to support production. The stronger form of sustainable development permits growth only under certain limited conditions: when it is designed to deal with necessary development, and when it is balanced by reduction in growth elsewhere. Strong forms view development over the long term and at the global level. Chapter 5 gives further consideration to this issue, and explores UN efforts to replace the emphasis on growth with a new concern for 'well-being'.

The top rung of the ladder represents the ideal approach to sustainable development. It offers a more profound vision aimed at structural change in society, the economy and the political systems. Some who hold this position reject the idea of sustainable development as formulated by Brundtland, but have gone on to modify the Brundtland position by injecting it with more radical, socialist considerations (Pepper, 1998).

For others, severe restraints on the consumption of the Earth's resources, on humankind's related economic activities and, most controversially, a reduction in human population, are all proposed. Here, promoting sustainable development is premised upon a radical change in our attitude to nature. Perhaps the most radical of these deep green positions is that of 'deep ecology', originally put forward by the Norwegian philosopher Arne Naess. The deep ecology position has three characteristics: attributing equal value to all life forms; seeking identification with non-human natural entities and systems; and advocating the development of policies that stress non-interference and the harmony of human life and nature (Naess, 1989). The substitution of human for natural capital is not allowed and gains in human welfare at the expense of radical transformation of the 'natural' environment are not tolerated (Sylvan and Bennett, 1994). Deep ecology rejects the assumption that human beings can and ought to manage the environment (Katz et al., 2000). As the management of the environment, albeit to different degrees, is an underlying assumption of all efforts to promote sustainable development, it thus rejects the sustainable development project. The environmental groups Earth First!, Sea Shepherd and the Animal Liberation Front represent examples of how deep green philosophy can influence environmental direct action.

So far, in discussing the ladder of sustainable development, the different approaches to nature and the natural world have been sketched along a continuum. The sustainable development project is rejected at both extremes, but for exactly the opposite reasons: for the pollution control approach, promoting sustainable development is seen as threatening economic growth by taking environmental considerations *too much* into account; at the opposite extreme, deep ecologists argue that sustainable development displaces considerations of nature, thus taking the environment *too little* into account (Jagers, 2002). Occupying the middle ground are a range of different understandings of sustainable development; each in turn can be associated with different policy imperatives. Implicit in the discussion so far has been that there is a link between the promotion of sustainable development and certain ideas about what constitutes right conduct (morals). Mention has been made, for example, of links with issues of justice and equity, which are discussed further in Chapter 6. These ideas have begun to permeate the discourse on sustainable development and, as a result, the promotion of sustainable development has now come to be associated with certain norms of behaviour. To complete the discussion on the concept of sustainable development, these normative principles are explored in greater depth.

The normative principles of sustainable development

As international engagement with the concept of sustainable development progressed from the time of the Brundtland Report, the term began to be associated with a number of normative principles. Normative principles are moral rules that specify what is good or bad, and thus shape how people ought to treat one another. In Brundtland, sustainable development was primarily associated with meeting human needs, especially the development needs of the poor and the protection of environmental resources. However, Brundtland also introduced other values into the discussion and this opened the way for a range of normative principles to come to be associated with the term.

Box 2.6

Normative principles of sustainable development

- Principle of Need
- Principle of Inter-generational Equity
- Principle of Intra-generational Equity
- Principle of Common but Differentiated Responsibilities
- Principle of Justice
- Principle of Participation
- Principle of Gender Equality.

Principle of need

As mentioned above, the Brundtland understanding of sustainable development gives priority to meeting the needs of both present and future generations. It holds that the satisfaction of human needs and aspirations is the major objective of development. However, the essential needs of vast numbers of people in developing countries for food, clothing, shelter and jobs are not being met. Brundtland also argues that, beyond meeting their basic needs, people have legitimate aspirations to an improved quality of life. From this perspective, sustainable development requires meeting the basic needs of all and extending to all the opportunity to satisfy their aspirations to a better life (WCED, 1987). In addition, living standards that go beyond the basic minimum are sustainable only if consumption standards everywhere have regard for long-term sustainability. In this context, sustainable development requires the promotion of values that encourage consumption standards that are within the bounds of the ecologically possible and to which all can reasonably aspire (WCED, 1987).

Inter- and intra-generational equity

According to the Brundtland Report, the promotion of sustainable development implies a concern for both intra-generational and inter-generational equity, especially with respect to resource use.

> 1 Intra-generational equity: refers to equity *within* our own generation.
> 2 Inter-generational equity: refers to equity *between* generations, that is, including the needs of future generations in the design and implementation of current policies.
>
> (WCED, 1987: 5–6)

1. Intra-generational equity

This highlights the importance of meeting the basic needs of present generations, given the widely uneven pattern of global development. The notion of equity within generations owes much to the work of John Rawls' *A Theory of Justice* (Rawls, 1971), although Rawls' earlier work pre-dates current concerns about the global environment. The principle of equity is fundamental to Rawls' theory of justice, in which he argues for equality in the distribution of basic social goods, such as liberty and opportunity, income and wealth, and social respect (Baker, 2012).

Present concerns about equity acknowledge the inequity in resource use between the North and the South, the rich and the poor, while at the same time seeing poverty as both a cause and a consequence of unsustainable behaviour. Poverty can lead to the over-exploitation of the resources of a local environment to satisfy immediate needs: 'Those who are poor and hungry will often destroy their immediate environment in order to survive' (WCED, 1987: 28). Poverty caused by a failure to address land reform, for example, can lead landless farmers to use ecologically damaging 'slash and burn' agricultural techniques, as is happening in the Amazon rainforest. Poverty can also lead to the growth of urban slums, which lack adequate infrastructure, especially for sewage and waste disposal, resulting in both health and environmental hazards. There is thus a relationship between poverty and exposure to the negative consequences of environmental degradation, such as polluted water. Such concerns led to the development of the environmental justice movement, particularly in the USA. This movement primarily addresses the negative impacts of environmental degradation on human health (Martinez-Alier, 1999). It began with an oppositional campaign in 1982 in a poor African American community in Warren Country, North Carolina, where locals campaigned against state plans to dispose of waste contaminated with polychlorinated biphenyls (PCBs) in a local landfill. The environmental justice movement is also typified by the actions of a local community in Love Canal in the USA in the 1970s, where low income housing was built on the site of a toxic dump, which subsequently led to severe health effects in the local community (Cutte, 2006). The movement spread over time, raising awareness of how the risks of environmental harm are unequally distributed, such that the poor bear environmental costs and have fewer environmental benefits. Furthermore, as we will see in Chapter 6, the movement is currently highlighting how global environmental change tends to impact more directly on the marginalised and the poor. The concern of the environmental justice movement, however, is narrower than those raised in the sustainable development agenda. The broader remit of the latter encompasses not just issues of health, and global equity and justice of access to and use of resources, but the maintenance of biodiversity and ecosystem health.

This links the promotion of sustainable development to questions of power and the removal of the disparities in economic and political relationships between the North and South. For Brundtland, there is a strong functional relationship between social justice and sustainable development because poverty is a major cause of environmental deterioration and the reduction of poverty is a precondition for environmentally sound development (WCED, 1987).

> Developing countries must operate in a world in which the resource gap between developing and industrial nations is widening, in which the industrial world dominates in the rule making of some key international bodies, and in which the industrial world has already used much of the planet's ecological capital. This inequality is the planet's main 'environmental' problem; it is also its main development problem.
>
> (WCED, 1987: 46)

However, Brundtland gives priority to the world's poor, independent of any poverty–environment relationship. This is because poverty is seen 'as an evil in itself' and sustainable development requires meeting the basic needs of all, thus extending to all the opportunity to fulfil their aspirations to a better life. The relationship between social and economic justice and physical sustainability is not just functional; that is, it does not merely serve a particular practical and efficiency purpose, but it is also normative, that is, it is based upon ethical considerations (Langhelle, 2000).

Making the link between poverty and environmental harm is not to deny that many communities in the developing world have devised sustainable coping strategies, as discussed in Chapter 11. In addition, it is not only the poor who over-exploit environmental resources but the rich as well, so that the alleviation of poverty may not necessarily lead to an end to environmental degradation. Poverty relief needs to be combined with other policies if environmental degradation is to be halted (Dobson, 1998). We will also see, in Chapter 7, that the relationship between the different strands of sustainable development, in particular those that relate to the social pillar and the need to address poverty, and to the ecological pillar, especially biodiversity protection, is not necessarily and in all cases a positive one.

2. Inter-generational equity

The idea of inter-generational equity dates as far back as the political philosophy of Kant, who developed the idea of posterity benefiting from the works of their ancestors. The philosopher Edmund Burke (1729–1799) also wrote about the idea of inter-generational partnership (Ball, 2000). However, in considering future generations, Brundtland developed quite the opposite perspective to that of previous thinkers, arguing that we borrow environmental capital from future generations and that 'Our children will inherit the losses' that this brings (WCED, 1987: 8). Brundtland argued that today's society might compromise, in many different ways, the ability of future generations to meet their essential needs (WCED, 1987). Rather than focusing upon the ways in which the actions of the present generation may help those of the future, Brundtland focused upon how today's unsustainable behaviour can narrow the options available for future generations. Promoting sustainable development requires foreclosing as few future options as possible (WCED, 1987: 46). Green theorists have developed Brundtland's ideas further, to suggest that our relations with other generations create obligations (Johnson, 2003). The exercise of obligations to future generations poses a problem, however: it is unclear how far into the future these obligations stretch. It would seem insufficient to restrict concern to the next generation only, as many environmental problems or processes work on a very long-term scale. A 'glacial' timescale, for example, applies to radioactive waste, with Plutonium 239 for example having a half-life of approximately 24,000 years. This means that after 24,000 years half of the radioactivity contained in the plutonium will have decayed. However, the hazardous life of radioactive waste is at least ten times the half-life, therefore these wastes will have to be isolated from the environment for 240,000 years (Greenpeace USA, 2014).

Considerations of inter-generational equity also raise another very difficult political issue, namely how future generations can be given some form of voice or consideration in present policy making. Not least among these problems is how to find out what the interests or needs of future generations will be. In addition, environmental management tasks, such as planning, monitoring and evaluation, typically do not fit in with the longer-term period needed to take account of future generations. As such, considerations of inter-generational equity require considerable extension of the timescale of current planning and policy making models and practices, as well as new practices of participation, discussed in more depth in Chapters 3 and 11.

Other theorists have argued that the principle of inter-generational equity brings with it more stringent requirements. Dobson in particular has argued that the principle means that future generations' human needs have to take *precedence over* the present generation's human wants. He argues that it

> would be odd for those who argue for the sustaining of ecological processes to put the wants of the present generation of human beings (which might threaten those processes) ahead of the needs of future generations of human beings (who depend upon them).
>
> (Dobson, 1998: 46)

It has also been argued that once the interests of future generations are taken into account, then concern for many features and aspects of the non-human natural world can be generated. This is not least because the loss of plant and animal species can greatly limit the options of future generations. Concern for other species thus becomes one essential prerequisite for future generations to meet their needs.

Common but differentiated responsibilities

The 1992 Stockholm Declaration (see Chapter 5) proclaimed the responsibility of governments to protect and improve the environment for present and future generations. After the Stockholm Conference, several states recognised in their constitutions or laws the right of their citizens to an adequate environment as well as the obligation of the state to protect the environment. This idea of environmental responsibility was further elaborated by the Brundtland Report, which called upon all governments to take responsibility for the environment, because the promotion of sustainable development involves guarding the *common fate* of humanity. However, in pursuing this responsibility, account has to be taken of the fact that not all countries have contributed in the same way, or to the same extent, to the current environmental crisis. Moreover, countries have different capacities for taking effective actions to deal with, or prevent further, environmental deterioration, such as in relation to climate change adaptation and mitigation, discussed in Chapters 6 and 7. The principle of 'common but differentiated responsibilities' provides a way of distributing the responsibilities and tasks associated with the promotion of sustainable development more fairly among developing and industrial countries. The principle acknowledges that industrial countries have been the main contributor to environmental problems through their patterns of resource exploitation, production and high consumption. It also recognises the unequal burden of implementing international environmental laws and agreements. Further, it takes account of the different capacity, including financial and technical capacity, available within countries to address the problem. In short, the use of the principle of common but differentiated responsibilities is driven by equity considerations.

The principle was subsequently to influence the Rio Declaration, the agreement that was agreed at the Rio Earth Summit in 1992, as discussed in Chapter 5.

> In view of the different contributions to global environmental degradation, States have common but differentiated responsibilities. The developed countries acknowledge the responsibility that they bear in the international pursuit of sustainable development in view of the pressures their societies place on the global environment and of the technologies and financial resources they command.
>
> (Principle 7 of the Rio Declaration)

The principle has two main components. The first is the common responsibility we have as a result of our common heritage and the common concern of humankind and which needs to be reflected in the duty of states to equally share the burden of environmental protection of common resources (Ostrom, 1994). The second component is the differentiated responsibility, which addresses issues in relation to substantive equality or, more specifically, the unequal distribution of material, social and economic situations across states. In particular, it acknowledges the different historical contributions made by different states to global environmental problems and the variation in their financial, technological and structural capacity to tackle them. In this sense, the principle establishes a conceptual framework for an equitable allocation of the costs of global environmental protection (De Lucia, 2012), as discussed in Chapter 6.

Prior to the Rio Earth Summit many international agreements had already recognised the different capabilities and needs of different states, and provided for differential treatment provisions. This was reflected in the Convention on the Long-range Transport of Air Pollution (1978) and the Montreal Protocol on Ozone Depleting Substances (1988), to give just two examples. However, what was particularly important about the Rio Declaration was that it established links between past economic exploitation of the global commons and the responsibility to carry out actions that remedy or mitigate the consequences of such exploitation. It thus anchors responsibility on past harm done, or culpability.

Differential obligations can take several different forms. It can oblige industrialised countries, for example, to transfer technology or resources, including finance and expertise, to developing countries. It can also lead to variations in the quantified targets and goals set in different countries through international agents. The Kyoto Protocol, an international agreement designed to address climate change, obliges industrialised countries to reduce their carbon emissions according to a negotiated scale, while developing countries have, as yet, no reduction targets to meet. Countries can also be given different timescales for implementing their obligations, as well as compensation, funding and resources to help implementation, as discussed in Chapter 6. Historical responsibility, however, is subject to intense negotiations in international forums where nations meet to tackle common environmental, economic and social issues. The Rio Declaration also called for the diffusion of *differential obligations* through international environmental law as a way of putting the principle into practice. These issues are discussed further in Chapter 6.

While drawing upon considerations of equity, the use of the principle of common but differentiated responsibilities has a strong functional logic. It is often used as a means of ensuring that developing countries sign up to and continue to participate in international environmental management regimes, such as the climate change regime mentioned above. Developing countries may be more motivated to implement conventions that acknowledge their vulnerability in the face of an environmental crisis that they did not primarily cause. When international conventions are faced with pervasive, multi-causal problems that traverse national boundaries, such participation is highly valued (Iles, 2003). Use of the principle thus helps to ensure that efforts to promote sustainable development have a more global reach.

Despite its widespread use, however, the principle continues to be contested in international environmental negotiations. Acrimonious debates also took place over whether the principle should be included in the Johannesburg Declaration released at the end of the 2002 World Summit on Sustainable Development (WSSD), discussed in Chapter 5. The fact that the United States failed in its repeated attempts to have the phrase omitted from the Declaration has been portrayed as perhaps the greatest accomplishment of the WSSD (Iles, 2003).

There is nevertheless a problematic side to the use of the principle of common but differentiated responsibilities. Developing countries, particularly during negotiations on international environmental treaties, often argue that environmental protection measures can interfere with their economic development strategies. The use of the principle may thus help perpetuate the perception that a trade-off exists between the environment and development, despite the fact that the model of sustainable development is designed to break such a perception. There is also the possibility that differential obligations can reinforce environmental degradation by permitting developing countries to continue polluting, destroying habitats or overusing their resources. The irony is that this behaviour can run down a county's natural resource base and destroy livelihoods, thus helping to perpetuate the very differences that were used to justify a country's special treatment in the first place (Iles, 2003). Thus, care needs to be taken in the way in which the principle is put into practice if it is to help promote sustainable development at the global level.

The principle of participation

There are both normative and functional reasons why the participation of non-state actors is an essential condition for the promotion of sustainable development. Taking a normative perspective, it can be argued that participation in decisions that shape one's life is considered a hallmark of democratic practice. Promoting sustainable development involves making difficult decisions about, for example, reducing consumption levels, or introducing taxes on goods that have negative environmental impacts, or prohibiting or placing restrictions on certain forms of behaviour, such as on the way of disposing of household waste. It is only though increased participation that society can construct 'a shared public basis' on which to ground the legitimacy and acceptance of such restrictions and corrections (Acterberg, 1993). As Brundtland has argued:

> Making the difficult choices involved in achieving sustainable development will depend on the widespread support and involvement of informed public and non-governmental organisations, the scientific community, and industry.
>
> (WCED, 1987: 21)

A second, normative argument is that participation is necessary because promoting sustainable development raises issues of an essentially 'subjective and value-laden' character (Paehlke, 1996). These include, for example, the value one attributes to nature, or to the possession of material goods. Given these value differences, agreement on the objectives of policy is unlikely unless these objectives are reached through participatory practices.

Functional arguments build upon Brundtland's belief that 'effective participation in decision making processes by local communities can help them articulate and effectively enforce their common interest' (WCED, 1987: 47). To this functional reason is added the argument that participation is the only approach to policy making that can incorporate the needs of all segments of society, future generations and other species (Dryzak, 1992; Pepper, 1998). This claim is somewhat contentious, however, and is discussed in Chapter 11, when several of the problems associated with participation are explored in greater depth.

Gender equality

Prompting sustainable development without consideration of the needs of the female half of the world's population is an empty gesture (Dobson, 1996). At a minimum it breaches the principles of inter- and intra-generational equity. This means that account has to be taken of the fact that environmental degradation affects men and women differently. Sustainable development policy has also to address the fact that women and men traditionally take on different societal tasks, have different functions and roles in relation to reproduction, and experience unequal treatment, including in relation to their access to and exercise of power.

Equitable participation of women in environmental decision making is also required, not least so as to give women space to discuss and identify their needs. In addition, the insights, experience and knowledge that women bring can help identify new policy solutions to better promote sustainable development. The links between women and the environment are explored by feminist environmentalism, a position discussed in Chapter 11. Attention is also paid to a somewhat different position, namely the argument that women and men differ in their relationship to nature, in their historical contribution to the environmental crisis and in the type of responsibility they bear for overcoming the legacy of past behaviour. This eco-feminist argument is also discussed in more detail in Chapter 11.

To date in this chapter we have explored the Brundtland formulation of sustainable development in considerable detail, while also discussing conceptual developments arising since the original WCED report was published. There are many who argue that since the Rio Earth summit and inspirited by the Brundtland Report, states have engaged with the promotion of sustainable development by promoting ecological modernisation. Indeed, so prevalent is the view that ecological modernisation is a particular expression or approach to sustainable development, and in particular that it has helped advance understandings of the economic dimensions of sustainable development, that some attention has to be given to this claim (Barry, 2010).

Ecological modernisation

Background

Work on ecological modernisation grew out of the belief that decoupling of economic growth from environmental destruction may be a feature of certain advanced industrial economies. The founder of ecological modernisation theory, Joseph Huber, argued that *super industrialisation*, that is, the development and application of more sophisticated technologies, could address environmental problems (Huber, 2004). Industrial society, he argued, has gone through three phases: industrial breakthrough (1789–1848); construction of industrial society (1848–1980) and a current phase of ecological modernisation (1980–). The current phase involves reconciling the impacts of human activity with the environment. These empirical observations gave rise to the theory of ecological modernisation, which holds that economic and environmental goals can be integrated within a framework of industrial modernity, thus over-coming the belief that there is a trade-off between economic growth and environmental protection (Moll, 1996; 1999; 2000). The work of Sim-monis (1989) and later work of Jänicke and Klaus (2012) shifted emphasis from technology to macro-economic structural change. They argued that ecological modernisation involves the restructuring of the technologies *and* the sectoral composition of national economies. In other words, ecological modernisation is part of a shift, in advanced economies, from energy- and resource intensive industries towards service- and knowledge intensive industries (Hajer, 1995). This has ena-bled growth of GDP to be decoupled from energy and resource use.

As the theory developed, attention was also turned to the role of the state in supporting ecological modernisation policy (Gouldson and Mur-phy, 1996). Stress is placed on the role of the state in steering busi-nesses towards eco-efficient practices, practices that do not undermine 'competitiveness'. On the contrary, these applications are seen as poten-tially creating new markets, employment, investment opportunities and technological advances (Barry, 2010). As such, ecological modernisa-tion discourse is explicit about not attempting to limit overall levels of consumption or production (Barry, 2010).

The use of the strategy of environmental policy integration is a hallmark of this approach. The Netherlands, Germany and Japan are the leaders in this field, as is the EU, as discussed in Chapter 3. As states engaged with ecological modernisation, they have also sought to build a new relationship with industry, as witnessed by the use of a range of new policy instruments, such as voluntary agreements with industry, to man-age the environment, matters discussed in Chapter 3. Ecological mod-ernisation also takes place through sector-specific activity, particularly in the industrial and energy sectors, where it involves the invention, innovation and diffusion of new technologies and techniques (Murphy, 2000). This further strengthens synergy between environmental protec-tion and economic growth, because policies to protect the environment can enhance efficiency and accelerate innovation, thereby providing an engine for further economic development. The restructuring of the Dutch chemical industry is the prime example. The pollution prevention schemes introduced by the American company 3M is another example. Ecological modernisation is also discussed in Chapter 4, when issues of production efficiency are explored in greater detail. Because it under-stands the environmental predicament as a problem of efficient resource allocation and use, the key strategy of ecological modernisation is to achieve eco-efficiency. This perspective regards growth as part of the solution to environmental problems, not as part of the problem.

Recent research by Mol builds upon sociological theory of risk to argue that ecological modernisation is the reflexive reorganisation of industrial society in the face of risk. Some scholars also went on to distinguish between a technocorporatist/weak and reflexive/strong ecological modernisation (Hajer, 1995; Christoff, 1996).

There is now a substantial body of literature on ecological modernisation and, as is so often the case in the social sciences, this has led to a widening and loosening of the theory's core ideas and to a disparate collection of empirical approaches. There are also disputes over the significance of findings, particularly in relation to other theories of social change, including those put forward by Beck (1992). In addition, as attention focused on what can be achieved through ecological modernisation, the boundaries blurred between ecological modernisation as a theory of social change and as a normative proposition. The study of ecological modernisation has thus become analytical and descriptive as well as normative. Much of the literature is concerned about what the state and industry *should* do in the pursuit of environmental protection, which will improve economic competitiveness at the micro (within firms) and macro levels.

However, the claims that the environmental problem can be, and indeed has been, addressed through the structures and processes of modernity have been heavily criticised (Blühdorn and Welsh, 2008). The empirical basis of the theory, that is, the extent to which European states or industrial sectors have undergone ecological modernisation, remains open to question. The theory also fails to clarify adequately the causal connections between ecological modernisation and the macro-economic structural changes that occurred in Western Europe as part of the shift from energy- and resource intensive industries towards service- and knowledge intensive industries. It has also a predominantly Western European focus, despite recent efforts to investigate non-European cases (Mol and Sonnenfeld, 2000). The rise of global environmental problems, such as climate change and biodiversity loss, also undermines the empirical claim that Western countries have been able to address the environmental consequences of growth.

The focus on efficiency gains is often seen as wildly optimistic, given that efficiency gains per unit of consumption are usually outstripped by overall increases in consumption, a phenomenon known as the rebound effect.

Efficiency rebound effects

Improvements in resource efficiency encourage greater use of the services that the resources helped to provide. While rebound effects vary widely in size, in some cases they may be sufficiently large to lead to an overall increase in resource consumption – an outcome that has been termed 'backfire'.

(*Source*: Sorrell, 2007)

Conceptually, the theory has been criticised for presenting over-simplified assumptions about the role of the state in inducing ecological transformation and for under-theorising the role that capitalist eco-efficiency and rationalisation can play in environmental reform. The argument that, through ecological modernisation, industrial society has begun to address the environmental problem is also seen as reductionist. It reduces the environmental problem to concerns about resource inputs, waste and pollutant emissions. This neglects the 'emancipatory' aims that featured in the environmental debates of the 1970s and 1980s in Europe (Blühdorn and Welsh, 2008). Environmentalism points to the aesthetic and moral values that ground our relationship with nature, nat-ural systems and environment, and calls for an ecological transformation of society based on major value changes. Focusing on the industrial nature of modernity means that the wider dimensions of the problem are not addressed. In short, ecological modernisation theory remains silent on issues of social justice, the distribution of wealth and on society–nature relations.

Despite disagreement about the merits of ecological modernisation as a theory, ecological modernisation *as a strategy of reform* offers distinc-tive advantages to advanced industrial society. To many policy makers it offers the hope that the environmental problem can be addressed with-out having to redirect the course of societal development (Blühdorn and Welsh, 2008; Spaargaren et al., 2000: 64). This is because it places stress on steering businesses towards eco-efficient practices, which do not undermine 'competitiveness'. This has allowed ecological moderni-sation to become the policy approach of choice in interpreting and implementing the goal of sustainable development (Barry, 2010). It has thus been adopted in government programmes in Germany, the

Netherlands, Sweden, Japan and the European Union (Baker, 2007; Dryzek, 1997).

As such, it has a 'seductive appeal', because belief in an efficiency response minimises the degree of social and cultural change that is necessary, especially in the high consumption societies of the West (Von Weizäcker et al., 1997). In this sense, it has been called 'a discourse of reassurance' (Dryzek, 1997; Blühdorn, 2007). It provides reassurance at several levels (Blühdorn, 2000). First, it supports the notion of rational progress and the continuity of our established patterns of social organisation and societal development. Second, it restores confidence in the power of the established political, economic and administrative system to respond effectively and efficiently to the negative 'externalities' of its economic model. Finally, it promises the continuity of modernity, with its principles of growth, profit and consumerist views of human welfare (Baker, 2007; Blühdorn, 2008). Unlike the reforms envisaged in Brundtland, or required in the stronger models of sustainable development, the strategy of ecological modernisation attempts to pre-empt any fundamental ideological conflict over modernity.

Distinguishing sustainable development and ecological modernisation

There is a tendency in the literature to approach the classic Brundtland formulation of sustainable development as if it were an empty conceptual shell, to be filled by whatever characteristics or variables are deemed appropriate to the political, social and cultural context within which it is applied. This tendency to conceptual pragmatism has allowed the term ecological modernisation to be used as if it were synonymous with the term sustainable development.

However, the strategy of ecological modernisation stands, we argue, in sharp contrast to the discourse on sustainable development, especially that proposed by the classic Brundtland formulation. To begin with, the wide-ranging tasks associated with the promotion of sustainable development, as detailed in Agenda 21, are reformulated under the ecological modernisation strategy as the technical, managerial task of 'decoupling' through eco-efficiency. Further, ecological modernisation still frames nature as a 'standing reserve' of exploitable resources. This does not challenge the Western economic development model either to limit growth or change existing patterns of high consumption. This fails to address the basic ecological contradiction in capitalism – that it requires constant expansion of consumption in a world characterised by finite resources. As the Brundtland Report has forcefully argued, promoting sustainable development in an ecological system characterised by finite resources requires a reduction in growth in the high consumption societies of the North, in order to make way for 'ecologically legitimate' development in the South. As such, Brundtland offers a strong model of sustainable development, premised on the belief that the industrialised world, viewing development primarily in terms of ever increasing material consumption, consumes in excess of its basic needs. This excess threatens the planet's ecological resource base and planetary system health. While the Brundtland understanding of sustainable development is open to different interpretations, we argue that at its core, it challenges the industrialised world to keep consumption patterns within the bounds of the ecologically possible and set at levels to which all can reasonably aspire. This requires changes in the understanding of well-being and what is needed to live a good life. Such change allows for necessary development in the South, in order to bring positive environmental, as well as social and economic, benefits (WCED, 1987: 51).

Furthermore, ecological modernisation places almost exclusive emphasis on technology and economic entrepreneurs as determinants of social change (Christoff, 1996). Social change, especially for Brundtland, is a process involving a wider set of actors and the promotion of sustainable development involves the engagement with a deeper set of principles. These include the normative principles of inter- and intra-generational equity. The social justice aspects of sustainable development are ignored by ecological modernisation (Langhelle, 2000). The result is that ethical considerations are sidelined by attention to efficiency procedures.

Thus, the promotion of sustainable development cannot be achieved by seeking to promote an ecologically modernised economy. Ecological modernisation, rather than promoting sustainable development, seeks to encourage growth as a solution to the planet's ecological crisis. As such, the two discourses are incompatible. In short, while the Brundtland formulation shares a belief in the advantages of technological development with the ecological modernisation strategy, the former envisages a common future built on fundamental processes of change. The Brundtland formulation of sustainable development requires changes not just at the technological and institutional levels, but also demands more fundamental social, economic, cultural and lifestyle changes, particularly in the high consumption societies of the West.

Conclusion

The popularisation of the discourse on sustainable development following the publication of the Brundtland Report should be understood in the larger context of growing ecological awareness, widening disparities between the North and South, especially the debt burden that has trapped many developing countries in poverty, and the rising opposition to the negative consequences of economic growth in some advanced capitalist nations. Because it offers a way of reconciling economic development and ecological protection, while being flexible enough to allow governments to take account of different political cultures, policy contexts and socio-economic needs, the Brundtland formulation of sustainable development has been able to become *the* guiding principle of international environmental negotiation and governance practice. However, not all environmentalists have endorsed the concept of sustainable development. Some reject the project on philosophical grounds, claiming that its underlying motives are too anthropocentric; for others, it is rejected on political grounds, either because it is not radical enough, or because it is too far too radical altogether.

The Brundtland approach is built upon a belief in the common heritage of humankind, trust in our technology and optimism about our willingness to engage collectively in the protection of our common future. At its core is a belief in the limits to growth, a belief bolstered by new research on planetary limits and ecological overshoot. The normative principles that have come to be associated with sustainable development have led to the elaboration of specific rights and obligations for states and have acted as guidelines for international and national environmental regulations and laws. These normative principles have widened the scope of those to whom environmental obligations are owed beyond states and beyond present generations. They also place obligations upon the individual, especially as a consumer.

These normative dimensions stretch their demands into the policy making system, or system of governance, to require that policy making processes become more inclusive and more gender sensitive, and that they facilitate the fuller participation of societal actors in decision making that affects their future. This points us to the task of the next chapter, which is to discuss the governance dimensions of sustainable development.

Summary points

- Sustainable development is about the long-term transformation of basic aspects of the present industrial economic system. Promoting sustainable development is about the construction of a new development paradigm, framed within the ecological limits of the planet.
- Sustainable development is a development paradigm based on principles of justice and equity and on limits to growth.
- There has been a proliferation in meaning and in application of the term, making the search for a precise definition a frustrating affair. More importantly, this is also a mistaken endeavour in that it misunderstands the function of political concepts.
- The policy imperatives associated with promoting sustainable development can be seen in terms of a ladder, the rungs of which range from a weak to an ideal form.
- At either end of such a continuum or ladder, the sustainable development project is rejected, but for entirely opposite reasons.
- Sustainable development has come to be associated with several normative principles that now guide environmental management practices and international law, but increasingly stretch into other issue areas.
- Promoting sustainable development also requires new governance practices.

Further reading

Atkinson, G., Dietz, S. and Neumayer, E. (eds) (2007), *Handbook of Sustainable Development* (Cheltenham: Edward Elgar).

Dobson, A. (1998), *Justice and the Environment: Conceptions of Environmental Sustainability and Dimensions of Social Justice* (Oxford: Oxford University Press).

Dresner, S. (2008), *The Principles of Sustainability* (London: Earthscan, 2nd edn).

Lafferty W. M. and Meadowcroft, J. (eds) (1996), *Democracy and the Environment: Problems and Prospects* (Cheltenham: Edward Elgar).

Rogers, P. P., Jalal, K. F. and Boyd, J. A. (2008), *An Introduction to Sustainable Development* (London: Earthscan).

World Commission on Environment and Development (WCED) (1987), *Our Common Future* (Oxford: Oxford University Press).

References

Acterberg, W. (1993), 'Can Liberal Democracy Survive the Environmental Crisis? Sustainability, Liberal Neutrality and Overlapping Consensus', in A. Dobson and P. Lucardie (eds), *The Politics of Nature: Explorations in Green Political Theory* (London: Routledge), 62–81.

Arrow, K., Bolin, B., Costanza, R., Dasgupta, P., Folke, C., Holling, C. S., Jansson, B-O., Levin, S., Maler, K-G., Perrings, C. and Pimentel, D. (1995), 'Economic Growth, Carrying Capacity, and the Environment', *Science*, 268 (April): 520–521.

Baker, S. (2007), 'Sustainable Development as Symbolic Commitment: Declaratory Politics and the Seductive Appeal of Ecological Modernisation in the European Union', *Environmental Politics*, 16 (2): 297–317.

Baker, S. (2012), 'Climate Change, The Common Good and the Promotion of Sustainable Development', in J. Meadowcroft, O. Langhelle and A. Ruund (eds), *Governance, Democracy and Sustainable Development* (Cheltenham: Edward Elgar), 249–271.

Baker, S., Kousis, M., Richardson, R. and Young, S. (eds) (1997), *The Politics of Sustainable Development: Theory, Policy, and Practice within the European Union* (London: Routledge).

Ball, T. (2000), 'The Earth Belongs to the Living: Thomas Jefferson and the Problem of Intergenerational Relations', *Environmental Politics*, 9 (2): 61–77.

Barry, J. (2010), 'Towards a Model of Green Political Economy: From Ecological Modernisation to Green Security', in L. Leonard and J. Barry (eds), *Global Ecological Politics* (Bingley, Yorkshire: Emerald Group), 109–128.

Beck, U. (1992), *Risk Society: Towards a New Modernity* (London: Sage).

Blühdorn, I. (2000), *Post-Ecological Politics: Social Theory and the Abdication of the Ecological Paradigm* (London: Routledge).

Blühdorn, I. (2007), 'Sustaining the Unsustainable: Symbolic Politics and the Politics of Simulation', *Environmental Politics*, 16 (2): 251–275.

Blühdorn, I. and Welsh. I. (eds) (2008), *The Politics of Unsustainability: Eco-Politics in the Post-Ecologist Era* (London: Routledge).

Christoff, P. (1996), 'Ecological Modernisation, Ecological Modernities', *Environmental Politics*, 5: 476–500.

Cutte, S. L. (2006), *Hazards, Vulnerability and Environmental Justice* (London: Routledge).

Daly, H. E. (1977), *Steady-State Economics* (San Francisco: Freeman).

Daly, H. E. and Cobb, J. B. (1990), *For the Common Good* (London: Green Print).

Dobson, A. (1996), 'Representative Democracy and the Environment', in W. M. Lafferty and J. Meadowcroft (eds), *Democracy and the Environment: Problems and Prospects* (Cheltenham: Edward Elgar), 124–139.

Dobson, A. (1998), *Justice and the Environment: Conceptions of Environmental Sustainability and Dimensions of Social Justice* (Oxford: Oxford University Press).

Dresner, S. (2008) [2002], *The Principles of Sustainability* (London: Earthscan, 2nd edn).

Dryzak, J. S. (1992), 'Ecology and Discursive Democracy: Beyond Liberal Capitalism and the Administrative State', *Capitalism, Nature, Socialism*, 10: 18–42.

Dryzek, J. S. (1997), *The Politics of the Earth: Environmental Discourses* (Oxford: Oxford University Press).

Gouldson, A. and Murphy, J. (1996), 'Ecological Modernisation and the European Union', *Geoforum*, 27 (1): 11–21.

Greenpeace USA (2014), 'Radioactive Waste: Nuclear Reactors Create Radioactive Waste That Will Remain Hazardous for 240,000 Years', available online at www.greenpeace.org/usa/en/campaigns/nuclear/safety-and-security/radioactive-waste/.

Hajer, M. (1995), *The Politics of Environmental Discourse: Ecological Modernization and the Policy Process* (Oxford: Oxford University Press).

Huber, J. (2004), *New Technologies and Environmental Innovation* (Cheltenham: Edward Elgar).

Iles, A. (2003), 'Rethinking Differential Obligations: Equity Under the Biodiversity Convention', *Leiden Journal of International Law*, 16: 217–251.

International Union for the Conservation of Nature and Natural Resources (IUCN) (1980), *World Conservation Strategy: Living Resources Conservation for Sustainable Development* (Gland, Switzerland: IUCN).

Jacobs, M. (1991), *The Green Economy: Environment, Sustainable Development and the Politics of the Future* (London: Pluto Press).

Jacobs, M. (1995), 'Justice and Sustainability', in J. Lovenduski and J. Stanyer (eds), *Contemporary Political Studies* (Belfast: Proceedings of the Political Studies Association of the UK, vol. 3), 1470–1485.

Jagers, S. C. (2002), 'Justice, Liberty and Bread – for All? On the Compatibility between Sustainable Development and Liberal Democracy', Department of Political Science, Gothenburg University, Gothenburg Studies in Politics, paper no. 79.

Jänicke, M. and Klaus, J. (2012), *Environmental Governance in Global Perspective: New Approaches to Ecological and Political Modernisation* (Shandong University Press).

Johnson, L. E. (2003), 'Generations and Contemporary Ethics', *Environmental Values*, 12: 471–487.

Katz, E., Light, A. and Rothenberg, D. (2000), *Beneath the Surface: Critical Essays in the Philosophy of Deep Ecology* (Cambridge, MA: MIT Press).

Lafferty, W. M. (1995), 'The Implementation of Sustainable Development in the European Union', in J. Lovenduski and J. Stanyer (eds), *Contemporary Political Studies* (Belfast: Proceedings of the Political Studies Association of the UK, vol. 1), 223–232.

Lafferty, W. M. and Meadowcroft, J. (2000), 'Introduction', in W. M. Lafferty and J. Meadowcroft (eds), *Implementing Sustainable Development:*

Strategies and Initiatives in High Consumption Societies (Oxford: Oxford University Press), 1–22.

Langhelle, O. (2000), 'Why Ecological Modernisation and Sustainable Development Should Not Be Conflated', *Journal of Environmental Policy and Planning*, 2: 303–322.

Lélé, S. (1991), 'Sustainable Development: A Critical Review', *World Development*, 19 (6): 607–621.

De Lucia, V. (2012), 'Common but Differentiated Responsibility', *Encyclopaedia of the Earth*, available online at www.eoearth.org/view/article/151320/.

Martinez-Alier, J. (1999), 'The Socio-Ecological Embeddedness of Economic Activity: The Emergence of a Transdisciplinary Field', in E. Becker and T. Jahn (eds), *Sustainability and the Social Sciences: A Cross-Disciplinary Approach to Integrating Environmental Considerations into Theoretical Reorientation* (London: Zed Books), 112–140.

Mcmanus, P. and Haughton, G. (2006), 'Planning with Ecological Footprints: A Sympathetic Critique of Theory and Practice', *Environment and Urbanization*, 18: 113–127.

Meadowcroft. J. (1999), 'Planning for Sustainable Development: What Can Be Learnt from the Critics?' in M. Kenny and J. Meadowcroft (eds), *Planning Sustainability* (London: Routledge), 12–38.

Meadows, D. H., Meadows, D. L., Randers, J. and Behrens, W. W. (1972), *The Limits to Growth: A Report for the Club of Rome's Project on the Predicament of Mankind*. Club of Rome Report (London: Pan Books).

Meadows, D. H., Randers, J. and Meadows, D. L. (2004), *The Limits to Growth: The 30-year Update* (London: Routledge).

Mol, A. (1996), 'Ecological Modernisation and Institutional Reflexivity: Environmental Reform in the Late Modern Age', *Environmental Politics*, 5 (2): 302–323.

Mol, A. P. J. (1999), 'Ecological Modernization and the Environmental Transition of Europe: Between National Variations and Common Denominators', *Journal of Environmental Policy and Planning*, 1: 167–181.

Mol, A. P. J. (2000), 'The Environmental Movement in an Era of Ecological Modernization', *Geoforum*, 31 (1): 45–56.

Mol, A. and Sonnenfeld, D. (eds) (2000), *Ecological Modernisation around the World: Perspectives and Critical Debates* (London: Frank Cass).

Murphy, J. (2000), 'Ecological Modernisation', *Geoforum*, 31 (1): 1–8.

Naess, A. (1989), *Ecology, Community and Lifestyle: Outline of an Ecosophy*, translated, edited and with an introduction by David Rothenberg (Cambridge: Cambridge University Press).

Ordway, S. (1953), *Resources and the American Dream, Including a Theory of the Limit of Growth* (New York: Ronald Press Company).

O'Riordan, T. (1981), *Environmentalism* (London: Pion Press, 2nd edn).

O'Riordan, T. (1985), 'What Does Sustainability Really Mean? Theory and Development of Concepts of Sustainability', proceedings of the conference

'Sustainable Development in an Industrial Economy', held at Queens' College, Cambridge, 23–25 June (Cambridge: UK Centre for Economic and Environmental Development).

Osborn, F. (1948), *Our Plundered Planet* (London: Faber and Faber).

Osborn, F. (1953), *The Limits of the Earth* (Boston: Little, Brown).

Ostrom, E., Gardner, R. and Walker, J. (1994) *Rules, Games, and Common-pool Resources* (Ann Arbor: University of Michigan Press).

Paehlke, R. C. (1996), 'Environmental Challenges to Democratic Practice', in W. M. Lafferty and J. Meadowcroft (eds), *Democracy and the Environment: Problems and Prospects* (Cheltenham: Edward Elgar), 18–38.

Paehlke, R. (2001), 'Environmental Politics, Sustainability and Social Science', *Environmental Politics*, 10 (4): 1–22.

Panayotou, T. (1993), *Empirical Tests and Policy Analysis of Environmental Degradation at Different Stages of Economic Development* (Geneva: International Labour Office, Working Paper WP238, Technology and Employment Programme).

Pearce, D. (1994), *Blueprint: Measuring Sustainable Development* (London: Earthscan).

Pearce, D. (1995), *Blueprint 4: Capturing Global Environmental Value* (London: Earthscan).

Pearce, D. and Barbier, E. B. (2000), *Blueprint for a Sustainable Economy* (London: Earthscan).

Pearce, D., Markandya, A. and Barbier, E. B. (1989), *Blueprint for a Green Economy* (London: Earthscan).

Pepper, D. (1993), *Eco-Socialism: From Deep Ecology to Social Justice* (London: Routledge).

Pepper, D. (1996), *Modern Environmentalism: An Introduction* (London: Routledge).

Pepper, D. (1998), 'Sustainable Development and Ecological Modernization: A Radical Homocentric Perspective', *Sustainable Development*, 6: 1–7.

Prezzey, J. (1989), *Definitions of Sustainability*, UK Centre for Economic and Environmental Development, Working Paper no. 9.

Rawls, J. (1971), *A Theory of Justice* (Cambridge, MA: Harvard University Press).

Rockström, J., Steffen, W., Noone, K., Persson, A., Chapin, F. S., Lambin, E., Lenton, T. M., Scheffer, M., Folke, C., Schellnhuber, H., Nykvist, B., De Wit, C. A., Hughes, T., van der Leeuw, S., Rodhe, H., Sörlin, S., Snyder, P. K., Costanza, R., Svedin, U., Falkenmark, M., Karlberg, L., Corell, R. W., Fabry, V. J., Hansen, J., Walker, B., Liverman, D., Richardson, K., Crutzen, P. and Foley, J. (2009), 'Planetary Boundaries: Exploring the Safe Operating Space for Humanity', *Ecology and Society*, 14 (2): 32, www.ecologyandsociety.org/vol14/iss2/art32/

Roseland, M. (2000), 'Sustainable Community Development: Integrating Environmental, Economic, and Social Objectives, *Progress in Planning*, 54: 73–132.

Schaefer, F., Luksch, U., Steinbach, N., Cabeça, J. and Hanauer, J. (2006), 'Ecological Footprint and Biocapacity: The World's Ability to Regenerate Resources and Absorb Waste in a Limited Time Period' (Luxembourg: Office for Official Publications of the European Communities), available online at http://epp.eurostat.ec.europa.eu/cache/ity_offpub/ks-au-06-001/en/ks-au-06-001-en.pdf

Simmonis, U. (1989), 'Ecological Modernization of Industrial Society: Three Strategic Elements', *International Social Science Journal*, 121: 347–361.

Simon, J. L. and Kahn, H. (1984), *The Resourceful Earth: A Response to Global 2000* (Oxford: Blackwell).

Sorrell, S. (2007), *The Rebound Effect: An Assessment of the Evidence for Economy-Wide Energy Savings from Improved Energy Efficiency*. A report produced by the Sussex Energy Group for the Technology and Policy Assessment function of the UK Energy Research Centre, October, available online at www.ukerc.ac.uk/Downloads/PDF/07/0710ReboundEffect/0710ReboundEffectReport.pdf

Sourcewatch (2009), 'Heidelberg Appeal', available online at www.source-watch.org/index.php?title=Heidelberg_Appeal

Spaargaren, G., Mol, A. P. J. and Buttel, F. H. (eds) (2000), *Environment and Global Modernity* (London: Sage), 209–228.

SRC (Stockholm Resilience Centre) (2009), *The Nine Planetary Boundaries* (Stockholm: Stockholm Resilience Centre), available online at www.stock-holmresilience.org/21/research/research-programmes/planetary-boundaries/planetary-boundaries/about-the-research/the-nine-planetary-boundaries.html

Sylvan, R. and Bennett, D. H. (1994), *The Greening of Ethics: From Anthropocentrism to Deep Green Theory* (Cambridge: White Horse Press).

Wackernagel, M. (2001), 'Advancing Sustainable Resource Management: Using Ecological Footprint Analysis for Problem Formulation, Policy Development and Communication, Redefining Progress', Oakland, CA: Redefining Progress. Available online at http://ec.europa.eu/environment/enveco/waste/pdf/wackernagel.pdf

Wackernagel, M. (2014), 'Ecological Footprint Policy? Land Use as an Environmental Indicator', *Journal of Industrial Ecology*, 18 (1): 20–23.

Wackernagel, M. and Rees, W. (1996), *Our Ecological Footprint: Reducing Human Impact on the Earth* (Gabriola Island, BC: New Society Publishers).

Von Weizäcker, M., Lovins, E., Hunter, A. B. and Lovins, L. (1997), *Factor Four: Doubling Wealth – Halving Resource Use* (London: Earthscan).

World Commission on Environment and Development (WCED) (1987), *Our Common Future* (Oxford: Oxford University Press).

The governance of sustainable development

Key issues

- Concept of governance
- Governance styles
- Sustainable development and governance
- Reflexive governance
- Normative requirements.

Introduction

Sustainable development has been described as a political concept replete with governance questions (Farrell et al., 2005: 143). This is because its pursuit raises a series of complex questions about how society can be steered towards an overarching objective – promoting sustainable development – that embraces social, economic and political goals now and into the future (ibid.). In putting the commitment to sustainable development into practice, inputs from an array of societal actors and institutions are required. This chapter explores the role that governance processes play in managing these inputs, and thus in steering society along sustainable development trajectories.

The concept of governance has proved very important for understating how society addresses the challenges of sustainable development. First, scholars have investigated the consequences of using different styles of governance, in particular markets and networks, for the delivery of sustainable development objectives (Jordan, 2008). Second, attention has been paid to identify how society ought to be governed in pursuit of sustainable development. We examine both of these approaches in this chapter. The chapter begins by discussing the concept of governance in general terms. This is followed by a more in-depth investigation of what specific governance requirements are needed in pursuit of sustainable development, introducing the reader to the notion of reflexive governance. The chapter then investigates claims as to the governance requirements that ought to be introduced if society is to move along a more sustainable development trajectory.

The concept of governance

Governance deals with managing, steering and guiding action in the realm of public affairs, especially in relation to public policy decision making (Baker and Eckerberg, 2008a). As such, it is usefully employed in a wide range of discussions about the responsibilities of public authorities in modern society (Meadowcroft, 2007a). Governance involves steering and control from the centre, whereby the state defines goals and sets priorities for public action (Pierre, 2000). However, the state does not act alone, and the term 'governance' seeks to capture the interactions between government, other public bodies, and private sector and civil society actors aimed at addressing public policy issues (Meuleman, 2008).

> Broadly speaking, 'governance' deals with managing, steering and guiding actions in the realm of public affairs.
>
> (Pierre, 2000)

Governance is typically grouped into three 'ideal types': hierarchies, markets and networks. Hierarchies involve traditional forms of top-down command and control by the state through, for example, legislation and regulation; fiscal and monetary measures (macroeconomic policy); strategic planning; and political brokerage. Governance can occur through market-based forms of resource allocation. It can also involve various forms of public–private collaboration, operating through organised networks (Pierre and Peters, 2000). Participation of actors drawn from civil society is also seen as an important element.

Countries differ in the ways in which they combine these different styles of governance and in how they co-ordinate activities across these styles. Analysis of state steering through these ideal types focuses on the extent to which a particular state or international agency has the political and institutional capacity to steer, the tools that it uses and whether the interests of different stakeholders are taken into account in policy decisions (Pierre, 2000).

Box 3.1

Governance: the ideal types

1. **Hierarchy**: top-down control and regulation by the state, with legal rules backed by (often criminal) sanctions.
2. **Market governance**: the use of a range of markets tools, such as prices, taxes, subsidies or permits, to provide incentives for action.
3. **Network governance**: various forms of public–private collaboration in policy making and delivery.

(*Sources*: Pierre and Peters, 2000; Black, 2008)

Sustainable development and governance

Sustainable development has been promoted using different styles of governance. Each style brings advantages and disadvantages, and in this section, attention is turned to discussing these, including through drawing upon a range of illustrative examples.

Hierarchical steering

The promotion of sustainable development requires state steering, or what can be referred to as 'goal directed intervention' by government or other state-like actors, such as the EU or the UN. Such actors have several governance functions and chief among these is their engagement in a continuous process of making, amending and enforcing regulation and legislation. Regulations, the classic instrument of hierarchical governance, prescribe both the means and ends of sustainable development policy. They provide a mechanism for formalising and codifying strategies and policies for sustainable development. Furthermore, enforcement of that legislation is critical for ensuring the delivery of strategies. Agenda 21 recognises this role when it calls upon states to develop and enforce integrated, sanction-obliging and effective laws and regulations that comply with well conceived social, ecological, economic and scientific principles (Lafferty and Eckerberg, 1998), as discussed in Chapter 8.

In addition to this 'command and control' task, the state needs to stimulate and facilitate a process of societal change in pursuit of sustainable development, including in relation to the adoption of new patterns of behaviour, for example, in matters of consumption. The state thus acts as a key agent of social change. This involves, for example, encouraging householders to adopt energy efficiency measures within the home through the provision of public subsidies and grants; or helping to reduce dependency upon cars through the enhancement of public transport provision. Governments have also to play a welfare function, such as supporting groups in society that are particularly vulnerable, for example in the face of climate change. This means ensuring that the health care system enhances its capacity to address the health risks associated with climate change; addressing problems associated with sea level rise, such as through the provision of flood defences or ensuring access to home insurance when the commercial sector is no longer willing to protect householders against the damages caused by flooding. Indeed, it has been argued that the continued existence, even pre-eminence, of a pro-active state *is essential* if the inter-generational and intra-generational aspects of sustainable development are to be respected (French, 2002).

The state also has broader-reaching and longer-term tasks, including in relation to the reordering of social institutions in ways that help the pursuit of sustainable development. This includes, for example, ensuring that the education system facilitates the emergence of new sustainable development values and norms. In addition, the rise of global environmental problems, such as biodiversity loss and climate change, also means a heightened role for the state in global environmental governance regimes, such as those operating through the UN. These regimes rely heavily upon the state acting as a cornerstone of international negotiations and playing a key role in implementing policies, such as those aimed at climate change mitigation and adaptation.

The state has also to intervene in the face of market failure, that is, where markets do not allocate scarce resources efficiently or in ways that achieve the highest possible social welfare outcomes. The market, left on its own without appropriate institutional direction, cannot sufficiently achieve public goals, such as poverty eradication and environmental conservation. It thus falls to the state to provide the legal and financial framework for the provision of public goods, such as clean water and a healthy environment (Stiglitz, 2000). Furthermore, the market cannot meet wider societal objectives, such as the equity aspects of sustainable development (French, 2002).

Economic activity also leads to so-called 'negative externalities', that is, by-products of production activities that are not reflected in price, such as environmental pollution. The fact that private economic activity produces negative externalities, while society as a whole has to struggle with the damaging impact of these externalities, serves to heighten the role of the state (Bleischwitz, 2003). This includes, for example, its role in enforcing the polluter pays principle and strict environmental standards. The state has also to deal with adaptations and information deficits (Stiglitz, 2000). Adaptation deficits refers to the speed at which markets and firms adapt to new circumstances, and this is particularly apt for our discussion. Processes of change are sometimes extremely slow and economic actors are often reluctant to risk being the 'first mover' lest it bring competitive disadvantage. Here governments are needed to 'kick start' such adaptive responses, for example in relation to climate change, including in the energy sector. The state, for example, played a key role in establishing LA21 processes and in facilitating information sharing across municipalities in order to help them devise local sustainable development actions (Lafferty and Eckerberg, 1998), as discussed in Chapter 8.

When efforts to promote sustainable development are viewed through the lens of market failure, it becomes clear that markets need hierarchical governance in order to function properly (Bleischwitz, 2003: 13). As Baker and Eckerberg have argued:

> The failure of the market to protect the global environmental commons still remains a solid ground for government intervention. Market failure is particularly evident as society pursues the promotion of sustainable futures, because markets work on short-term horizons, deal with profits and promote individual advancement. The pursuit of sustainable development requires transition to a society whose policy processes guard the future, promote equity and pursue the common good.
>
> (Baker and Eckerberg, 2008b: 227)

It is for this complex variety of reasons that many governments have adopted a proactive approach to sustainable development planning. We see this in particular in the formulation and implementation of national sustainable development strategies.

In summary, we can see that the governance of sustainable development gives a heightened and positive role for the state in directing societal

Box 3.2

National sustainable development strategies

UN encouragement

The development of national sustainable development strategies (NSDS) has been encouraged by the UN ever since the Rio Earth Summit of 1992, as reflected in Chapter 8 of Agenda 21. Many states have adopted such NSDS as part of their commitment to the Johannesburg Plan of Implementation. The EU has also followed this trend, as discussed in Chapter 9.

The 2002 World Summit on Sustainable Development reiterated a call to all countries to make progress in the formulation and elaboration of national strategies for sustainable development.

National SD strategy

A strategy is not simply a planning document. It is a continuing and adaptive process of strategic and co-ordinated action. It involves ongoing monitoring, learning and improvements.

Why a strategy?

Traditional plans are seen as lacking the flexibility needed to promote sustainable development. Unlike plans, a strategy rarely implies initiating a completely new or stand-alone strategic planning project. Strategies are considered particularly apt for efforts to promote sustainable development. This is because they encourage innovation and adaptive change, in relation to both structures (such as institutions) and processes (such as planning).

Evidence

A study of nineteen countries found four main types of national sustainable development strategies: comprehensive and multi-dimensional (e.g. Germany); cross-sectoral (e.g. Cameroon); sectoral (e.g. Canada); and integration of sustainable development into existing planning processes (e.g. Mexico).

The survey found increased public participation in the formulation and implementation of NSDS. It also found that countries use a mix of policy initiatives, although the use of economic policy initiatives is still in its infancy.

However, findings also show that strategies often pursued the path of least political resistance and they were often poorly monitored. Furthermore, the choice of strategy type was at times dictated by capacity constraints. NSDS continue to be peripheral in national budgetary priorities and are only weakly linked to local sustainable development strategies. Significant challenges remain in terms of building trust among stakeholders, providing sufficient time for the participatory process and strengthening the capacity of civil society to engage.

(*Sources*: OECD, 2009; Swanson et al., 2004; Meadowcroft, 2007b)

change, even if this steering remains open-ended and, as we will see below, increasingly collaborative in nature.

Challenges to traditional governance

Despite the obvious importance of hierarchical steering, traditional methods of governance have come under challenge from the increasingly complex, dynamic and interdependent nature of contemporary policy making (Kooiman, 1993; 2000). This has meant that no single actor, including the state, has the knowledge and resource capacity to tackle problems unilaterally (Kooiman, 2000). A good example is provided by the emergence of complex, global environmental problems, such as climate change, as discussed in Chapter 6. Globalisation and Europeanisation have also contributed to the changing context of state policy making (Kohler-Koch and Rittberger, 2006). These processes have resulted in actors, agencies and processes operating at the supranational levels having growing influence on domestic policy, as discussed in Chapter 5 on the UN and Chapter 9 on the EU. Europeanisation has also seen many member states adopting new programmes and plans, including financial packages, aimed at enhancing the capacity of sub-national, regional and local authorities to act in pursuit of sustainable development, as discussed in Chapter 8.

Traditional methods of governance were also strongly challenged during the neoliberal wave of reforms that swept many industrialised states in the latter half of the twentieth century. The spread of neoliberalism brought new approaches to public service provision and delivery, including privatisation and the increased use of public–private partnerships (Kooiman, 2000: 15). As such, governments 'offloaded' functions, either to the market or to civil society groups and even to the individual. Many states also make increased use of 'softer' steering instruments, such as voluntary agreements. These changes continue to be driven by ideological beliefs as to the limited efficiency and effectiveness of the state. The state's capacity to intervene was also restricted by a questioning public and their scepticism in the face of technocratic, state-centred approaches to environmental risk regulation and management (Gunningham, 2009). This often erupted into civil action to overturn public policy decisions; for example, in relation to nuclear power, road building and airport extensions, or to demand more environmentally sensitive regulation of toxic by-products of industrial production.

From this discussion, we can see that the term 'governance' is used to capture important changes in contemporary policy making. However, we also see that term is used as a normative prescription, to describe the type and style of steering that should be adopted to achieve a preferred societal goal (Kohler-Koch and Rittberger, 2006). As a normative prescription, governance can be used in a neoliberal sense by those seeking to reduce the part played by governments in favour of enhancing reliance upon market mechanisms and corporate interests (Kemp et al., 2005: 17).

However, matters are not as simple as that: during the latter half of the twentieth century, the term governance came to be associated with the call for the promotion of democratic pluralism, especially greater participation of societal actors in policy making, as discussed in Chapter 11.

While this range of different influences has acted to different degrees on different states, the result of these various pressures is such that policy making has now come to involve the enhanced engagement of economic and social actors in policy making and delivery (Lafferty, 2004). Governance now takes place in the context of the displacement of political and institutional policy capacity *downwards* in the political system to regional and local authorities, *outwards* to social and economic agencies and NGOs and *upwards* to trans-national institutional systems such as the EU and the UN (Painter and Pierre, 2005: 1). As a result of these changes the term 'governance' is employed in contemporary discussions to capture the use of governing mechanisms that do not rest on the authority and sanctions of government alone (Gouldson and Bebbington, 2007: 7). We must be careful, however, to continue to acknowledge the important role of the state in the pursuit of sustainable development, even if that role is now enacted in conjunction with other actors. The state remains a key player in initiating and co-ordinating the sustainable development planning processes. This includes establishing framework legislation, developing strategies, initiating funding mechanisms, and spurring the sub-national levels, as well as economic actors and civil society, to engage in appropriate policies and actions.

Market governance for sustainable development

The latter half of the twentieth century saw significant increases in the range of market-based instruments involved in environmental management, and they now extend from subsidies through to emission charges and tradable permits (OECD, 2003). Their use within the environmental policy arena is now a key feature of contemporary sustainable development governance (Jordan, 2008; Wurzel et al., 2012), as discussed in Chapter 6. Their growing use emphasises the closer co-operation that now exists between public and private actors in the formulation and implementation of sustainable development policy.

Market-based environmental policy instruments use financial incentives, such as subsidies, taxes, price differentials or market creation, to bring about change in technology, behaviour or products to promote positive environmental outcomes.

Market-based instruments aim to ensure that producers and consumers take account of the environmental implications of their actions. They are seen as offering more freedom and incentives to adapt behaviour, when compared to command and control approaches. Using the tax system, for example, to address environmental externalities can be an effective way of achieving behavioural change. First, consumers will be encouraged to forsake higher priced goods or services for lower priced alternatives that are less environmentally damaging. Second, businesses will be encouraged to move away from using environmentally damaging production methods and resources. Furthermore, as polluters are required to pay tax for pollution, this encourages technological developments.

The use of market instruments is also supported by the argument that they can relieve regulatory burden by reducing the overall volume of legislation affecting economic actors. Here, the overlap can be clearly seen between, on the one hand, the rise of neoliberalism that emphasises belief in the efficiency of markets and their associated economic instruments when compared to interventionist measures; and on the other hand, the growth in use of new forms of market governance (Lenschow, 1999: 40–41).

Another example of the use of market-based instruments is the growing use of payments for ecosystem services (PES), discussed in Chapters 7 and 12. The use of market-based instruments is closely tied to the use of voluntary agreements, which are commitments undertaken by industry to pursue actions leading to environmental improvements (see OECD, 2003). Their popularity stems principally from the perception that they provide flexible and inexpensive ways to manage environmental problems. They stress the virtues of co-operative governance. Voluntary agreements first appeared in Japan in the 1960s and then later in France

Box 3.3

Market-based instruments

Advantages

- Cost effective, as they offer lowest cost to society to control pollution;
- Decrease externalities, or make the decision makers bear the cost;
- Provide incentives for technological innovation and diffusion;
- Generate revenue for government.

Disadvantages

- Institutional constraints, such as underfunding, inexperience, unclear jurisdiction and lack of political will, limit effective implementation;
- Administrative intensity, as they require monitoring, legal design, public consultation, enforcement and collection;
- They are potentially regressive through increasing the cost of goods and services; firms could pass on charges to consumers through higher prices;
- They could have perverse effects, e.g. water pollution charges may cause more pollution if based on the effluent concentrations and not on total pollution load.

A good example of the positive use of market-based instruments for environmental governance is provided by the introduction of a tax on plastic carrier bags in Ireland.

and are in growing use in the EU and the USA. One important form of voluntary agreement is so-called negotiated agreements, which are formal 'contracts' between industry and public authorities aimed at addressing particular environmental problems. These are negotiated between environmental regulators and industry associations in specific polluting sectors and/or geographic areas (Bailey, 2007). They deal with a diverse range of matters, such as climate change mitigation and adaptation, reduction of air pollutants and the use of recycled product and materials.

Despite strong arguments in favour of voluntary agreements, concerns have been raised about whether this governance tool can adequately protect the environment, let alone promote sustainable development. Research has found that reliance on voluntary agreements can lead to reduced levels of environmental quality, relative to what might have

Case study 3.1

The Irish plastic bag tax

Background

Plastic bags take between 450 and 1,000 years to degrade and thus form long-lasting litter. They are the most common man-made item seen by sailors at sea. Plastic bags have a major negative impact on wildlife, including ingestion and entanglement that leads to infection, amputation, starvation and death.

Every year 800,000 tonnes of single-use plastic bags are used in the European Union. During the Beachwatch Big Weekend 2011 in the UK, an average of thirty-eight plastic bags for every kilometre of coastline was found.

The issue in Ireland

Plastic bag litter was a visible and persistent component of litter pollution in Ireland, including throughout the countryside and along the coastline. Two characteristics of Ireland are germane here: first, Ireland is a country of frequent high winds, and bags released in the open may travel long distances. Second, Ireland is a country of hedgerows, with relatively small fields enclosed by shrubs, hedges, clumps of trees and ditches, in which discarded plastic bags are trapped and accumulate. This accumulation becomes particularly evident in winter, when the masking effect of deciduous vegetation is absent.

Policy

In March 2002, the Irish government introduced a Plastic Bag Environmental Levy of 15 cents per plastic bag provided to shoppers at point of sale in retail outlets. Retailers must pass on the full amount of the levy as a charge to customers at the checkout. The current levy of 22 cents was introduced in July 2007.

The primary purpose of the levy was to reduce the consumption of disposable plastic bags by influencing consumer behaviour. All levies are remitted into the Environment Fund and are used to support waste management, anti-litter and other environmental initiatives.

The levy on plastic shopping bags has a strong anti-litter and anti-pollution emphasis. The regulations do not distinguish between biodegradable plastic bags and other plastic bags. Biodegradable bags still take a considerable amount of time to degrade.

Outcome

The levy had an immediate effect on consumer behaviour, with a decrease in plastic bag usage from an estimated 328 bags per capita to 21 bags per capita overnight. It cut the use of plastic bags in supermarkets by 90 per cent in just 5 months.

However, such progress was not to endure, as the use of plastic carrier bags began to increase again, resulting in the introduction of an increased levy in 2007.

Alternatives to disposable plastic shopping bags, such as reusable bags, were made available in shops. Consumers have now switched to using these alternatives.

Policy success

The example shows how the introduction of a price signal, in this case a product tax, can influence consumer behaviour significantly.

Research found a strong perception amongst householders surveyed that the effect on the environment has been positive. Many report feeling guilty when they forget to bring their own long-life bag and have to pay the levy! 'We are not aware of another tax that induces such an enthusiasm and affection from those who are liable to pay it.' The retail industry has been very positive in terms of implementation and acceptance.

The Irish case also shows the need for ongoing monitoring and, where necessary, policy adjustments to ensure continued success.

Other examples

Many countries have a levy on plastic bags, while others have an outright ban.

(*Sources*: Convery et al., 2007; Irish Department of Environment, Community and Local Government, 2007; Irish Citizens Information Board, 2011)

been achieved legislatively (Segerson and Miceli, 1998). Their use has also been heavily criticised by environmental NGOs, who argue that the risk of non-compliance is high and that the existing monitoring requirements are based on self-monitoring and hence prone to bias (Euractive, 2008).

Using networks and public–private partnerships

Contemporary governance can also be distinguished from the more hierarchical approach by the enhanced role of policy networks in policy making (Rhodes, 2007). Network governance involves a 'more-or-less fluid set of state and societal actors linked together by specific interest and resource relationships' (Howlett, 2002: 235). Networks consist of actors that have a relatively stable relationship and share a policy interest and are therefore willing to co-operate, including through sharing

Box 3.4

Advantages of networks for the governance of sustainable development

Policy formulation

- Promote consensus building around policy solutions;
- Provide a forum for development of innovative policy solutions;
- Offer opportunities for deliberation on value related issues;
- Help create synergy between different competences and sources of knowledge;
- Support policies that are attuned to local needs and better adapted to local conditions.

Policy implementation

- Prove more cost effective because they combine resources from public, private and voluntary sectors;
- Bring more effective implementation by getting a range of groups on board.

Policy reflection and learning

- Can accommodate change because membership, goals and structures are adaptable;
- Help avoid institutional apathy, typically in political settings with many actors with conflicting interests, especially at global level;
- Facilitate institutionalised learning.

Wider value

- Democratise policy process by making it inclusive and open to wider stakeholder views.

resources, in order to influence policy outcomes. They tend to form around specific issue-areas.

> Policy networks are ... characterised by interdependence between actors, patterned resource exchange, and informal rules or norms in specific policy sectors.
>
> (Peterson and O'Toole, 2001: 303)

Networks can include membership drawn from NGOs, the private sector, business associations, trade unions, non-government and voluntary bodies, and various government agencies, scientific networks and international institutions. Networks are seen to bring particular advantages for the governance of sustainable development that stretch across the various stages of the policy process.

Box 3.5

The UN Global Compact

The Global Compact is the largest voluntary corporate responsibility initiative in the world. It works with a range of stakeholders, including UN agencies, governments, corporate actors, civil society, labour, and other non-business interests.

It is a strategic policy initiative for businesses that are committed to support and enact a set of core values in the areas of human rights, labour standards, the environment and anti-corruption. The initiative also supports broader UN goals, including the Millennium Development Goals.

Aims and objective

'The initiative seeks to combine the best properties of the UN, such as moral authority and convening power, with the private sector's solution-finding strengths, and the expertise and capacities of a range of key stakeholders.'

The aim is to engage business, as a primary driver of globalisation, to help ensure that markets, commerce, technology and finance advance in ways that benefit economies and societies.

The ten principles

Members are asked to align their operations and strategies with ten principles:

Human rights

Principle 1 Businesses should support and respect the protection of internationally proclaimed human rights;

Principle 2 Businesses should make sure that they are not complicit in human rights abuses.

Labour

Principle 3 Businesses should uphold the freedom of association and the right to collective bargaining;

Principle 4 Elimination of all forms of forced and compulsory labour;

Principle 5 Abolition of child labour;

Principle 6 Elimination of discrimination in respect of employment and occupation.

Environment

Principle 7 Businesses should support a precautionary approach;

Principle 8 Undertake initiatives to promote greater environmental responsibility;

Principle 9 Encourage development and diffusion of environmentally friendly technologies.

Anti-Corruption

Principle 10 Businesses should work against corruption in all its forms.

Membership

The Global Compact has over 8,700 corporate participants and other stakeholders from over 130 countries. It also has Local Networks that link business, many in developing and emerging markets.

Activities

The Compact provides a practical framework for the development, implementation, and disclosure of sustainability policies and practices. It offers participants a wide spectrum of work plans, management tools and resources to help advance sustainable business models and markets. It also shares best practices.

The UN Global Compact has been involved in major UN events, including the UN Conference on the Least Developed Countries (2011), and Rio+20 (2012).

(Source: UN, 2013)

There is a vast array of sustainable development networks now in operation, working across the entire governance scale from the international down to the local levels (Robinson and Keating, 2005). The UN Global Compact, a network that combines multiple stakeholders, including representatives drawn from governments, the private sector and the NGO community, provides a good example.

Some networks promote voluntary approaches and thus their proliferation supports the development of market governance. Such actions often

Case study 3.2

Access and benefit-sharing in genetic resources

Background

Voluntary codes of conduct arose amidst concern about the failure of states and international organisations to regulate for the conservation of plant and animal genetic resources and for their sustainable use, and to do so in ways that adequately protect plant and animal genetic resources. This concern led to self-regulation and the creation of collective norms of management by associations of biological resource users, such as botanical gardens and private corporations. As a result, several groups have established ethical codes of conduct or voluntary mechanisms for protection and benefit-sharing of genetic resources.

Examples

(i) MOSAICC

In 1997 the international code of conduct MOSAICC for the *ex situ* collections of microbial cultures was launched. MOSAICC stands for Micro-Organisms Sustainable use and Access regulation International Code of Conduct. This is a voluntary code of conduct to support the implementation of the CBD at the microbial level. The MOSAICC network involves twelve partners, including representatives from the North and South, NGOs, and the commercial sector, including the OECD, the IUCN (World Conservation Union), the International Mycological Institute, and the Agricultural Research Council, Republic of South Africa.

(ii) Kew Gardens common principles on access and benefit-sharing

The Royal Kew Gardens in Britain has also formed a network of botanic gardens working to implement the access and benefit-sharing provisions of the CBD. Two voluntary approaches have been developed: the Principles on Access to Genetic Resources and Benefit-Sharing (the Principles) and the International Plant Exchange Network (IPEN). IPEN established a system of exchange for a network of gardens that have signed up to a Common Code of Conduct.

In addition to these initiatives, some companies have also created ethical codes on an individual basis, in the belief that this will improve their reputation as reliable suppliers of genetic material.

Critique

It is difficult to determine how much these different initiatives of self-regulation have helped support greater compliance with international regimes, such as the FAO, UNEP and WTO, as there is a lack of research in the area.

Voluntary agreements also encounter problems of co-ordination between the different regimes, each of which have their own sets of requirements and reporting protocols.

(*Sources*: Dedeurwaerdere, 2005; BCCM, 2011; BGCI n.d.)

Box 3.6

Problems with network governance

Networks are inefficient

- As actors in networks have diverse interests, the search for consensus and co-operation takes time.
- Networks may be indecisive, inefficient and may impede effective policy delivery.

Networks lack legitimacy

- Networks include non-elected actors.
- They can be unrepresentative, populated by elites, and exclude ordinary citizens.
- Democracy may thus be undermined by network governance.

Networks promote self-interest

- Non-elected actors who claim to represent community interests may instead be motivated by self-interest.
- Networks may serve private vested interests rather than the public good.

Power and conflict are inherent in networks

- There are power imbalances within networks and therefore asymmetric relationships between actors.

- Some may have access to better expert knowledge, more control over information and communication resources, and greater financial resources.

- Asymmetrical distributions of power within networks mean some groups can benefit from network opportunism.

Networks increase fragmentation

- The use of networks can result in fragmented service delivery, and this can lead to reduced control over implementation.

Networks pose problems of accountability

- Networks can obscure accountability and result in confusion about who is accountable to whom, for what, and who takes responsibility for outcomes.

- The presence of networks can increase the demands on the state for monitoring and co-ordination.

take place when governments fail to act and can be led by environmental champions, that is, those with a strong commitment to act in support of our collective future.

As we can see from this example and the other examples discussed in this chapter, while the benefits afforded by network governance can be considerable, there are also disadvantages related to their use. These drawbacks bring potential for governance failure.

Problems with network governance can be illustrated by the case of waste management in Ireland. In Ireland, neoliberal-inspired reforms in the late twentieth century and into the early twenty-first century saw a rise in the use of public–private partnership agreements, including for the management of waste. In the waste sector, however, this diminished the control that local authorities could now exercise over the strategic management of waste. In addition, as privatisation saw a rise in charges to domestic consumers for waste disposal, this led to an increase in illegal dumping. In turn, the commercialisation of waste has turned waste into a profitable commodity, a process that does little to encourage waste reduction (Connaughton et al., 2008). As this case shows, network governance arrangements can weaken the role of environmental administrators, especially when the administration loses competence over implementation (Holzinger et al., 2006).

Faced with these concerns, Baker and Eckerberg (2008b) have pointed out that the use of new forms of governance of sustainable development must continue to go hand-in-hand with the central and formal role of governments. Hierarchical governance has a place in initiating actions, co-ordinating responses, legitimising and entrenching decisions and monitoring performance and unintended consequences, and must remain a central component of governance if we are successfully to steer society along more sustainable development trajectories.

The 'right' governance mix

From our discussion to date, we can see that there is a variety of appropriate ways of undertaking the practices of governance to respect the commitment to the promotion of sustainable development. Each governance style has its own role to play: network governance can lay the basis of consent and of long-term support; market governance can stimulate entrepreneurship and self-regulatory responsibility (Meuleman and in't Veld, 2009) and hierarchical governance can kick-start action and ensure the transparency, accountability and legitimacy of governance in pursuit of sustainable development. The mix of governance styles needs to reflect the social, cultural and political characteristics of a country or a region, as well as the particular sustainable development challenge that policy is trying to address. As such, a situational mixture of network governance, market governance and hierarchical governance is needed (Meuleman, 2008: 73). So far in this chapter we have explored the use of the three main styles of governance, namely hierarchies, markets and networks, and their role in the governance of sustainable development. There are some, however, who claim that the promotion of sustainable development represents such a distinctive, all-embracing and challenging public policy objective that it requires a unique governance approach, one that goes beyond the mere mixing of appropriate governance styles.

Distinctive governance requirements

Several authors have highlighted the distinctive governance challenges involved in the pursuit of sustainable development. Two characteristics of sustainable development are given particular attention: the scale of social transformation required and the character of the steering logic involved in promotion efforts. In the section that follows, we explore this work and identify its significance.

The scale of transformation

Practical experiences of linking environment and development policy, particularly since the 1990s, has led to a new understanding of the complexities of social, economic and ecological interactions. This has brought insight into the scale of the transformation necessary to promote sustainable development, that is, a form of development that mutually reinforces societal, economic and ecological well-being over time. This includes understanding that sustainable development is both a dynamic concept as well as a multi-dimensional one and it requires action across a variety of temporal and spatial scales, for example taking account of future generations in present policy, and taking action at both the local and global levels in both the design and implementation of policies. In addition, it is a very ambitious project, as it requires decoupling economic development from both environmental harm and social exclusion and inequality.

The governance challenges presented by these characteristics go beyond the need to promote sound practices of environmental and resource management. Rather, they require changes in both production patterns (what is produced) and consumption levels (how much is consumed and by whom), which are premised upon steering society towards new cultural values that replace those of consumerism, as discussed in Chapter 4. It also necessitates transformation of key economic sectors, including energy, transport, agriculture, manufacturing and construction, as discussed in Chapter 7. All of these changes have to be underpinned by technological development taking place alongside profound shifts in the organisation of society. In addition, if they are to be successful, both behavioural change and systems innovation demand strong societal engagement and cannot simply be imposed from the top (van Zeijl-Rozedma et al., 2008). Added to this complexity there is also a very distinctive steering logic involved in the promotion of sustainable development.

The steering logic

Steering society in the pursuit of sustainable development is faced with distinctive features that present major steering challenges and also have important implications for how steering is best conducted.

Steering in the context of uncertainty and ambivalence

Promoting sustainable development is not about implementing a blueprint, nor is it about helping society reach an identified or identifiable end state, or about establishing static structures or facilitating fixed qualities of social, economic or political life now or into the future. Rather, it is an open-ended process, one that is not usefully conceived of as aiming towards a specified or specifiable target (Kemp et al., 2005: 16). As an ongoing process, its desirable characteristics change over time, across space and within different social, political, cultural and historical contexts. This means that promotion efforts are undertaken in the context of goals that are inherently open and subject to change. In short, governance has to cope with the complexity and the indeterminacy of sustainable development as a steering objective (Kemp et al., 2005; Meadowcroft, 2007a).

Promotion efforts also have to take place amidst a profound lack of knowledge about the processes, relationships and impacts involved. In particular, knowledge of the complex and dynamic interactions between society, economic development, technology and nature remains limited. The ecological system, including both the climate and ecosystem systems, is poorly understood, and forecasts for the future have large margins of error. This uncertainty can present problems because traditionally policy making is based on the belief that effective policy requires clear goals and a good understanding of the relevant causal relationships (Kemp et al., 2005). When faced with uncertainty, the typical response of policy makers is to ignore the problem and simply assume certainty exists. However, there is increased recognition that efforts to promote sustainable development are best served when policy makers acknowledge inherent policy ambivalence and uncertainty and set about devising new and innovative ways in which to steer society under conditions of uncertainty (Kemp et al., 2005). At the same time, parts of the requirements for the promotion of sustainable development do have a known technical component and many technical solutions are already being put to use (Kemp et al., 2005). This includes, for example, the adoption of resource-efficient, ecological modernisation procedures in production processes or of new technological innovations in the energy sector. Governance for sustainable development has thus to take place in a dual context: acting now in technical and often sub-optimal ways while at the same time steering towards more transformative approaches, which must themselves remain open-ended. With such steering, care has to be taken to ensure that the choices made at present do not unduly foreclose options for, or visions of, our sustainable future.

Steering for the long term

Sustainable development, as we saw from Chapter 2, has a strong inter-generational element. This means that steering has to adopt a long-term horizon and that policy processes have to find ways to put future problems on the agenda and work with them at the present time. It also requires institutional flexibility to enable institutions to both anticipate and handle unforeseen problems or opportunities (Meuleman and in't Veld, 2009). However, adopting long-term perspectives in decision making is very challenging. First, public policy typically takes place within a very short time frame, often dictated by electoral cycles. Lengthening the policy time frame would not only be unpopular among elected politicians, but it adds to uncertainty as the further into the future we project, the weaker becomes our knowledge base. Governments have often only limited incentives to learn these new skills, as future problems are not popular with politicians, have limited issue salience and do not always win votes at elections (Meuleman and in't Veld, 2009).

Steering through multiple levels and sectors

The important role played by sub-national, regional and local authorities in the promotion of sustainable development is increasingly acknowledged, as discussed in Chapter 8. In addition, Brundtland stressed how promoting sustainable development is a steering activity that cuts across functional and administrative boundaries and established territorial jurisdictions. The recent growth of governance arrangements for the environmental management of natural resources that cut across jurisdictional boundaries, such as seas and rivers, is a case in point. Examples include river basin authorities dealing with the River Danube, the Black Sea and the Mediterranean Sea. Increasingly, these complex multi-level governance structures and processes have to engage with economic actors, as well as civil society organisations, such as environmental NGOs, that also operate across multiple levels.

Tensions and struggles within and between the authorities and actors operating at these different levels are typical of modern political processes. This presents challenges not only for policy co-ordination and for the delineation of responsibilities, but also in ensuring transparency and accountability and in monitoring compliance. Indeed, as pointed out in our earlier discussion, part of the function of hierarchical forms of governance is to help ensure such co-ordination.

More specifically analysis has identified two co-ordination needs that are crucial for the promotion of sustainable development: horizontal and vertical co-ordination. The 'vertical' dimension refers to the linkages between higher and lower levels of government, including their institutional, financial, and informational aspects. Vertical co-ordination addresses the need for coherence, compatibility and complementarity between the policies and actions that operate across the different levels. Ensuring compatibility and complementarity between national policies for different sectors and those of the EU, or between EU policies and those of the UN, is required to achieve vertical co-ordination. Horizontal co-ordination focuses on the co-ordination of action within each level. This requires coherence between different sectoral policies within a state, for example, between its policies on sustainable development, transport, agriculture and energy.

In short, when we look at the steering logic involved in the governance of sustainable development, several distinctive challenges emerge. Governance of sustainable development has to take account of the need to steer in the context of uncertainty, to take account of the long term, and to promote both horizontal and vertical policy co-ordination (see Nilsson and Eckerberg, 2007; Meadowcroft, 2007a). These challenges are not easy to meet.

Reflexive governance is believed by many researchers to offer the best way to deal with the steering logic of sustainable development, particularly as it relates to the problem of governance in the face of uncertainty (Voβ and Kemp, 2006). The literature on reflexive governance draws from systems theory, in particular the latter's focus on complexity, uncertainties and nonlinear processes of change (Loorbach, 2010).

Reflexive governance

Research on reflexive governance started from the realisation that governments are increasingly devoting time and resources to dealing with problems that are themselves a result of governing. As such, governance becomes ever more preoccupied with repairing the unintended consequences of prior attempts at shaping societal development. Reflections seek to identify these side effects and either pre-empt them, or address them at an early stage in policy making. Reflexive governance involves not just societal self-reflection, but also the development of new kinds of strategies, processes and institutions that emerge under this condition of 'self-confrontation'.

In the reflexive governance literature, sustainable development is understood as a particular kind of problem framing, one that emphasises the interconnectedness of different problems and scales, as well as the long-term and indirect effects of actions. Rather than attempting to ignore or minimise the ambiguity of goals, uncertainty about cause and effect relations and the feedbacks that occur between sustainable development steering activities, reflexive governance seeks to find ways to address and handle this ambiguity. In doing this, sustainable development comes to be seen to be more about the organisation of processes than about particular outcomes. Thus the approach recognises the limitations of rigid methods aimed at achieving predetermined outcomes, and of steering efforts that delimit problems and treat them in linear ways. In this view, the promotion of sustainable development becomes a constant process in which further adjustments are made as environmental, social and economic conditions change, changes that are, in part, the outcome of previous interventions. Reflexive governance also links well with the democratic and participatory dimensions of sustainable development, as it requires participatory goal formulation and interactive processes between societal actors for the elaboration of sustainable development strategies, as discussed in Chapter 11.

Voß and Kemp (2006) have derived cornerstone strategies to handle governance problems of sustainable development. These strategies offer a practically orientated framework for reflexive governance. They take into account the need for governance to be more anticipatory, orientated towards the long term, to make use of visions of sustainability and to be

---

Case study 3.3

Reflexive governance in the Netherlands

Transition management in the Netherlands is seen as an example of reflexive governance for sustainable development, particularly as applied in the energy sector.

The Netherlands has developed an approach to energy policy that involves exploring different policy options and trajectories and employing steering methods that maintain flexibility, especially through variation and selection, instead of rigid planning.

This transition management is seen as innovative in two ways: first, it offers a prescriptive approach to governance that can be used as a basis for operational policy models; second, it explicitly adopts a normative model by taking sustainable development as its long-term goal.

(*Source*: Voß and Kemp, 2006)

more concerned with learning, innovation and adaptation. This also involves reassessing the place of instrumental reason in public life and abandoning linear models of planning. As such, it moves beyond the modernist modes of rationality towards more self-conscious, holistic and communicative forms of reason.

Reflexive governance, as both a model of governance and as a policy practice, is not without its critics. The idea that we can actively manage moves towards sustainable development could be seen to represent a reassertion of the intent to 'socially engineer' (Walker and Shove, 2007). In defence of reflexive governance, it could be argued that this approach differs from the modernist project in that its methods of engineering are no longer, at least not exclusively, conceived of in modernist terms. Others argue, however, that the notion of reflexive governance is not even needed. Meadowcroft (2007a), for example, has argued that the key elements of reflexivity, in particular as they relate to reconsidering existing practices, critically appraising current institutions and exploring alternative futures, were embedded in the concept of sustainable development from the outset under the Brundtland formulation.

In addition, the reflexive governance literature has been criticised for downplaying the political dimensions of reflexive processes and for its silence on issues of power and the ways in which the distribution of power shapes steering outcomes (Walker and Shove, 2007). Assumptions that the world can be characterised by co-operation, collaboration and consensus building have also been questioned. It also leaves unaddressed the question of who wins and who loses out as transitions are steered and managed in one direction but not another. The approach has also been criticised for holding questionable assumptions about the way contemporary policy making is undertaken, in particular its tendency to downplay the role of the state in policy making and for ignoring how institutional configurations often hinder the capacity of actors to engage in reflexive steering (Hendriks and Grin, 2010).

Further adjustments

At the beginning of this chapter, we mentioned that in addition to the use of the term 'governance' to describe empirical phenomena, that is, to capture important changes in contemporary policy making processes, the term is also used in another, albeit related way. The concept 'governance' can be understood not only as a description of changing patterns in the way in which society is steered but as a normative prescription, that is, it can be used to describe the type and style of steering that *should* be adopted to achieve a preferred societal end point (Kohler-Koch and Rittberger, 2006). As a normative goal, governance for sustainable development refers to processes of socio-political governance oriented to steering societal development among more sustainable lines (Farrell et al., 2005). We have discussed this issue throughout the chapter, but in this section attention is turned to an examination of calls for further adjustments to practices of governance – changes that *ought* to occur in order to ensure that social development progresses along a sustainable development trajectory.

These calls arise both from awareness of the limitations of existing governance practices and from realisation that the promotion of sustainable development raises many new issues for society to ponder. They have led to a set of normative goals that outline what governments as well as agencies and organisations ought to promote and how this should be done. Several actors and agencies have elaborated on the governance changes they believe are needed if society is to respond to the urgent development needs of our time without compromising the ability of future generations to meet their own needs. These issues include how to steer Western society towards reductions in consumption, and how to steer in the context of ecological limits but alongside unprecedented population growth. As a normative project, the governance *for* sustainable development becomes a goal-orientated activity, that is, one that seeks to achieve certain (desired) societal outcomes and to avoid other (less promising) social futures (Meadowcroft, 2007a). It requires adjusting the structures that regulate societal interactions so that they can encourage positive developmental adaptation (Meadowcroft et al., 2005: 5). For high consumption societies, instigating governance processes that encourage and facilitate radical transformations of patterns of production and consumption could be one such step.

Turning attention to this matter, Meadowcroft et al. (2005) have identified a series of wide ranging changes that need to be introduced in order to promote our sustainable futures. These can be seen as an indicative list of the normative requirements for governance *for* sustainable development:

In order to put these governance changes into practice, new knowledge is required. This includes developing a better understanding of ecological processes and of social–ecological interactions, and of the possibilities

---

Box 3.7

Normative requirements: governance for sustainable development

- Developing appropriate political frameworks for iterative rounds of 'future visioning', goal identification, policy design and implementation, performance monitoring and policy adjustment;

- Adopting a long-term focus, reaching into several generations;

- Integrating different kinds of knowledge into decision processes, and establishing an appropriate balance between lay and expert knowledge;

- Structuring engagement as a learning process, so that governments and other social actors acquire experience, experiment with options, draw lessons and learn how more sustainable social practices can be generated;

- Evolving complex systems of multi-level governances, where decision makers remain responsible to citizens, communities and stakeholders;

- Maintaining political support for long-term adjustment, despite fluctuating short-term preoccupations of politicians and electorates;

- Incorporating sustainable development into education and cultural practices, individual codes of conduct and popular morality;

- Recognising that governance for sustainable development will require difficult choices, because not all social goods are commensurate and win/win scenarios are not always practical;

- Recognising that serious failures, setbacks and disappointments are involved, not least because these processes involve great uncertainty and indeterminacy.

(*Source*: Adapted from Meadowcroft et al., 2005: 6–8)

---

of, and limits to, their conscious adjustment. However, several, deep-rooted changes are also needed. These include integrating the economic, social and environmental dimensions of decision making across society, a call for integration that was discussed in Chapter 2. This integration also calls for transforming unsustainable practices embedded in core economic sectors. Such changes will bring significant struggles within and between societies over the orientation of development, the distribution of costs and benefits, the identification of public interest and the definition of the appropriate role of government in orienting and adjusting to change (Meadowcroft et al., 2005).

Conclusion

This chapter has explored the issue of governance, examining to what extent existing governance styles help to promote sustainable futures. Empirically, this chapter focused on the growing array of practices used to steer society's future, providing critical reflections on these practices.

Traditionally, governments have drawn heavily upon regulations and sanctions as steering instruments for policy, so-called 'command and control' steering. This sees the state pass legislation governing the environmental consequences of economic activity. The state ensures compliance with international agreements, sets regulatory standards in relation to pollution prevention and control, oversees the application of environmental law, negotiates voluntary agreements with industry, monitors the impacts of its policies, deals with their distributional consequences, and addresses the relationship between environmental and other policy fields, such as transport, energy and trade.

The governance of sustainable development requires acceptance of the premise that no one actor can tackle this complex task. Increasingly, non-state actors, including civil society and economic actors, are sharing responsibility with the state. These partnerships have been institutionalised through a variety of administrative arrangements and have been operationalised using a variety of new policy tools. These efforts have been driven, in part, by the desire of governments and international organisations to foster the internalisation of sustainable development norms into business practices and consumer behaviour. As a result, the governance of sustainable development now involves complex multi-centred, multi-layered and diffuse inputs from a range of public and private actors. These developments are important, because they provide new ways to connect private interests to public objectives.

From our discussion, it is clear that the governance of sustainable development gives a heightened and positive role for the state in directing societal change, even if this steering remains open-ended and increasingly needs to be collaborative in nature. As the chapter progressed, discussion focused on the specific characteristics of sustainable development as a steering objective, and on how these characteristics add both complexity and ambiguity to governance. There is no 'best form' of governance for the promotion of sustainable development. Governance structures and processes have to be context-specific and open-ended; they have to embrace uncertainty, yet set intermediate, realisable goals for policy action and delivery. This need for diversity in the mixture of governance approaches has led to recent experiments that seek to make governance adaptability and flexibility an integral part of new, reflexive forms of sustainable development governance (Voβ et al., 2006). The chapter finished by exploring what adjustments ought to be made to governance structures and processes to help society in its pursuit of sustainable development.

All these approaches have in common the belief that intervention in pursuit of normative goals, like those of sustainable development, is possible and potentially effective (Walker and Shove, 2007: 219). Yet our discussion pointed to the far-reaching and radical demands that the promotion of sustainable development poses for governance processes, as well as the aims and objectives of governance regimes.

Gouldson and Bebbington argue that we can look at the potential influence of new and emerging forms of governance in an optimistic or a pessimistic manner. The optimistic view sees new forms of governance as providing opportunities or spaces for communicative action and social learning. This is achieved by enhancing transparency and creating opportunities for engagement and accountability, which can, in turn, lead to more legitimate decision making processes and more socially desirable outcomes. As such, new approaches to governance have transformative potential. The challenge is to ensure that this potential can be realised. The more pessimistic view on governance points to the prevailing power relations, and the ways in which particular roles, responsibilities and levels of influences are assigned, in ways that enable the more powerful actors to use governance processes to legitimise their activities without obliging them to transform those activities (Gouldson and Bebbington, 2007: 10–11).

Summary points

- There is a growing array of practices used to steer society towards sustainable development;
- These include combining different governance styles, that is, markets, hierarchies and networks;
- Hierarchical governance remains central for the steering of societal change;
- The use of markets and networks continues to pose problems with respect to equity, transparency and accountability;

- New forms of reflexive governance are proposed, given the wide-ranging challenges presented by the pursuit of sustainable development;
- Deeper changes in governance practice are also envisaged as necessary for the pursuit of sustainable futures.

Further reading

Baker, S. (2010), 'The Governance Dimensions of Sustainable Development', in P. Glasbergen and I. Niestroy (eds), *Sustainable Development and Environmental Policy in the European Union: A Governance Perspective* (The Hague, Netherlands: Open University Press), 21–59.

Baker, S. (2014), 'Governance', in Carl Death (ed.), *Critical Environmental Politics* (London Routledge), 100–110.

Farrell, K. N., Kemp, R., Hinterberg, F., Rammel, C. and Ziegler, R. (2005), 'From "for" to Governance for Sustainable Development in Europe: What Is at Stake for Future Research?' *International Journal of Sustainable Development*, 8 (1–2): 127–151.

Jordan, A. (2008), 'The Governance of Sustainable Development: Taking Stock and Looking Forwards', *Environment and Planning C: Government and Policy*, 26 (1): 17–23.

Lafferty, W. M. (ed.) (2004), *Governance for Sustainable Development: The Challenge of Adapting Form to Function* (Cheltenham: Edward Elgar).

Meadowcroft, J. (2007), 'Who Is in Charge Here? Governance for Sustainable Development in a Complex World', *Environmental Policy and Planning*, 9 (3–4): 193–212.

Wurzel, R., Zito, A. and Jordan, A. (2012), *Environmental Governance in Europe: A Comparative Analysis of New Environmental Policy Instruments* (Cheltenham: Edward Elgar).

References

Baker, S. and Eckerberg, K. (2008a), 'In Pursuit of Sustainable Development at the Sub-National Level: The New Governance Agenda', in S. Baker and K. Eckerberg (eds), *In Pursuit of Sustainable Development: New Governance Practices at the Sub-National Level in European States* (London: Routledge, 2008), 1–26.

Baker, S. and Eckerberg, K. (2008b), 'Combining Old and New Governance in Pursuit of Sustainable Development', in S. Baker and K. Eckerberg (eds), *In Pursuit of Sustainable Development: New Governance Practices at the Sub-National Level in European States* (London: Routledge), 208–227.

Bailey, I. (2007), 'Market Environmentalism, New Environmental Policy Instruments and Climate Policy in the United Kingdom and Germany', *Annals of the Association of American Geographers*, 97 (3): 530–550.

BCCM (Belgian Co-ordinated Collections of Micro-organisms) (2011), 'MOSAICC, Micro-Organisms Sustainable Use and Access Regulation International Code of Conduct', available online at http://bccm.belspo.be/projects/mosaicc/

BGCI (Botanic Gardens Conservation International) (n.d.), 'Kew Developing a Policy on Access and Benefit Sharing', available online at www.bgci.org/resources/abs_policy/

Black, J. (2008), 'Constructing and Contesting Legitimacy and Accountability in Polycentric Regulatory Regimes', *Regulation and Governance*, 2: 137–164.

Bleischwitz, R. (2003), *Governance of Sustainable Development* (Wuppertal, Germany: Wuppertal Institute for Climate, Environment and Energy).

Connaughton, B., Quinn, B. and Rees, N. (2008), 'Rhetoric or Reality: Responding to the Challenge of Sustainable Development and New Governance Patterns in Ireland', in S. Baker and K. Eckerberg (eds), *In Pursuit of Sustainable Development: New Governance Practices at the Sub-National Level in Europe* (Abingdon/New York: Routledge/ECPR studies in European Political Science), 145–168.

Convery, F., McDonnell, S. and Ferreira, S. (2007), 'The Most Popular Tax in Europe? Lessons from the Irish Plastic Bags Levy', *Environmental and Resource Economics*, 38 (1): 1–11.

Dedeurwaerdere, T. (2005), 'The Contribution of Network Governance to Sustainable Development', Université catholique de Louvain, Institute for Sustainable Development and International Relations, paper no.13, available online at http://uclouvain.academia.edu/tomdedeurwaerdere/Papers/984675/The_contribution_of_network_governance_to_sustainable_development

Euractive (2008), 'Environmental Voluntary Agreements', available online at www.euractiv.com/climate-environment/environmental-voluntary-agreemen-linksdossier-188217?display=normal

Farrell, K. N., Kemp, R., Hinterberg, F., Rammel, C. and Ziegler, R. (2005), 'From "for" to Governance for Sustainable Development in Europe: What Is at Stake for Future Research?' *International Jouranal of Sustainable Development*, 8 (1–2): 127–151.

French, D. A. (2002), 'The Role of the State and International Organizations in Reconciling Sustainable Development and Globalization', *International Environmental Agreements: Politics, Law and Economics*, 2: 135–150.

Gouldson, A. and Bebbington, J. (2007), 'Corporations and the Governance of Environmental Risk', *Environment and Planning C: Government and Policy*, 25: 4–20.

Gunninghan, N. (2009), 'Environment Law, Regulation and Governance: Shifting Architectures', *Journal of Environmental Law*, 21 (2): 179–212.

Hendriks, C. and Grin, J. (2010), 'Contextualizing Reflexive Governance: The Politics of Dutch Transitions to Sustainability', *Journal of Environmental Policy and Planning*, 9 (3–4): 333–350.

Holzinger, K., Knill, C. and Schäfer, A. (2006), 'Rhetoric or Reality? "New Governance" in EU Environmental Policy', *European Law Journal*, 12: 403–420.

Howlett, M. (2002), 'Do Networks Matter? Linking Policy Network Structure to Policy Outcomes: Evidence from Four Canadian Policy Sectors 1990–2000', *Canadian Journal of Political Science*, 35(2): 235–267.

Irish Citizens Information Board (2011), 'Plastic Bag Environmental Levy', available online at www.citizensinformation.ie/en/environment/waste_management_and_recycling/plastic_bag_environmental_levy.html

Irish Department of Environment, Community and Local Government (2007), 'Plastic Bags', available online at www.environ.ie/en/Environment/Waste/PlasticBags/

Jordan, A. (2008). 'The Governance of Sustainable Development: Taking Stock and Looking Forwards', *Environment and Planning C: Government and Policy*, 26 (1): 17–23.

Kemp, R., Parto, S. and Gibson, R. B. (2005), 'Governance for Sustainable Development: Moving from Theory to Practice', *International Journal of Sustainable Development*, 8 (1– 2): 13–30.

Kohler-Koch, B. and Rittberger, B. (2006), 'The Governance Turn in EU Studies', *Journal of Common Market Studiess*, 44 (S1): 27–49.

Kooiman, J. (ed.) (1993), *Modern Governance: New Government–Society Interactions* (London: Sage).

Kooiman, J. (2000), 'Societal Governance: Levels, Modes, and Orders of Social-Political Interaction', in J. Pierre (ed.), *Debating Governance* (Oxford: Oxford University Press), 138–166.

Lafferty, W. M. (ed.) (2004), *Governance for Sustainable Development: The Challenge of Adapting Form to Function* (London: Edward Elgar).

Lafferty, W. and Eckerberg, K. (eds) (1998), *From the Earth Summit to Local Agenda 21: Working Towards Sustainable Development* (London: Earthscan).

Lenschow, A. (1999), 'Transformation in European Environmental Governance', in B. Kohler-Koch and R. Eising (eds), *The Transformation of Governance in the European Union* (London: Routledge), 39–60.

Loorbach, D. (2010), 'Transition Management for Sustainable Development: A Prescriptive, Complexity-Based Governance Framework', *Governance: An International Journal of Policy, Administration, and Institutions*, 23 (1): 161–183.

Meadowcroft, J. (2007a), 'Who Is in Charge Here? Governance for Sustainable Development in a Complex World', *Environmental Policy and Planning*, 9 (3–4): 193–212.

Meadowcroft, J. (2007b), 'National Sustainable Development Strategies: Features, Challenges and Reflexivity', *European Environment*, 17: 152–163.

Meadowcroft, J., Farrell, K. N. and Spangenberg, J. (2005), 'Developing a Framework for Sustainability Governance in the European Union', *International Journal of Sustainable Development*, 8 (1/2): 3–11.

Meuleman, L. (2008), *Public Management and the Metagoverance of Hierarchies, Networks and Markets: The Feasibility of Designing and Managing Governance Style Combinations* (The Hague, Netherlands: Physica-Verlag).

Meuleman, L. and in't Veld, R. J. (2009), *Sustainable Development and the Governance of Long-term Decisions*. Advisory Council for Spatial Planning, Nature and the Environment, Preliminary studies and background studies, number V.17 (The Hague, Netherlands: RMNO), available online at www.eeac-net.org/download/EEAC_WG_GOV_SDGovernanceLong-termDecisions_1009_final.pdf

Nilsson, M. and Eckerberg, K. (eds) (2007), *Environmental Policy Integration in Practice: Shaping Institutions for Learning* (London: Earthscan).

OECD (Organisation for Economic Co-operation and Development) (2003), *The Use of Economic Instruments for Pollution Control and Natural Resource Management in EECCA*, CCNM/ENV/EAP(2003)5 (Paris: OECD).

OECD (2009), 'Strategies for Sustainable Development', available online at www.oecd.org/document/40/0,3343,en_2649_34421_2670312_1_1_1_1,00.html

Painter, M. and Pierre, J. (2005), 'Unpacking Policy Capacity: Issues and Themes', in M. Painter and J. Pierre (eds), *Challenges to State Policy Capacity: Global trends and Comparative Perspectives* (Basingstoke: Palgrave Macmillan).

Peterson, J. J. and O'Toole, L. J. (2001), 'Federal Governance in the United States and the European Union: A Policy Network Perspective', in K. Nicolaidis and R. Howse (eds), *The Federal Vision. Legitimacy and levels of governance in the United States and the European Union* (Oxford: Oxford University Press).

Pierre, J. (2000), 'Introduction: Understanding Governance', in J. Pierre (ed.), *Debating Governance* (Oxford: Oxford University Press), 1–12.

Pierre, J. and Peters, B. G. (2000), *Governance, Politics and the State* (London: Macmillan).

Rhodes, R. A. W. (2007), 'Understanding Governance: Ten Years On', *Organisational Studies*, 28 (8): 1243–1264.

Robinson, L. and Keating, J. (2005), 'Networks and Governance: The Case of LLENs in Victoria ARC Linkage Project'. Occasional paper 3, Centre for Post Compulsory Education and Lifelong Learning, University of Melbourne, available online at www.edfac.unimelb.edu.au/cpell/documents/ARCPaper3.pdf

Segerson, K. and Miceli, T. J. (1998), 'Voluntary Environmental Agreements: Good or Bad News for Environmental Protection?' *Journal of Environmental Economics and Management*, 36: 109–130.

Swanson, D., Pintér, L., Bregha, F., Volkery, A. and Jacob, K. (2004), *National Strategies for Sustainable Development, Challenges, Approaches and Innovations in Strategic and Co-ordinated Action, A 19-Country Study* (Winnipeg, Canada: International Institute for Sustainable Development and Deutsche Gesellschaft für Technische Zusammenarbeit (GTZ) GmbH), available online at www.iisd.org/pdf/2004/measure_nat_strategies_sd.pdf

Stiglitz, J. (2000), *Economics of the Public Sector* (London: W. W. Norton, 3rd revised edn).

UN (2013), *Overview of the Global Compact*, available online at www.unglobalcompact.org/AboutTheGC/index.html

van Zeijl-Rozema, R., Cörvers, R., Kemp, R. and Martens, P. (2008), 'Governance for Sustainable Development: A Framework', *Sustainable Development*, 16: 410–421.

Voβ, J-P. and Kemp, R. (2006), 'Sustainability and Reflexive Governance: Introduction', in J-P. Voβ, D. Bauknecht and R. Kemp (eds), *Reflexive Governance for Sustainable Development* (Cheltenham: Edward Elgar), 3–28.

Voβ, J-P., Bauknecht, D. and Kemp, R. (eds) (2006), *Reflexive Governance for Sustainable Development* (Cheltenham: Edward Elgar).

Walker, G. and Shove, E. (2007), 'Ambivalence, Sustainability and the Governance of Socio-Technical Transitions', *Journal of Environmental Policy and Planning*, 9 (3–4): 213–225.

Wurzel, R., Zito, A. and Jordan, A. (2012), *Environmental Governance in Europe: A Comparative Analysis of New Environmental Policy Instruments* (Cheltenham: Edward Elgar).

Sustainable production and sustainable consumption

Key issues

- Sustainable consumption and production
- Efficiency and Factor 4/10
- Corporate social responsibility
- Population
- Lifestyles and voluntary simplicity
- Instrumental versus moral rationality.

Introduction

Many of today's consumption and production patterns are unsustainable (MEA, 2005). These patterns are driving resource depletion, which is compounded by a burgeoning in the generation of waste, both of which are putting pressure on the Earth's fragile resources and on planetary systems. Following from the Rio Earth Summit, there have been several initiatives on what has become known as sustainable consumption and production (SCP). SCP initiatives seek to decouple economic growth from environmental degradation. This is done through the development of production processes that use fewer resources and generate less waste, including hazardous substances, while yielding environmental benefits and frequently productivity and economic gains. They also encourage sustainable patterns of resource consumption, and of their related goods and services.

For many, achieving SCP patterns will bring improvements in economic development while also bringing business, employment and export opportunities. This in turn can help support development opportunities for poorer countries. For others, the promotion of sustainable development through the lens of SCP weakens efforts, because it fails to address production and consumption levels. The fault lines between these two perspectives are explored in this chapter.

The chapter begins by examining the UN approach to SCP, explaining its origins, its relationship to the Millennium Development Goals (MDGs) and detailing several SCP initiatives. Promoting sustainable production and consumption is best viewed as an interrelated task, especially in an interconnected world where supply chains are global. Resource extraction, the production of intermediate inputs, distribution, marketing, waste disposal and post-consumer reuse takes place across, and also links, the world's national economies (Royal Society, 2012).

While recognising this linkage, the chapter also deals with sustainable production and sustainable consumption separately. There are several reasons for this. The drive towards efficiency is more marked in efforts to promote sustainable production than in relation to the pursuit of sustainable consumption. Sustainable production targets the behaviour of producers of goods and services. Although producers consume resources in the production of goods and services, for the sake of simplicity we treat consumption as the end product of production, an activity that brings in citizens as consumers and that therefore raises separate issues in relation to consumer behaviour and consumer values. We thus treat sustainable production as addressing matters in relation to the supply side of economic activity, whereas sustainable consumption addresses the demand side. Furthermore, the drive to promote sustainable consumption has given rise to an array of actions and to the development of new values that reveal sharp contrasts with the sustainable production agenda. Throughout the chapter, examples and case studies are presented to illustrate these differences.

Sustainable consumption and production

Relationship to sustainable development

The 1992 Rio Earth Summit stressed that the transformation of consumption and production patterns in both developed and developing countries is an essential step in the promotion of sustainable development. The achievement of SCP is one of the Rio Declaration Principles.

> To achieve sustainable development and a higher quality of life
> for all people, States should reduce and eliminate unsustainable
> patterns of production and consumption and promote appropriate
> demographic policies.
>
> (Principle 8 of the Rio Declaration)

At the Summit, states committed themselves to two objectives:

1 Seeking production patterns which reduce consumption of energy
   and non-renewable resources and limit pollution, particularly the production of environmentally harmful waste.
2 Adopting harmonious consumption patterns which could be satisfied
   by natural resources on a sustainable basis.

SCP is seen to support the continued use of goods and services that
respond to basic needs, especially for the world's impoverished people,
and bring a better quality of life by minimising the use of natural
resources, toxic materials and emissions of waste and pollutants over
the life cycle (Baker, 1996).

Sustainable consumption and production (SCP) focuses on the sustainable
and efficient management of resources at all stages of value chains of goods
and services.

(UNEP, 2012a)

Matters in relation to SCP were also addressed in several parts of
Agenda 21, notably those dealing with population, energy, transportation and waste, and in the chapters on economic instruments and
the transfer of technology. Chapter 4 of Agenda 21 stresses that
'action is needed to promote patterns of consumption and production
that reduce environmental stress and will meet the basic needs of
humanity' (UN, 1992). Emphasis in particular is placed upon achieving SCP by addressing high consumption lifestyles in developed
countries.

> the major cause of the continued deterioration of the global environ-
> ment is the unsustainable pattern of consumption and production,
> particularly in industrialised countries, which is a matter of grave
> concern, aggravating poverty and imbalances.
>
> (Agenda 21, Ch. 4.3 [UN, 1992])

SCP is also seen as a key way in which to support the MDGs.

Box 4.1

SCP and the MDGs

Links

Developing countries face increasing commodity prices, especially of food and fossil fuels, alongside continuing declines in the extent, quality and productivity of their eco-systems. This reinforces the importance of the two key messages of SCP: the need to dramatically improve the resource efficiency of consumption and production, and to reduce the pressure on increasingly scarce resources by promoting their sustainable use.

MDG 7 and MDG 1

The most obvious connection between the SCP agenda and MDGs is MDG 7, the achievement of environmental sustainability. Attaining MDG 7 requires the production and consumption of more goods and services to meet the basic needs and aspirations of the world's poorest, while keeping within the limits of ecosystems.

The 2011 *Human Development Report* (HDR) made projections on how environmental degradation associated with current consumption and production patterns would impact the development path of developing countries. The projections showed that these impacts could prevent any further increase in the Human Development Index in some countries from 2030 onwards. The MDG 1 goal of ending hunger and poverty will not be met unless there is also significant progress towards the MDG 7 goal of environmental sustainability.

(*Source*: UNEP, 2012b)

The WSSD in Johannesburg acknowledged that SCP is a prerequisite for sustainable development, as discussed in Chapter 5. As part of the Johannesburg Plan of Implementation, countries were asked to develop a '10 Year Framework of Programmes' (10YFP) to accelerate the shift towards SCP. This resulted in the so-called Marrakech Process, a UN-led international collaboration involving multiple stakeholders. Its mission is to establish task forces and engage in dialogues with different stakeholders, in order both to refine the concept of SCP and to show how it could be made operational in very different countries, economic sectors and cultural contexts. The work undertaken through the Marrakech Process provided a major input into the 10YFP (2012–2020) adopted at the UNCSD (Rio+20) in June 2012.

Box 4.2

The 10YFP adopted at Rio+20

Aim

The 10YFP aims to promote social and economic development within the carrying capacity of ecosystems by addressing and, where appropriate, decoupling economic growth from environmental degradation.

This is to be achieved by improving efficiency and sustainability in the use of resources and production processes and reducing resource degradation, pollution and waste.

Vision

The 10YFP is expected to support new market opportunities for products and technologies, in particular from developing countries. The belief is that, by stimulating competitiveness, countries can develop an inclusive economy delivering full and productive employment.

The 10YFP is also expected to support the achievement of the MDG and the implementation of targets and goals agreed under relevant multilateral environmental agreements.

Actions

All countries are expected to take action, but with developed countries taking the lead, especially through the provision of financial and technical assistance and through supporting capacity building. Plans are to target:

(a) Consumer information
(b) Sustainable lifestyles and education
(c) Sustainable public procurement
(d) Sustainable buildings and construction
(e) Sustainable tourism, including ecotourism.

(*Source*: UN, 2012)

As will be discussed in Chapter 5, the approach to sustainable development adopted at Rio+20 differs in key ways from the original Brundtland formulation and the principles of the Rio Earth Summit. This difference is crystallised in the 10YFP, where an instrumental rationality stresses the need for resource efficiencies to promote competitive economies. Although the UN links SCP to wider welfare and equity goals, eco-efficiency remains a key goal of SCP strategies. Limits to growth or the potential for efficiency gains to be lost through the rebound effect are not considered. Decoupling is also restricted to 'appropriate' contexts. By not addressing limits to consumption and production, actions agreed at Rio+20 restrict the SCP agenda and are more akin to weaker models of sustainable development and their related pursuit of ecological modernisation, as discussed in Chapters 2 and 3.

> Eco-efficiency: the delivery of competitively priced goods and services which satisfy human needs and produce quality of life while progressively reducing ecological impacts and resource intensity, through the life cycle, to a level at least in line with the earth's estimated carrying capacity.
>
> (WBCSD, 2000)

Achieving eco-efficiency in the production of goods and services forms part of a wider Rio+20 agenda of green growth, which aims at decoupling environmental pressures from economic growth. Decoupling refers to the process of reducing the resource intensity of, and environmental damage relating to, economic activities (OECD, 2003: 13).

> Decoupling: reducing the amount of physical emissions or natural resource use per unit of economic output. This can be achieved either through increased efficiency, 'eco-efficiency', stemming from technological change, or a shift to less environmentally damaging products.

Decoupling is seen as a way of overcoming barriers to growth-orientated development arising as a result of the finite nature of the Earth's resources, and, as discussed in Chapter 3, can be out of kilter with sustainable development.

SCP policies and initiatives

Under the auspices of the UN, international organisations and regional bodies, alongside business and industry groups, have engaged in a broad range of initiatives under the remit of SCP (see Table 4.1). The aim of many of these activities is to promote cleaner production in industrial manufacturing. Cleaner production results from conserving raw materials, water and energy, eliminating toxic and dangerous raw materials, and reducing the quantity and toxicity of all emissions and wastes during the production process. Waste management is also an important component, involving waste minimisation and disposal, including through recycling.

Table 4.1 SCP policies and initiatives

Objectives	Strategic targets	UN activities	EU/regional initiatives	National level engagement	Governance and regulation	Management tools	Corporate engagement	Civil society engagement
Promote SD Attain MDG Maintain ecosystem services	Energy and resource efficiency Cleaner production Waste management	Round tables Global Compact Global reporting initiatives Marrakech Process Rio+20 10YFP Collaborating centre on SCP MEAs UNEP strategic approach to international chemicals management UNEP Bali strategic plan for technology support and capacity-building	EU SCP/ sustainable industrial policy (SCP/SIP) Action Plan 2008 EU integrated product policy EU 2020 strategy EU SPREAD sustainable lifestyles 2050 Regional framework programmes on SCP, e.g. Africa, Asia	National SCP programmes and action plans Action plans on environmental technologies National cleaner production centres Green public procurement, e.g. US federal requirements for the procurement of energy-efficient products Green growth initiatives Sector specific approaches, e.g. energy, food	Guidelines on unfair commercial practices Regulations on product energy efficiency EU Eco-design of energy using Products Directive 2009 EU Registration, Evaluation, Authorisation and Restriction of Chemical substances (REACH) Directive 2007 EU packaging and packaging waste directives, 1994 USA National Action Plan for Energy Efficiency, Vision for 2025	Eco labelling Energy labelling Eco Management Audit Systems (EMAS) ISO standards e.g. ISO 14001 Product environmental footprint Corporate environmental footprint Environmental technology verification Cradle-to-grave waste management	CSR Initiatives Voluntary agreements Codes of practice Sustainable value chains WBCSD SCP policies	SCP roundtables/ retail forums (e.g. European Food SCP round table) Reduce, recycle, repair (3Rs) Collaborative consumption (sharing, swapping, trading, etc.) Co-producers (e.g. growing own food) Lifestyle changes: Efficient living (wasting less) Different living (focus on high quality goods and services) Sufficient living (reducing consumption) Social innovation: Voluntary simplicity Fair-trade Transition towns

Many SCP actions use market- and information-based policies and schemes that are voluntary. At the regional level, a number of intergovernmental bodies have established SCP frameworks, such as the EU's Sustainable Consumption and Production and Sustainable Industrial Policy (SCP/SIP) Action Plan (CEC, 2008). This contains a number of proposals for improving the environmental performance of products and increasing the demand for more sustainable goods and production technologies. It also seeks to encourage European industry to take advantage of the opportunities to innovate. The Resource-Efficient Europe initiative under the Europe 2020 Strategy provides another example, discussed in Chapter 9. The EU has also established the European Food SCP Roundtable and the EU Retail Forum to support SCP. At the EU member state level, dedicated national SCP action plans have been developed by the Czech Republic, Finland, Poland, and the United Kingdom, to give some examples. Similar regional level initiatives can be found in Central America.

The Arab region has also developed SCP strategies with the support of the Marrakech Process, while in Latin America and the Caribbean a Regional Council of Government Experts on SCP was set up in 2003 to support the implementation of an SCP regional strategy. There is also an Asia and the Pacific Roundtable on SCP. The Mercosur countries (Argentina, Brazil, Paraguay, Uruguay, and Venezuela) have also developed a Policy for Promotion and Co-operation on SCP, which focuses on encouraging co-operation on SPP and sustainable consumption, stimulating eco-innovation and promoting education (UNEP, 2012a).

In Africa, a regional 10YFP has spurred a number of sub-regional, national and local SCP programmes. For example, pilot projects for mainstreaming SCP in national and city level development policies and action plans on SCP have been conducted in Tanzania and Cairo in Egypt. There is also an African Roundtable on SCP, established in 2002. Similarly, in Asia and the Pacific, the Green Growth Initiative has been adopted by countries as a way to reconcile tensions between poverty reduction and environmental sustainability. This initiative promotes SCP, including through green tax reform.

Links between SCP and poverty alleviation are also reflected in the Arab Regional Strategy for SCP, which was endorsed in 2009 by an array of regional intergovernmental organisations. It encourages the use of products and services that conserve water and energy while contributing to poverty eradication and sustainable lifestyles.

These examples reveal the array of actors involved in the promotion of SCP, operating across international, regional and national scales. These activities

Case study 4.1

The Central American Commission for Environment and Development (CCAD) initiative on sustainable public procurement

Background

Governments are among the largest single consumers within any given market. The CCAD launched an initiative to promote sustainable public procurement (SPP) at the regional level and by national governments.

Policy

A Regional Policy on Public Procurement identifies opportunities for more efficient use of materials, resources and energy. The aim is to contribute to the protection of human health and foster the development of a regional market for sustainable and innovative goods and services.

The policy addresses four specific areas:

1 Institutional: ensuring that relevant information and methodologies are adapted to the specific country context;
2 Legal: getting SPP included in member country legislation;
3 Technical: giving support to the providers of goods and services in the shift towards more sustainable production practices;
4 Information and capacity-building: developing the technical skills for implementing SPC in both the public and the private sectors.

Stakeholders

Stakeholders included representatives of the ministries of Environment, Economy, Agriculture, Tourism, and Labour, national procurement authorities and civil society organisations.

Implementation

Green procurement guidelines for implementation have been developed by the Centro de Gestión Tecnológica e Informática Industrial, Costa Rica, and via the Marrakech Task Force on Sustainable Public Procurement.

(Source: UNEP, 2012a)

can support plans of action designed to implement UN-level commitments while addressing specific needs across scale. However, as mentioned in the introduction to this chapter, it is also important to explore sustainable production separately from sustainable consumption initiatives.

Sustainable production: the business perspective

What is sustainable production?

Efforts to promote sustainable production are built around the belief that the utilisation of inappropriate technologies, especially in manufacturing, is resource- and energy inefficient, risks workplace hazards, yields high levels of waste and results in pollution (ISO, 2000). From this perspective, it is more rational to prevent unsustainability from the very start of the production process rather than attempting to deal with the consequences.

> Sustainable production is the creation of goods and services using processes and systems that are non-polluting, that conserve energy and natural resources, that are economically efficient, safe and healthy for workers, communities and consumers, and are socially and creatively rewarding for all working people.
>
> (EEA, n.d.)

While the definition of sustainable production provided by the European Environment Agency is broad in its scope and ambitions, most efforts involve redesigning production processes so that they use materials and energy more efficiently. Several research institutes have suggested that OECD countries will need to cut their per capita pollution and resource intensities by a factor of ten or more over the next half century if they are to reduce the burden they place on the global environment to sustainable levels.

The concepts of eco-efficiency and factor 4/10 targets were included in the conclusions to the Earth Summit+5 in 1997. The World Business Council for Sustainable Development has also adopted the factor 4/10 concept. This eco-efficiency approach exhibits what we may call an 'instrumental rationality': it instructs agents to use those means that are necessary in relation to their given ends, in this case resource efficiency in production. It does not question the nature of these ends, in this case the nature or extent of what is produced and for whom. It is strongly linked to the pursuit of ecological modernisation, discussed in Chapter 3.

Box 4.3

Efficiency gains: factor 4/10

Factor 4

The goal of being twice as productive with half the resources (material and energy), leads to a factor 4 improvement in efficiency.

Factor 10

This means ten times as much productivity from the same inputs (ranging to the same productivity with one tenth of the resources). Factor 10 equates to a 90 per cent decrease in resource usage.

Many experts are convinced that without radical dematerialisation in advanced countries, sustainability cannot be reached.

(*Source*: von Weizsacker et al., 1998)

Case study 4.2

Leather production in Viet Nam

TanTec produces leather goods for overseas companies from tanneries located in China and Viet Nam. It has achieved significant cost savings by implementing energy-efficiency and waste management practices.

The Saigon TanTec uses on average only 33 mj of energy per square metre of leather, compared with a leather industry standard of approximately 52 mj, as calculated by the British Leather Technology Centre. These energy reductions have been achieved through a host of improvements, including the installation of continuously controllable compressors and pumps, the retrofitting of dryers, the installation of energy-efficient re-tanning drums, and the shift from oil to liquified gas as a heating source. In addition, the factory has invested in energy efficient lighting systems and translucent plastic panels that allow sunlight to penetrate the workspace. Bamboo walls also allow for a natural ventilation of the factory. Use is also made of reed grasses and 'wetland' methods for wastewater management and post-manufacture purification.

The company has now achieved a 40 per cent reduction in its energy consumption, and has reduced CO_2 emissions by 2,700 tons per year. In addition, TanTec has reduced water and chemical consumption by 50 per cent and 15 per cent, respectively.

(*Source*: UNEP, 2012a)

Business reporting on commitments to sustainable production is also becoming increasingly common. The Global Reporting Initiative (GRI) sets out reporting principles and performance indicators that organisations can use to measure and report on their operations. In an effort to create a globally accepted reporting framework, the GRI and the Prince's Accounting for Sustainability Project formed the International Integrated Reporting Committee, which brings together financial, environmental, social and governance information in a consistent and comparable format (www.accountingforsustainability.org/international_network).

The International Organisation for Standardisation (ISO) has also been instrumental in establishing standards for environmental management systems (EMS) and tools to guide companies around the world on the conduct of Lifecycle Assessments (LCAs).

> The sustainability life cycle approach deals with product life cycles from resource extraction, material production, manufacturing, use (or service delivery), to reuse, recovery, end-of-life treatment, and disposal of remaining waste.

The ISO 14000 standards enable an organisation to identify and manage the environmental impacts of its activities, establish environmental performance objectives and targets, and adopt a life cycle perspective in managing those impacts. Reporting activity, especially when it is undertaken through a standardised rubric, adds legitimacy to companies' green claims, while at the same time enabling them to highlight their green credentials in an effort to attract the growing body of ethical consumers. Many of the concepts discussed in Chapter 3 in relation to ecological modernisation find their practical expression within these initiatives. Among the best known, and some would argue the most integrated attempts to promote SP across the life cycle of its products is provided by the firm PUMA. PUMA is one of the world's leading 'sport lifestyle' companies, producing footwear, sports clothing and accessories. It also provides an example of good practice in relation to the exercise of Corporate Social Responsibility (CSR), discussed in the next section.

Businesses have also formed platforms for information-sharing on wider sustainable development issues, such as the World Business Council for Sustainable Development and the Africa Corporate Sustainability Forum. Many are based on principles of corporate social and

Case study 4.3

PUMA

Background

PUMA has a long history of engagement in CSR activities, stretching back to 1993 when it included a Code of Conduct in its supply contracts.

PUMA believes that it has both the 'opportunity and the responsibility to contribute to a better world for the generations to come'. This long-term sustainability vision has been embedded into the company's operations and product cycle. It also holds regular dialogues to discuss its sustainability strategy with relevant stakeholders.

Programmes

Puma operates several programmes, including PUMA.safe, which focuses on environmental and social issues; PUMA.peace, supporting global peace; and PUMA.creative, supporting artists and creative organisations. PUMA.safe has a sub-project on PUMA. safe ecology. This focuses on environmental issues, including implementation of the Restricted Substances List amongst all its suppliers; the collection of Environmental Key Performance Indicators, and improving energy and water efficiency as well as waste management. PUMA.safe ecology also works to develop more sustainable product materials.

Carbon neutrality

PUMA is involved in the UNEP Climate Neutral Network. In 2006, it signed the UN Global Compact and endorsed the 'Seal the Deal!' campaign in support of a binding international agreement on climate change under the Kyoto Protocol. It also operates a fair trade policy. PUMA also participates in other initiatives such as the Carbon Disclosure Project, and the Cotton Made in Africa initiative that supports sub-Saharan African farmers' efforts to secure sustainable livelihoods.

At the Business for the Environment Summit in Seoul, Korea in 2010, PUMA announced that it will completely offset its own global CO_2 emissions, to become the first carbon neutral company within the sport lifestyle industry.

Implementation

In 2002, PUMA published its first environmental and social report. In 2009, it conducted a CO_2 inventory based on the internationally accepted Greenhouse Gas Protocol guidelines. The inventory focused on properties or operations that PUMA owns or has direct control over, such as their offices, stores and warehouses, as well as the transportation of goods and staff business travel. This data was collected at a national level in forty-six countries and consolidated in a central environmental management database. The calculations show that PUMA produced approximately 107,000 tons of CO_2 company-wide in 2008. Several changes were subsequently introduced, including a shift to green electricity providers, installation of solar power, and new energy conservation measures. Targets and timetables for the reduction of GHG emissions were also set.

Similarly, PUMA discovered that the carbon footprint of a T-shirt could be reduced by cutting out one step in the dyeing process – and at the same time generating savings for the supplier. Other carbon footprint product studies were also conducted for a pair of shoes and PUMA's first fair trade football. In 2012 it opened a green store in Bangalore that uses solar PV panels, an air tunnel for cooling, shoe cartons-cum-carry bags and organic cotton for apparels. The surface layout of the store has been designed to ensure optimum use of daylight. Materials utilised in the store are locally sourced and recycled.

Exemplar

The company has received considerable international attention for its product life-cycle assessments and its CSR activities. In 2003 it was awarded a prize from the German Network of Business Ethics; it is listed in FTSE4GOOD (an index of CSR standards) and in the Dow Jones Sustainability Index. It also has Oekom sustainability ranking, a world leading sustainability rating agency. Its 2007 sustainability report received A+GRI status, has been rated by Greenpeace as 'green' and has achieved complete Fair Labour Association accreditation, which seeks fair working conditions in the supply chain.

Some limitations

PUMA's carbon offsetting does not include CO_2 emissions through transportation of products. Policies also presume an accurate record of the company's environmental impacts. Extracting information from smaller players further down its supply chain involves some calculated guesswork. The producers of its raw materials account for over a third of its indirect GHG emissions and more than half of its water consumption.

PUMA's much publicised 2011 Environmental Profit and Loss accounts were also criticised over their methodology, in particular how they actually valued the various environmental impacts. Most impacts on the environment, such as GHG emissions, water depletion, air pollution and so on, are unvalued by the market. In addition, PUMA's policies are not aimed at consumption reduction.

(*Sources*: PUMA, 2010; 2013; Wilson, 2010)

environmental responsibility (CSR). The principle has come to play a particularly important role as a driver of resource efficiency, cleaner production and environmental management programmes.

Corporate social responsibility

Corporate social responsibility emerged against a background of growing public distrust in companies following major environmental disasters, such as in Love Canal in the USA in the 1970s, toxic deaths in Bhopal, India in 1984, the Exxon Valdez oil spill in Alaska in 1989, and the Nike scandal in 1997, when workers in one of its contract factories in Vietnam were found to be exposed to toxic fumes up to 177 times the Vietnamese legal limit. Widely reported corporate scandals in the 1990s, such as fraud in ENRON, one of the world's leading suppliers of electricity and natural gas, and money laundering in the global food company Parmalat added to public concerns. This led to public information campaigns against many companies and to consumer boycotts. The 1990s were also marked by the emergence of social mobilisation, as seen for example in the growth of the anti-globalisation and the global democracy movements, epitomised by the Seattle protests at the World Trade Organisation's ministerial conference in 1999. These movements are leading the push for greater accountability and transparency in international trade agreements and in the behaviour of multinational companies (Roach and Toffel, 2008).

In the last two decades, a number of international initiatives have tried to rebuild trust in businesses by improving their environmental and social profiles and impacts. These are, at least in part, driven by fear of new regulation or backlash from groups that might harm the organisation, for example through consumer boycotts. They have given rise to the notion of corporate social responsibility, although several other terms are used, such as corporate responsibility, corporate sustainability or even business ethics (Goodwin and Bartlett, 2008). The classic definition of CSR provided by Wood continues to be widely quoted in the literature.

Corporate social responsibility is a business organisation's configuration of principles of social responsibility, processes of social responsiveness, and policies, programmes, and observable outcomes as they relate to the firm's societal relationships.

(Wood, 1991)

The International Organisation for Standardisation's voluntary guidance on social responsibility published in 2008 (the so-called ISO 26000) stresses the need to integrate CSR into all business processes so as to realise the potential contribution of business to sustainable development. According to this standard, an organisation is responsible for the impacts of its decisions and activities on society and the environment, and should therefore engage in transparent and ethical behaviour (ISO, 2008). What is central to the idea of CSR is that a company acts as a 'moral agent' in society, taking some degree of responsibility beyond that traditionally given to its shareholders (Martinuzzi et al., 2011).

Box 4.4

Multi-dimensional CSR

Economic	Environmental
Pursue sound corporate governance practices;	Support the protection of air, water, land biodiversity;
Ensure transparency through economic, social and environmental reporting;	Minimise amount of toxic substances, emissions, sewage and waste;
Engage in fair competition;	Conserve natural resources, apply renewable energy and reduce usage of raw materials;
Foster innovation;	Engage in climate protection;
Combat bribery and corruption;	Boost innovation for efficiency;
Employ socially responsible investment;	Consider the whole product life-cycle, facilitate reusability and recyclability of products.
Protect intellectual property rights;	
Offer safe and high-quality products/ services;	
Foster sustainable consumption and production;	
Implement sound risk management systems.	

Social	Global
Engage in fair and efficient human resource management;	Raise stakeholders' awareness of social and environmental topics;
Guarantee safety, occupational health and security;	Practise sound stakeholder management;
Respect freedom of association;	Facilitate sustainable supply chains;
Abandon discrimination and encourage diversity;	Respect human rights;
Respect consumer interests.	Engage in poverty reduction;
	Participate in development of public policies.

(*Sources*: Modified from Martinuzzi et al., 2010 and 2011)

The idea of gaining competitive advantage is the core of the business case for CSR. However, whether or not firms with elevated social or environmental performances also perform better financially is uncertain (Roach and Toffel, 2008). Furthermore, despite the many incentives and initiatives, SCP has not yet become a core criterion in financial decision making, presenting major obstacles to the replication and scaling-up of good practices (UNEP, 2012a). Whether or not efforts to address sustainable consumption are similarly limited is addressed in the next section.

Sustainable consumption

The 1998 UN Human Development Report on *Consumption for Human Development* (UNDP, 1998) pointed out that the growth in consumption, especially during the latter half of the twentieth century, was unprecedented in its scale and diversity, but badly distributed, resulting in gaping inequalities. While the world's dominant consumers are overwhelmingly concentrated in the developed world, the environmental damage from consumption falls most severely on the poor. Furthermore, rising pressures for conspicuous consumption can turn destructive, reinforcing exclusion, poverty and inequality (UNDP, 1998).

Consumption levels are closely related to population levels and both play key roles in the pursuit of sustainable development (Royal Society, 2012). At present there are no well charted ways for 10 billion people to achieve lifestyles like those enjoyed in the developed world, because further economic growth will collide with the Earth's finite sources (Royal Society, 2012). However, the relationship between population and consumption is not straightforward.

While technology can partly help address the need to supply the basic needs of the growing population of humans on the planet, without socio-political change the challenge of promoting sustainable development cannot be addressed.

Box 4.5

Population and consumption

Every member of society must consume a minimum amount of goods and services to survive. Each additional person added to the population will necessarily result in an increase in total or aggregate consumption. If per capita consumption is constant, each person added to the population increases total consumption by the same amount.

But per capita consumption is not constant, either over time or between richer and poorer segments of a population. Moreover, the age structure of a population, that is, the proportion of working age in the population relative to the number of young and old dependents, influences consumption and economic growth. The more people of working age there are in a population, the more potential producers there are, therefore there is more potential for economic growth

But again, life is not simple. There must be jobs for the people of working age, and the productivity of people at work depends heavily on their education, health, and the availability of local investment and credit to create productive jobs for them.

The more economic growth there is, the more consumption there may be, but not necessarily. Gross domestic product, for example, may grow because exports are increased, leading to little or no growth in consumption in the producing country.

Age structure affects consumption, as each age group's behaviour is governed differently by tastes and perceptions about needs, but is constrained by general standards of living. Increasing life expectancy increases the amount that people are able to save and therefore may increase their lifetime consumption levels.

(*Source*: Royal Society, 2012)

Defining sustainable consumption

The Oslo Roundtable on Sustainable Production and Consumption in 1994 proposed a working definition of sustainable consumption, known as the Oslo Definition, which has since become widely accepted.

Sustainable consumption is 'the use of goods and services that respond to basic needs and bring a better quality of life, while minimising the use of natural resources, toxic materials and emissions of waste and pollutants over the life cycle, so as not to jeopardise the needs of future generations'.

(*Source*: IISD, 1995)

Box 4.6

Characteristics of sustainable consumption

- Shared: ensuring basic needs for all;
- Strengthening: building human capabilities;
- Socially responsible: the consumption of some must not compromise the well-being of others;
- Sustainable: consumption must not mortgage the choices of future generations.

(*Source*: UNDP, 1998)

When viewed from the perspective of promoting global sustainable development, sustainable consumption means raising the consumption levels of more than a quarter of humanity that is currently unable to meet its basic needs. It also means discouraging patterns of consumption that have a negative impact on society and that reinforce inequalities and poverty. This requires, in turn, that efforts are made to achieve more equitable international burden-sharing in reducing and preventing global environmental damage and in reducing global poverty. The UNDP has encapsulated the requirements in four key actions: consumption must be shared, strengthening, socially responsible and sustainable.

Food policy highlights the variety of issues surrounding sustainable consumption. In addition to the ecological, social and economic aspects of food consumption, public health concerns are also an integral factor in ensuring the sustainable development of the food sector. The links between consumption and production are also visible. Rapidly increasing food and commodity prices, in part driven by increased fuel prices, reflect further the inter-linkages. Considering a projected population of nine billion in 2050, feeding the world will be a major challenge, given current consumption trends.

In the face of population growth, global environmental change and the rise of new consumption economies, for example in China and India, putting in place a more sustainable food supply system is clearly an urgent need. However, getting the world onto a sustainable consumption trajectory is a complex task, as seen from the following case study.

Case study 4.4

Sustainable food

Policy issues

There are clear links between food policy and other policy sectors, such as agriculture, transport, land use policy, health policy and environmental policy.

Many people live in food poverty and the world is plagued by periodic, major famines. In developed countries, high fat, sugar and salt content in food is contributing to poor health and there are substantial amounts of food waste. There are also growing concerns about food security, especially given biodiversity loss, climate change and human population growth.

Agricultural intensification remains a major plank of food production policy, including through the EU Common Agricultural Policy. This has negative impacts on the environment, including water and soil, and on biodiversity.

Food production is increasingly characterised by a dominant, highly concentrated agri-food sector. As a result, governments tend to restrict themselves to a marginal role and to soft policy tools, such as consumer information and education. Food regulations generally tend to concentrate on food safety issues and protecting consumers' health by setting standards of hygiene.

Sustainable food consumption

Sustainable consumption of food requires changes along the length of the food chain. The UK Sustainable Development Commission (2008) emphasises the need for safe, healthy and nutritious food, while ensuring decent living and working conditions of those working along the food chain, while also maintaining respect for animals.

Barriers to action

The achievement of sustainable food consumption requires policies that link policies on nutrition and food, environment, and health and social cohesion and those that address poverty alleviation.

However, individual choices and economic practices around food production and consumption are constrained by structural and institutional processes. Sustainable food consumption requires the availability of food that is produced in a sustainable way and which is affordable. The distance between production and consumption also needs to be narrowed. This would help to reduce food miles. An important contribution will come from the reduced consumption of meat and dairy products, which in turn requires value changes.

(*Source*: Reisch et al., 2011)

Complexity of task

Within the growth-orientated model of economic development, as countries increase their income and gross national product, so does their demand for goods and services rise. Expanding economies usually lead to higher consumption. Thus, as the economies and aspirations of emerging and Third World countries grow, so too does their consumption. In addition, during and after a global recession citizens are encouraged to spend rather than save, because under this model spending appears the only way to escape from the financial slow down (Jackson, 2009). This establishes a consumption momentum from which it is difficult to escape.

In addition, as international trade increases, the production of goods can become increasingly detached from direct consumption. Goods sold in one country are often made in another and, in turn, involve the use of natural resources from yet another country. This can be seen in the case of China, discussed in Chapter 2. This makes it more difficult for consumers of final goods to see the impact of their consumption, including on the natural environment and social conditions of other places (Royal Society, 2012). In addition, lock-in, particularly technological lock-in, can act as a barrier to the promotion of sustainable consumption, and more generally to system transition (Elzen et al., 2004). Central to the idea of lock-in is that technologies and technological systems follow specific paths that are difficult and costly to escape from. Lock-in is said to account for the continued use of a range of inferior technologies, ranging from the QWERTY keyboard to the internal combustion engine (Perkins, 2003). Developed economies are also locked into a complex of hydrocarbon-intensive technologies and infrastructures.

Technological lock-in refers to macro level forces that create systematic barriers to the diffusion and adoption of efficient and sustainable technologies.

(Carrillo-Hermosilla, 2013)

Addressing consumption also raises another issue: that of the social value associated with possessing goods and services. There is a long history of social science research that shows the links between consumption, social values and status. Veblen in *The Theory of the Leisure Class* (1899) coined the phrase 'conspicuous consumption' to capture acts of purchasing and using certain goods and services so as to identify oneself to others as having superior wealth and social standing. Horkheimer and

Adorno (1947) have also critically explored the ways in which culture and the creation of meaning in modern consumer societies is handed over to commercial interests and industrial processes. More recently, Douglas and Isherwood (1979) argued that practices of consumption communicate meaning and so make culture visible, with Douglas herself arguing in her famous quotation: 'An individual's main objective in consumption is to help create the social universe and to find in it a creditable place' (Douglas, 1976: 207).

In speaking about consumption from a sustainable development perspective, however, it is important to make a distinction between consumption and consumerism: consumption per se is not inherently unsustainable. Consumerism is best understood as a cultural condition in which economic consumption becomes a way of life (Miles, 1998). As such, consumerism is inherently unsustainable as it is a state of affairs in which more and more cultural functions are handed over to the activity of consumption, such that it colonises more and more aspects of human experience (Jackson, 2006).

Viewed through the lens of consumerism, the pursuit of sustainable consumption cannot be framed as a mere technical problem – one in which efficiency gains in resource productivity allow levels of consumption to remain the same or to increase (Evans and Jackson, 2008). This fails to address the ways in which consumption is embedded in modern ways of living (Jackson, 2004). From this perspective, the challenge for sustainable consumption is about addressing the social and cultural functions performed by consumerism.

> the transition to a sustainable society cannot hope to proceed without the emergence or re-emergence of some kinds of meaning structures that lie outside the consumer realm.
>
> (Jackson, 2004: 17)

This makes the promotion of sustainable consumption about the development of ways of living that maintain the satisfaction of needs and conditions of human flourishing in a manner that is less reliant on the economic consumption of goods and services. This process has been referred to as decommodification.

Decommodification: any political, social, or cultural process that reduces the scope and influence of the market in everyday life.

(Vail, 2010)

Decommodification forms part of the adoption of what is referred to as 'sustainable lifestyles' (Spaargaren, 2003).

Sustainable lifestyles

Several groups have argued that more radical approaches are needed, ones that move beyond the instrumental rationality embedded in the efficiency approaches to sustainable consumption and production. This involves addressing the causes of unsustainability, and often from the standpoint of moral reasoning. The European Environment Bureau, for example, has argued that a planned, intelligent response to SCP requires a sustainability transition approach, one which refocuses the lens of our societal objectives, economic activities and behaviour (Fedrigo and Tukker, 2009). This new view promotes human relationships, and community cohesion, rather than just growth in terms of money and material goods. Moving beyond GDP as a measure of progress brings in an economic agenda based on well-being. The promotion of well-being is also stressed by the UN, as discussed in Chapter 6. Sustainability transition also requires accepting planetary limits and bringing actions back within ecological carrying capacity (Fedrigo and Tukker, 2009).

The approaches that promote sustainable lifestyles address the citizen as a member of society, rather than as a mere consumer (Micheletti and Stolle, 2012). One way this is done is by considering the social and physical infrastructure in which production and consumption take place. For example, the public cannot be expected to drive cars less often if there are no efficient and affordable public transport systems available to them, or if it costs less to use more polluting forms of transport. There are also several social innovation practices that have tried to move the agenda even further. These include the WWF's 'One Planet Living' campaign that promotes low-impact living in the areas of housing, food and mobility. The Transition Town movement provides another example: a bottom-up movement to realise social and environmental goals such as carbon neutrality and the development of local food supplies. The Slow Food movement also promotes traditional, locally produced foods (Baker and Mehmood, 2014; Fedrigo and Tukker, 2009). Among the best known and most wide reaching social innovation projects in pursuit of sustainable futures is that of the Fair Trade movement.

Case study 4.5

Fair Trade

Background

Fair Trade (or 'Fairtrade') arose in response to the failure of conventional trade to deliver sustainable livelihoods and development opportunities to people in the poorest countries of the world. Fair Trade is not a charity but a trading partnership that seeks greater equity in international trade. Fair Trade organisations support producers, raise awareness and campaign for changes in the rules and practice of conventional international trade.

Links to sustainable development

Fair Trade contributes to sustainable development by offering better trading conditions to, and securing the rights of, marginalised producers and workers, especially in the Global South.

The Charter of Fair Trade Principles

In 2009, Fairtrade International (FLO) along with the World Fair Trade Organisation adopted the Charter of Fair Trade Principles. Its key principles include:

● Market access for marginalised producers;

● Sustainable and equitable trading relationships;

● Capacity building and empowerment;

● Consumer awareness raising and advocacy.

FLO

FLO is a non-profit, multi-stakeholder body responsible for the strategic direction of Fairtrade, and sets Fairtrade standards and supports producers. It co-ordinates labels for around fifteen product groups, from agricultural commodities to gold and sports balls. This helps consumers identify goods that have been produced under socially fairer and more environmentally friendly conditions. These are national organisations that market Fairtrade in their country. In 2009, FLO-certified product sales amounted to €3.4 billion worldwide. Thousands of products carry the Fairtrade mark.

There are two distinct sets of Fairtrade standards, which acknowledge different types of disadvantaged producers. One set applies to smallholders who are working together in co-operatives or other organisations with a democratic structure. The other standard applies to workers whose employers pay decent wages, guarantee the right to join trade unions, ensure health and safety standards and provide adequate housing where relevant.

Products have a Fairtrade Price, which is the minimum that must be paid to the producers. In addition, producers get an additional sum, the Fairtrade Premium, to invest in their communities.

Significance

Fair Trade can be seen as a 'social contract' in which buyers (including final consumers) agree to do more than is expected by the conventional market, such as paying fair prices, providing pre-finance and offering support for capacity building. In return for this, producers use the benefits of Fair Trade to improve their social and economic conditions, especially among the most disadvantaged members of their organisation.

(*Sources*: World Fair Trade Organisation and Fairtrade Labelling Organisations International, 2009; Fairtrade International, 2011)

Concern about the impacts of consumerism on other dimensions of human flourishing and needs, such as personal and social relationships, has also led to another movement, the Voluntary Simplicity movement. This movement promotes a deeper approach that reflects strong sustainable development values. Unlike many of the other movements and in sharp contrast to the SCP agenda of many international organisations, it is particularly concerned to reduce consumption volumes. It is built upon the rejection of consumerism.

Box 4.7

Voluntary simplicity

Background

Voluntary simplicity, or simple living, is a way of life that rejects the high-consumption, materialistic lifestyles of consumer cultures. It affirms 'the simple life' or 'downshifting'.

The rejection of consumerism arises from recognition that high consumption habits are degrading the Earth and are unethical in a world of great human need. It rejects the idea that the meaning of life consists in the consumption or accumulation of material things.

Core idea

The movement is built on the affirmation of simplicity. This holds that very little is needed to live well and that abundance is a state of mind, not a quantity of consumer products. Human beings can live meaningful, free, happy, and infinitely diverse lives, while consuming no more than a sustainable and equitable share of nature.

Duane Elgin, one of the main contemporary activists expounding voluntary simplicity, defines it as 'a manner of living that is outwardly simple and inwardly rich … a deliberate choice to live with less in the belief that more life will be returned to us in the process'.

David Shi defines voluntary simplicity as 'enlightened material restraint'.

This counter-cultural philosophy has deep philosophical roots and can be found in various religious traditions, including Celtic spirituality, Buddhism and the historical peace churches, such as the Amish. Deep ecology proponent Arne Naess also expounded voluntary simplicity, as did the ecologist Henry David Thoreau.

Approach

Sometimes called 'the quiet revolution', voluntary simplicity involves providing for material needs as simply and directly as possible, minimising expenditure on consumer goods and services, and directing progressively more time and energy towards pursuing non-materialistic sources of satisfaction and meaning. This means accepting a lower income and a lower level of consumption, in exchange for more time and freedom to pursue other life goals.

The primary attributes of the simple life include: thoughtful frugality; a suspicion of luxuries; a reverence and respect for nature; a desire for self-sufficiency; a commitment to conscientious rather than conspicuous consumption; a privileging of creativity and contemplation over possessions; an aesthetic preference for minimalism and functionality; and a sense of responsibility for the just uses of the world's resources.

Personal and social progress is measured by increases in the qualitative richness of daily living, the cultivation of relationships, and the development of social, intellectual, aesthetic, and/or spiritual potentials.

Voluntary simplicity does not mean living in poverty, becoming an ascetic monk, or indiscriminately renouncing all the advantages of science and technology.

Development

The Voluntary Simplicity movement is a diverse social movement made up of people who are resisting high consumption lifestyles and who are seeking, in various ways, a life of lower consumption but higher quality. It is spread throughout the globe, and runs such activities as National Downshifting Week (UK) and the Small House Movement (USA), to name but two.

Critique of voluntary simplicity

Voluntary simplicity can be misinterpreted as glorifying or romanticising poverty. There is also a risk that advocates may be understood to be downplaying the plight of those in the world who genuinely live lives oppressed by material deprivation.

It has also been criticised for being necessarily agrarian and for being regressive and anti-technology. However, while it does question the assumption that science and technology are always the most reliable paths to health, happiness, and sustainability, it also holds that a blind faith in science can itself be 'anti-progress'.

(*Sources*: Alexander, 2009; Elgin, 1998; 2000; Shi, 2007; http://simplicitycollective.com/start-here/what-is-voluntary-simplicity-2)

Despite the important contribution that these approaches have made to the pursuit of sustainable consumption, many remain the pursuit of a marginal minority. Whether or not these actions could be easily 'scaled up' and transferred to a society-wide level remains to be investigated.

Conclusion

To realise the Brundtland goal of increasing the levels of consumption of the world's poor while staying within the carrying capacity of the planet, both sustainable patterns of production and of consumption are needed. Agenda 21 recognises the great imbalances in global and national patterns of consumption and production. To address these imbalances, sustainable development requires improving the environmental quality and addressing the equity of both production systems and consumption patterns.

Change can be difficult to realise, because society is locked into unsustainable patterns which are difficult to alter: many such patterns are deeply embedded in technological systems and rooted in cultural habits. However, international organisations, governments and the business sector have recognised that improvements can be made to patterns of consumption and production in ways that raise the quality of life and enhance efficiency and competitiveness, while community-based organisations have supported action to move towards sustainable lifestyles.

However, a key dividing line remains. On the one hand, there are those who claim that necessary improvements in environmental quality and in equality can be achieved through the substitution of more efficient and less polluting goods and services: for example, using low waste, energy efficient technologies. On the other hand there are claims that the promotion of sustainable development requires reductions in the volumes of goods and services that are both produced and consumed. These positions map on to the lower and higher rungs, respectively, of the ladder of sustainable development. Addressing patterns of production and consumption has proved insufficient to promote sustainable development. On a planet characterised by finite resources and fragile planetary systems, stronger forms of sustainable development call for levels of production and their related consumption activities to be reduced.

Summary points

- Both sustainable production and sustainable consumption are key determinants of sustainable development.
- While interrelated, differences in approaches can be seen by exploring sustainable production and sustainable consumption separately.
- Sustainable production tends to focus on efficiency measures, which can decrease pressure on natural resources but can also be subject to the rebound effect.
- Efforts to promote sustainable consumption can be blocked by institutional, technical, and structural barriers and by societal values that result in an unwillingness to address levels of consumption.
- Alternative lifestyle approaches move from an instrumental rationality to moral reasoning, promoting the virtues of just and simple living.

Further reading

Alexander, S. (ed.) (2009), *Voluntary Simplicity: The Poetic Alternative to Consumer Culture* (New Zealand: Stead and Daughters).

Jackson, T. (2006), *The Earthscan Reader on Sustainable Consumption* (London: Earthscan).

UNEP (2012), *The Global Outlook on Sustainable Consumption and Production (SCP) Policies* (New York: UNEP), available online at www.unep.fr/scp/go/publications.htm

Web resources

CORPUS Knowledge Hub, 'Sustainable Consumption and Production', www.scp-knowledge.eu/

International Institute for Sustainable Development has document links and other online resources on sustainable consumption and production, at www.iisd.ca/consume/overview.html.

References

Alexander, S. (ed.) (2009), *Voluntary Simplicity: The Poetic Alternative to Consumer Culture* (New Zealand: Stead and Daughters).

Baker, S. (1996), 'Sustainable Development and Consumption, the Ambiguities: The Oslo Ministerial Roundtable Conference on Sustainable Production and Consumption', *Environmental Politics*, 5 (1): 93–100.

Baker, S. and Mehmood, A. (2014), 'Social Innovation and the Governance of Sustainable Places', *Local Environment*, doi: 10.1080/13549839.2013.842964.

Carrillo-Hermosilla, J. (2013), 'Technological Lock-in', *The Encyclopaedia of Earth*, available online at www.eoearth.org/article/Technological_lock-in

CEC (2008), *Sustainable Consumption and Production and Sustainable Industrial Policy Action Plan* (Brussels: Commission of the European Communities, COM(2008) 397 final), available online at http://ec.europa.eu/environment/eussd/pdf/com_2008_397.pdf

Douglas, M. (1976), 'Relative Poverty, Relative Communication', in A. Halsey (ed.), *Traditions of Social Policy* (Oxford: Basil Blackwell).

Douglas, M. and Isherwood, B. (1979), *The World of Goods: Towards an Anthropology of Consumption* (London: Penguin Books).

EEA (n.d.), 'Sustainable Consumption and Production' (Copenhagen: EEA, EnviroWindows, platform for knowledge sharing and development), available online at http://ew.eea.europa.eu/ManagementConcepts/scp/

Elgin, D. (1998), *Voluntary Simplicity: Toward a Way of Life That Is Outwardly Simple, Inwardly Rich* (New York: William Morrow, revised edn).

Elgin, D. (2000), *Promise Ahead: A Vision of Hope and Action for Humanity's Future* (New York: HarperCollins).

Elzen, B., Geels, F. W. and Green, K. (eds) (2004), *System Innovation and the Transition to Sustainability: Theory, Evidence and Policy* (Cheltenham: Edward Elgar).

Evans, D. and Jackson, T. (2008), 'Sustainable Consumption: Perspectives from Social and Cultural Theory' (Guilford: University of Surrey, RESOLVE Working Paper 05-08), available online at www.esrc.ac.uk/my-esrc/.../330b057b-d023-4d85-ac4c-2b5ef2a16254

Fairtrade International (2011), *What Is Fair Trade?* available online at www.fairtrade.net/what_is_fairtrade.html

Fedrigo, D. and Tukker, A. (2009), *Blueprint for European Sustainable Consumption and Production: Finding the Path of Transition to a Sustainable Society* (Brussels: EEB Publication no. 2009/07), available online at http://eeb.org/publication/2009/0905_SCPBlueprint_FINAL.pdf

Goodwin, F. W. and Bartlett, J. L. (2008), *Public Relations and Corporate Social Responsibility (CSR)* (Queensland University of Technology, Working Paper), available online at QUT Digital Repository, http://eprints.qut.edu.au/15427/1/15427.pdf

Horkheimer, M. and Adorno, T. W. (1947), *Dialectic of Enlightenment* (London: Verso).

IISD (1995), International Institute for Sustainable Development Digital Archive, Oslo Ministerial Roundtable Conference on Sustainable Production and Consumption, 6–10 February 1995, Oslo, Norway. Background papers for meeting: Elements for an International Work Programme on Sustainable Production and Consumption, Part 1: 'The Imperative of Sustainable Production and Consumption', available online at www.iisd.ca/consume/oslo004.html

ISO (International Organisation for Standardization) (2000), 'Conserving Our Environment: Sustainable Production and Consumption' (Geneva: ISO), available online at www.iso.org/iso/livelinkgetfile?llNodeId=22292&llVolId=-2000

ISO (2008), *The ISO Survey 2007* (Geneva: ISO).

Jackson, T. (2004), 'Consuming Paradise? Unsustainable Consumption in Cultural and Social-Psychological Context', in K. Hubacek, A. Inaba and S. Stagl (eds), *Driving Forces of and Barriers to Sustainable Consumption*, Proceedings of an International Workshop, Leeds, 5–6 March, 9–26, available online at http://portal.surrey.ac.uk/pls/portal/docs/page/eng/research/ces/cesresearch/ecological-economics/projects/fbn/paradise.pdf

Jackson, T. (2006), *The Earthscan Reader on Sustainable Consumption* (London: Earthscan).

Jackson, T. (2009), *Prosperity without Growth? The Transition to a Sustainable Economy* (London: Sustainable Development Commission), available online at www.sd-commission.org.uk/data/files/publications/prosperity_without_growth_report.pdf

Martinuzzi, A., Gisch-Boie, S. and Wiman, A. (2010), *Does Corporate Responsibility Pay Off? Exploring the Links between CSR and Competitiveness in Europe's Industrial Sectors* (Vienna: Vienna University of Economics and Business Administration, Research Institute for Managing Sustainability).

Martinuzzi, A., Krumay, B. and Pisano., U. (2011), *Focus CSR: The New Communication of the EU Commission on CSR and National CSR Strategies and Action Plans* (Vienna: ESDN European Sustainable Development Network, Quarterly Report no. 23, December), available online at www.sd-network.eu/quarterly%20reports/report%20files/pdf/2011-December-The_New_Communication_of_the_EU_Commission_on_CSR_and_National_CSR_strategies.pdf

MEA (2005), *Living Beyond Our Means: Natural Assets and Human Well-Being, Statement from the Board*, available online at www.maweb.org/documents/document.429.aspx.pdf

Micheletti, M. and Stolle, D. (2012), 'Sustainable Citizenship and the New Politics of Consumption', *Annals of the American Academy of Political and Social Science*, 644: 88–120.

Miles, S. (1998), *Consumerism as a Way of Life* (London: Sage).

OECD (2003), *Environmental Indicators: Development, Measure and Use* (Paris: OECD, reference paper), available online at www.oecd.org/environment/indicators-modelling-outlooks/24993546.pdf

Perkins. R. (2003), 'Technological "Lock-in"', *International Society for Ecological Economics Internet Encyclopaedia of Ecological Economics*, available online at http://isecoeco.org/pdf/techlkin.pdf

PUMA (2010), 'Puma Announces Carbon Neutral Plans for 2010', available online at http://safe.puma.com/us/en/2010/04/puma-announces-carbon-neutral-plans-for-2010/

PUMA (2013), 'Sustainability', available online at http://about.puma.com/sustainability/

Reisch, L. A., Lorek, S. and Bietz, S. (2011), 'Policy Instruments for Sustainable Food Consumption' (CORPUS Consortium, Discussion Paper 2), available online at www.scp-knowledge.eu/sites/default/files/Food_Policy_Paper.pdf

Roach, B. and Toffel. M. (2008), 'Social and Environmental Responsibility of Corporations', in Cutler J. Cleveland (ed.), *Encyclopaedia of Earth* (Washington, DC: Environmental Information Coalition, National Council for Science and the Environment), available online at www.eoearth.org/article/Social_and_environmental_responsibility_of_corporations

Royal Society UK (2012), *People and the Planet* (London: The Royal Society Science Policy Centre, Report, 01/12, available online at http://royalsociety.org/uploadedFiles/Royal_Society_Content/policy/projects/people-planet/2012-04-25-PeoplePlanet.pdf

Shi, D. E. (2007), *The Simple Life: Plain Living and High Thinking in American Culture* (Athens, GA: University of Georgia Press, 2nd edn).

Spaargaren, G. (2003), 'Sustainable Consumption: A Theoretical and Environmental Policy Perspective', *Society and Natural Resources*, 16: 687–701.

UN (1992), *Agenda 21* (New York: UN, Division for Sustainable Development), available online at http://sustainabledevelopment.un.org/content/documents/Agenda21.pdf

UN (2012), *A 10-year Framework of Programmes on Sustainable Consumption and Production Patterns* (Rio de Janeiro, Brazil: UN, A/CONF.216/5, 19 June), available online at www.unep.fr/scp/pdf/ACONF_216_5_10YFP.pdf

UNDP (1998), *Human Development Report 1998: Consumption for Human Development* (New York: UNDP), available online at http://hdr.undp.org/en/reports/global/hdr1998/

UNEP (2012a), *The Global Outlook on Sustainable Consumption and Production (SCP) Policies* (New York: UNEP), available online at www.unep.fr/scp/go/publications.htm

UNEP (2012b), 'Sustainable Consumption and Production for Poverty Alleviation' (New York: UNEP), available online at www.unep.org/pdf/SCP_Poverty_full_final.pdf

Vail, J. (2010), 'Decommodification and Egalitarian Political Economy', *Politics and Society*, 38 (3): 310–346.

Veblen, T. (1899), *The Theory of the Leisure Class* (London: Prometheus Books, 1998 edn).

WBSCD (World Business Council for Sustainable Development) (2000), *Eco-efficiency: Creating More Value with Less Impact* (Conches-Geneva: WBCSD), available online at http://oldwww.wbcsd.org/DocRoot/BugWjalu0wHL0IMoiYDr/eco_efficiency_creating_more_value.pdf

von Weizsacker, E., Lovins, A. B. and Lovins, L. H. (1998), *Factor Four: Doubling Wealth, Halving Resource Use: The New Report to the Club of Rome* (London: Routledge).

Wilson, B. (2010), 'Puma Thinks Outside the Box with Eco-packaging', *BBC News*, 19 April, available online at news.bbc.co.uk/1/hi/business/8619165.stm

Wood, D. J. (1991), 'Corporate Social Performance Revisited', *The Academy of Management Review*, 16 (4): 691–718.

World Fair Trade Organization and Fairtrade Labelling Organizations International (2009), 'A Charter for Fair Trade Principles', available online at www.fairtrade.net/fileadmin/user_upload/content/2009/about_us/documents/Fair_Trade_Charter.pdf

Part II
International engagement with sustainable development

Global governance and UN environment summits

Key issues

- Global environmental governance
- Principles of good governance
- UN summits and conferences
- UN Global Compact
- Private governance.

Introduction

The previous chapters discussed how the promotion of sustainable development is a global task, one that is directed towards the establishment of a more equitable relationship between the North and the South, the rich and the marginalised, the peoples of the present and the generations of the future. The Brundtland Report was clear that the promotion of sustainable development requires effective, international co-operation. It called for the strengthening of international institutions for global environmental governance, as well as changes in existing international agencies concerned with a range of policy areas, including development, trade regulations and agriculture.

> The objective of sustainable development and the integrated nature of the global environment/development challenges pose problems for institutions, national and international, that were established on the basis of narrow preoccupations and compartmentalized concerns
>
> (WCED, 1987: 9)

The Brundtland Report came at a time of growing concern about the inability of national level institutions to address newly emerging, global environmental problems. In turn, the Brundtland approach, with its focus on sustainable development as a quintessential global project, contributed to the need for new arrangements at the international level. This chapter examines the rise of global environmental governance through the United Nations (UN) system and the emergence of the principles of good governance to guide engagement. This is followed by a detailed examination of the major UN summits and conference events. The significance of these events is then explored, revealing both positive, but also highly negative assessments of the contribution of the UN to the promotion of sustainable development.

The development of global environmental governance

The UN has played a particularly important role in the development of a system of global governance for the promotion of sustainable development. There are now over thirty specialised UN agencies and programmes involved in the promotion of sustainable development at the global level. These include the Food and Agricultural Organisation, the UN Development Programme (UNDP) and the UN Environment Programme (UNEP). The UNEP is the overall co-ordinating environmental organisation of the UN. It is charged with the tasks of assessing global, regional and national environmental conditions and trends; developing international and national environmental instruments; and strengthening institutions for the wise management of the environment (UNEP, 2003). There are also a substantial number of international regimes, dealing with matters across the environmental, social and economic dimensions of sustainable development. These include regimes addressing environmental issues such as hazardous waste, ozone depletion, biodiversity loss and climate change; social matters, such as in relation to poverty alleviation; through to economic issues such as aid and trade. An international regime exists when there are agreed upon formal and informal institutional structures, principles, norms, rules and decision making procedures and action programmes to address a specific issue (Young, 1997). The task of such regimes is to secure negotiations, set standards of management, and find effective responses to challenges, including through adopting implementation strategies and funding mechanisms. Examples of such regimes include those developed under the auspices of the UN to deal with climate change (the UN FCCC) and to address biodiversity loss (the CBD), both of which are dealt with in subsequent chapters. Such conventions provide a general framework for action, while their associated technical protocols outline steps to address specific aspects of the problem. Regimes typically include multilateral environmental agreements (MEAs), that is, legally binding agreements between states relating to environmental matters. UNEP administers the growing array of MEAs, including the Vienna Convention's Montreal Protocol on Substances That Deplete the Ozone Layer, the Convention on International Trade in Endangered Species of Wild Fauna and Flora (CITES), the Basel Convention on the Control of Transboundary Movements of Hazardous Wastes and Their Disposal, the Convention on Prior Informed Consent, Procedure for Certain Hazardous Chemicals and Pesticides in International Trade and the Cartagena Protocol on Biosafety to the Convention on Biological Diversity, as well as the Stockholm Convention on Persistent Organic Pollutants. Other international regimes that lie formally outside the UN's remit, especially the WTO, which deals with global rules of trade, also play an important part in global sustainable development governance, an issue discussed in Chapter 3.

While the internationalisation of environmental governance is primarily built upon negotiations and agreements between states, non-state actors, including environmental non-governmental organisations (NGOs) and economic interest groups play an increasingly significant role. This collective activity, including the development of international environmental regimes, their formal and informal institutional arrangements, their norms and principles, their MEAs as well as their conventions and protocols, when it is combined with the involvement of civil society, has resulted in the emergence of what has been termed 'global environmental governance'.

Global environmental governance is the establishment and operation of a set of rules of conduct that define practice, assign roles and guide interaction so as to enable state and non-state actors to grapple with collective environmental problems within and across state boundaries

(Young, 1997)

The development of global environmental governance has led to the claim that the state no longer plays a central role in international environmental governance, a claim refuted in Chapter 3. However, a more balanced view would see that, while international environmental governance is undoubtedly undergoing some new and rather innovative developments, this has not been at the expense of the state. The state remains the primary actor in global environmental governance, even if it now plays that role in close collaboration with other actors (Vogel, 2010). According to Young, these efforts both reflect and affect significant developments in the character of international society:

> Although states remain central players in natural resources and environmental issues, nonstate actors have made particularly striking advances both in the creation of environmental regimes and in efforts to make these regimes function effectively once they are in place. Environmental concerns are clearly one significant force behind the rising interest in the idea of global civil society.
>
> (Young, 1997: 2)

The role of the state in the promotion of sustainable development, and how this role is increasingly combined with participation from civil society and the engagement of economic actors, is discussed in considerable detail in Chapter 11. The current chapter concentrates on the international level; and the role of economic actors and groups from civil society and the emergence of what has been referred to as 'private governance'. First, attention is turned to what may be referred to as the conduct of global governance.

Promoting good governance

During the 1980s, the notion of good governance gained prominence in development discourse, especially among donor countries. However, it was soon used in a broader political manner. By the 1990s, the benchmark of good governance had expanded to include governance practices whose legitimacy is derived from a democratic mandate, which followed the rule of law, promoted free market competition, and encouraged greater participation of civil society in policy making processes (Kohler-Koch and Rittberger, 2006: 29).

The development of principles of 'good governance' has provided a normative framework for the governance of sustainable development and has particular salience within the international arena. The Brundtland Report provided an important starting point for this by arguing that reconciling environmental and development goals hinges upon the fair and equitable distribution of bargaining power, so as to ensure that the influence and voice of the world's poor is heard and reflected in international decisions and outcomes. The Report also spoke of the need to address the imbalances in the current structures of global governance and to create more inclusive international governance systems.

Subsequently, the UN summits have led to a set of expectations about the conduct of global environmental governance regimes. This has led to the elaboration of a set of principles for good governance practice for the promotion of sustainable development. Some of these were elaborated at the Rio Earth Summit, where the Rio Principles outlined several of the core elements of good governance for sustainable development. The WSSD Plan of Implementation, stemming from the UN-sponsored Johannesburg Summit, discussed below, also supported the argument that 'good governance within each country and at the international level is essential for sustainable development' (UN, 2002a: 2). Following from the Rio Summit and subsequent summits and conferences, 'good governance' within the UN now refers to a diverse array of criteria including effectiveness and efficiency, rule of law, participation, accountability, transparency, respect for human rights, absence of corruption, tolerance and gender equality (UNDP, 1997).

Several other international organisations, including the EU, have also emphasised good governance, not only within their own policies and practices but also internationally. The OECD has also developed a similar checklist of what good governance for sustainable development might mean in practice for policy makers (OECD, 2002). By the 1990s, the World Bank had also become a prominent advocate of the idea of propagating sound fiscal management and administrative efficiency as a precondition for the promotion of sustainable development (World Bank, 2002). However, it has been argued that the Bank has adopted quite a restrictive approach to 'good governance', mainly confining itself to matters concerning the effectiveness of government, and the introduction of neoliberal economic reforms. This has been at the expense of addressing the quality of governance, in particular its democratic content, including in relation to the distribution of power and resources within a given society (Santiso, 2001).

Several of these principles of good governance are discussed in detail in this book. The precautionary principle is discussed in Chapter 7, while EU efforts to promote environmental policy integration are discussed in Chapter 9. Chapter 6 explores how the principle of common but differential responsibilities has shaped international environmental law, while Chapter 11 looks in some detail at the principles of participation and of

---

Box 5.1

UN principles of good governance

General principles of good governance

- Application of the rule of law;
- Transparency and accountability;
- Effectiveness and efficiency;
- Subsidiarity, that is, that actions should be taken at the appropriate level of government;
- Participation and responsiveness to the need of stakeholders;
- Gender equity.

Specific principles of good governance for sustainable development

- Precautionary principle;
- Principle of common but differentiated responsibilities;
- Ecosystems approach;
- The polluter pays principle;
- Principle of environmental policy integration.

---

gender equality. These explorations will reveal how the good governance principles have generated severe controversy among the states participating in the UN summits.

The United Nations environmental summits

Summits and conferences represent the public face of the UN's environmental engagement (Death, 2010). They have resulted in numerous declarations, plans of action and sector-specific environmental conventions or laws. Summits have become central events shaping current international sustainable development policy and, more broadly, they are having a major influence on international relations, especially in relation to trade policy, North–South relations, and, as discussed above, have been instrumental in the development of civil society at the global level.

The UN Conference on the Human Environment, held in Stockholm in 1972, marked the beginning of the new era of international co-operation on the environment. The Conference was organised around two basic beliefs: that poverty was a cause of environmental degradation and that environmental problems could be solved by the application of scientific and technological knowledge. The Conference was significant for several reasons. It helped legitimise the view that environmental problems are of a global nature and therefore require international engagement. This led to the establishment of the UNEP, with a mandate

> to be the leading global environmental authority that sets the global environmental agenda, that promotes the coherent implementation of the environmental dimensions of sustainable development within the United Nations system and that serves as an authoritative advocate for the global environment.
>
> (UNEP, 2003)

The Conference also stimulated governments in over 100 countries to establish environmental ministries and agencies. It provided the impetus for the subsequent and rather rapid development of international environmental law, especially in relation to marine pollution, depletion of the ozone layer and transboundary movement of hazardous waste. It also marked the beginning of an explosive growth in the number of environmental NGOs operating at the international level.

Ten years later, the Stockholm +10 Conference was held in Nairobi, Kenya. In contrast to the belief in the capacity of scientific and technological knowledge to solve environmental problems that marked the 1972 Stockholm Conference, the Nairobi Conference paid particular attention to the need to address the underlying economic and social causes of environmental degradation. This led to the establishment of the World Commission on Environment and Development (WCED), chaired by Gro Harlem Brundtland, discussed in Chapter 2. One of the concrete proposals made in the Brundtland Report was for the UN to hold an 'Earth Summit'.

The 1992 Rio Earth Summit

In 1989, the General Assembly of the UN called for a UN conference on environment and development, later to be known as the Earth Summit, held in Rio de Janeiro in 1992. The Summit was attended by the largest ever number of heads of state and of government. This indicates the importance that had by then become attached to the subject of environmental deterioration and to recognition of the need to find ways to reconcile environmental protection with economic development policies at the international level.

The Rio Earth Summit focused on two key issues: first, the link between environment and development; second, the practical issues surrounding the promotion of sustainable development, especially the introduction of policies that balance environmental protection with social and economic concerns, particularly in the developing world. The Earth Summit resulted in an ambitious programme for promoting sustainable development.

First, the Rio Declaration on Environment and Development was agreed, presenting twenty-seven principles of sustainable development. These include the normative principles of common but differentiated responsibilities and the equity principles (inter- and intra-generational equity), as discussed in Chapter 2. The Declaration also contains several principles of good governance, including the precautionary principle and the principle of subsidiarity. Many of these principles address development concerns, but the Declaration also dealt with principles concerning trade and the environment and the role of civil society and social and economic groups. The Rio Declaration also stresses the right to, and need for, economic development and poverty alleviation, especially in the developing world.

Second, Agenda 21 was agreed, which presented not only an astute analysis of the causes and symptoms of unsustainable forms of development, but an authoritative set of ideas on how to promote sustainable

---

Box 5.2

Agreements reached at the Rio Earth Summit, 1992

a. Rio Declaration on Environment and Development
b. Agenda 21
c. UN Framework Convention on Climate Change (UN FCCC)
d. UN Convention on Biological Diversity (CBD)
e. Forest Principles.

development in practice. Agenda 21 consists of forty chapters that out-line action plans across a wide range of areas. It stresses the importance of bottom-up participation, especially community-based approaches, through Local Agenda 21, discussed in Chapter 8 of this book. Chapter 8 of Agenda 21 also called for countries to adopt national strategies for sustainable development that build upon and harmonise the various sec-toral economic, social and environmental policies and plans that are operating in the country.

Third, two legally binding conventions were signed at Rio: the UN Framework Convention on Climate Change (UN FCCC) and the UN Convention on Biological Diversity (CBD), discussed in Chapters 6 and 7 respectively. The Convention to Combat Desertification also resulted from discussions held at Rio, but was not agreed until 1995. Fourth, the Forest Principles, known formally as the 'Non-legally Binding Authoritative Statement of Principles for a Global Consensus on the Management, Conservation and Sustainable Development of all Types of Forests' was also agreed. This outlined general principles of forest protection and sustainable management, while affirming the sov-ereign right of the state to exploit forests. It was subsequently to lead in 2007 to the 'Non-legally Binding Instrument on all Types of Forests' (UN, 2007). Some had hoped, however, for a legally binding forest convention, rather than the 'softer' forest principles. Agenda 21, the Rio Declaration and the forest principles are non-binding statements of intent, sometimes termed 'soft laws', which provide guidelines for future actions.

Fourth, the Rio Earth Summit led to the establishment of new institu-tions. Chief among these was the Commission on Sustainable Develop-ment (CSD), whose primary role was to monitor progress on the agreements reached at Rio. However, while it was initially envisaged as a high-level institution, it was linked to the Economic and Social Council rather than the General Assembly of the UN, a link that is seen as having limited its powers. The CSD had a very broad mandate and programme of work, which it pursued through annual sessions, less for-mal inter-sessional meetings, voluntary reporting from member states on implementation of Agenda 21 and through holding 'multi-stake-holder dialogues'. The CSD has also been charged with establishing sustainable development indicators, work also undertaken by the OECD. At the UN Conference on Sustainable Development (Rio+20) in 2012, member states agreed to establish a political forum to replace the CSD, as discussed below.

Conflicting views on the Rio Earth Summit

Despite the many agreements reached at Rio, and the range of outputs, conventions, laws organisations and activities that it has spawned, the Rio Earth Summit is not without its critics. The two Bush administrations in the US, for example, have repeatedly questioned the sustainable development principles that form the basis of the Rio Declaration, not least the upholding of the precautionary principle as a guide to international environmental law. Far from remaining an abstract debate, the US administration's rejection of the precautionary principle has proved highly influential, shaping its position on several key international environmental policy issues, including global climate change, as discussed in Chapter 6. The US administration has also held the Rio process to ransom, as it were, by consistently refusing to meet the financial obligations laid down at Rio. Similarly, Chapter 7 discusses ongoing issues arising from the broad and highly ambitious CBD, and Chapters 8 and 11 explore continuing problems in relation to securing action to address climate change under the UN FCCC.

Criticism has not only originated from within the US administration, or in relation to specific conventions. Some of the most scathing criticism of both the Rio Summit and the entire international governance process that it has spawned has come from within the radical green movement. The *Ecologist* magazine has argued that the very premise of the Earth Summit was flawed (*Ecologist*, 1993). By focusing attention on how the environment should be managed, it addressed the wrong question. The real question, it argues, is not how, but who will manage the environment and in whose interest. It raised the issue of how the 'global' environment was defined at Rio: designating certain issues as global, such as climate change and biodiversity loss, and others as local, such as desertification, was seen to reflect the interests of the politically and economically powerful nations of the industrialised world. The *Ecologist* asked whose common good is being protected by the Rio process. In reply, it argued that powerful states use institutions such as the UN to transform their own state interests into international agreed upon environmental norms and governance systems. Particularly in the negotiating stage of regime formation, states have a powerful interest in ensuring that considerations of costs, benefits or problems of domestic implementation remain dominant factors in shaping outcomes (Breitmeier, 1997). Radical greens argue that this is precisely what was happening at Rio and has continued since.

Box 5.3

The *Ecologist's* radical repudiation of the Rio Earth Summit

1   It is not poverty which is the root cause of environmental degradation, but the Western style of wealth.
2   Over-population is caused not cured by modernisation, as it destroys the traditional balance between people and their environment.
3   The 'open international economic system' of the Rio Declaration will extinguish cultural and ecological diversity.
4   The problem of pollution is not solvable by pricing the environment (market solutions), but by reversing the enclosure of the commons, that is, the process whereby the common resources upon which people have traditionally depended are being brought under the remit of commercial interests and commodified.
5   The call for more 'global management' constitutes another example of Western cultural imperialism.
6   The idea that developing countries urgently need the transfer of Western technology smacks of arrogance. It assumes that ignorance is characteristic of Third World people.

(*Source*: Pepper, 1996)

In addition to the radical green critique of Rio and its related developments, one that draws heavily upon a developing world perspective, there is another, more fundamental critique provided by radical environmentalists and green theorists. Because this critique is built upon a principled rejection of global governance regimes and therefore not specifically directed at the Rio Conference, it is discussed in a separate section at the end of this chapter.

The 1997 Earth Summit +5

In 1997, a United Nations General Assembly Special Session (UNGASS) was convened in New York to review the implementation of Agenda 21. This event is known as the Earth Summit 11, Earth Summit +5 or simply as UNGASS. The hope was that the Summit would revitalise commitments to sustainable development and help to prioritise actions for the post-1997 period (Osborne and Bigg, 1998). Reports on the state of the world's environment prepared for UNGASS painted a dismal picture. They showed that the global environment had continued to deteriorate at an alarming rate, with rising levels of greenhouse gas emissions, toxic pollution and waste, while renewable resources, notably fresh water, forests, topsoil and marine fish stocks, continued to be used at unsustainable rates (see Yamin, 1998/99 for a listing and review of reports). World leaders began the UNGASS meeting with the sobering realisation that five years after the Rio Earth Summit the planet's health was worse than ever.

Yet despite this knowledge, there were no major breakthroughs at New York. Discussions became bogged down in North–South differences on how to finance sustainable development. Pledges made at Rio by donor countries in the North to increase official development assistance (ODA) and make environment-friendly technologies available to developing countries had not been kept. Rather, this aid had declined from an average 0.34 per cent of donor country gross national product in 1991 to 0.27 per cent in 1995. Many in the South saw this as a breakdown of the global partnership for development declared at the Rio Summit.

The hotly contested debate on whether there should be a legally binding forest convention also undermined the Earth Summit +5. Canada and the EU strongly favoured a new treaty, but the US, Brazil, India and most major environment organisations were opposed. Many developing countries did not want to see a legally binding forest convention, fearing that it would hamper their development plans. In contrast, major environmental organisations opposed a convention fearing that it would at best offer only weak measures and, at worst, bring forests within the remit of global environmental management regimes that would, paradoxically, open up forests to commercial logging under the guise of introducing sustainable practices.

The final document adopted by delegates disappointed many. It contained few new concrete commitments on action.

> We are familiar with the tactics being played. Posturing, spinning declarations of intent, pointing the finger at others, pandering to interest groups, weighing short-term profits and immediate electoral gains, and emphasizing the need for clearer definitions, dialogue or information-gathering ... These prevent action plans from being operationalised into implementation programmes.
> (Ambassador Razali Ismail of Malaysia, President of the General Assembly and Chair of the Special Session, address to the UNGASS, 23 June 1997 [International Institute for Sustainable Development, 1997])

However, the UNGASS meeting was significant in that NGOs were able for the first time to deliver speeches to the plenary sessions and to have access to ministerial-level consultations. This reveals how the system of environmental governance promoted by the UN encourages the participation of private and voluntary groups (Elliott, 2002). Such engagement lends legitimacy to UN proceedings, while at the same time giving the UN access to vital information, as many NGOs have built up considerable expertise and local knowledge on a wide range of environmental and social issues. In addition, the participation of NGOs helps in the formation of implementation partnerships, as seen in the development of Type II partnerships at the WSSD, discussed below.

A parallel Rio+5 Forum was held at the same time as the Earth Summit +5, and organised by the Earth Council. The Earth Council is an environmental NGO set up by Maurice Strong, who chaired the negotiations at the original 1992 Rio Summit. This meeting pointed to the many areas of rapid environmental degeneration, to the rise in poverty and

---

Box 5.4

Outcome of the Earth Summit +5

- Confirmation of political commitment to the promotion of sustainable development;
- Frank assessment of progress;
- Recognition of the key role of the UN;
- Clarification of the role of specific institutions;
- Confirmation of UNCED targets and commitments for ODA;
- More focused programme of work for the CSD;
- New programme for the further implementation of Agenda 21;
- Continuation of work on forests and consideration of a possible legally binding instrument in the future;
- Beginning of intergovernmental process within CSD on freshwater and energy;
- Better understanding of issues around sustainable development and tourism, transport, information and changing production and consumption patterns;
- A number of new practical agreements in special areas, such as world-wide phase-out of lead from gasoline.

(*Source*: Osborn and Bigg, 1998)

inequality, and to the growing scarcity of many key environmental resources, especially water. It highlighted the need for industrial countries to instigate policies aimed at promoting new patterns of sustainable consumption and production.

In contrast to the official Summit, the alternative Forum struck a chord with ongoing debates about the moral, political and financial responsibilities of the industrial world to address its 'ecological debt', that is, the negative ecological legacies that it has bequeathed both to the South and to future generations. The Brundtland Report and the Rio Declaration had made it clear that attempts to promote sustainable development require that the North take these responsibilities seriously and act upon them. This belief is also reflected in the UN Millennium Declaration, and its related Millennium Development Goals designed to address the social pillar of sustainable development, as discussed in Chapter 11. The Forum also highlighted the gap between declaratory political commitments, on the one hand, and, on the other, engaging in actions necessary to bring about change in the international economic and political system to promote sustainable futures.

The 2002 World Summit on Sustainable Development

The third summit, the World Summit on Sustainable Development (WSSD) was held in Johannesburg in 2002. The WSSD set the twin goals of providing a ten-year review of the 1992 Earth Summit and of reinvigorating global commitment to sustainable development. When judged in terms of both its scope and its complexity, the WSSD was the biggest event of its kind organised by the UN to date. There was also a concurrent Global Peoples Forum, organised by and for civil society. However, many argued that the Summit was doomed to failure even before it started, because national representatives and civil society organisations came to Johannesburg with very different and sometimes incompatible agendas. The result was a dauntingly long 'wish list' which Johannesburg could only but fail to meet.

Several processes fed into the WSSD, including the outputs of the UN environmental conferences held up to that date, and a series of UNCED documents and reports, in particular, the Millennium Declaration, the WTO Doha Declaration and the Monterrey Consensus. Like the Rio conferences, the WSSD was preceded by four preparatory meetings, known as 'PreComs'. Twenty-two reports on the implementation of

Agenda 21 were presented at the first PreCom. They highlighted the limited achievements of global efforts in the years since 1992. They showed a variety of problems, ranging from a fragmented approach; lack of progress in addressing unsustainable patterns of production and consumption; inadequate attention to the core issues of water and sanitation, energy, health, agriculture, biodiversity protection and ecosystem management (issues known under the acronym WEHAB); lack of coherence between international policies on finance, trade, investment, technology and sustainable development; and insufficient financial resources; to an absence of mechanisms for technology transfer. These reports helped focus the attention of the Summit on identifying practical ways to implement the previous declarations of commitment to sustainable development. (For a summary of these reports, see UN, 2002b.) As a result, there were several agreements reached at the WSSD.

The Summit resulted in the Johannesburg Declaration on Sustainable Development, which declared the need to promote sustainable development through multi-level policy actions, adopting a long-term perspective and encouraging broad participation. However, the Declaration is more in the nature of a political compromise, desperately needed to lend weight and legitimacy to the closing moments of the Summit. The Declaration lacked the intellectual sophistication or authority that the Rio Declaration still commands (Hens and Nath, 2003). Unlike the Rio Declaration, the Johannesburg Declaration was also seen as unlikely to lead to new international negotiations or legal conventions.

The second output, the Joint Plan of Implementation, is the core outcome of the WSSD, and describes how existing commitments and targets might be met. One of these is the commitment to achieve sustainable harvesting practices in the world's fisheries by 2015. The Plan prioritises WEHAB initiatives. Agreements in relation to

---

Box 5.5

Agreements reached at the WSSD

a. Johannesburg Declaration on Sustainable Development
b. Johannesburg Plan of Implementation
c. WEHAB initiatives
d. Type II partnerships.

renewable energy use and addressing biodiversity loss were also reached, but are regarded as particularly weak as no new commitments were made to increase aid or deal with debt. The links between poverty, biodiversity and aid are discussed in Chapter 7. More generally, the Plan was seen to lack innovative thinking, especially the section that dealt with climate change. Similar concerns have been raised about the failure to address the links between globalisation, trade and sustainable development, where the Plan only calls for 'an examination' of the relationship between trade, environment and development. It failed to invoke the precautionary principles in dealing with potential problems at the interface between the environment, trade and development, nor did it present any new insights into the mechanisms linking trade, environment and sustainable development (Hens and Nath, 2003). In the words of one critic:

> This Summit has failed the poor and vulnerable peoples of the world. It has not reached agreement on the radical action, with clear timetables and targets, needed to tackle the world's environmental problems, from climate change and renewable energy to forest and species loss. The world's Governments must agree to meet again and determine that next time they will do better.
>
> (Ricardo Navarro, chair of FoE International, FoE UK [FoE, 2002a])

The most innovative, but potentially most problematic, implementation strategy agreed in the Plan are the so-called Type II partnerships. While governments negotiate Type I actions, Type II partnerships are where groups of countries, their national or sub-national governments, the private sector, especially the business community, and civil society actors work in partnership to support specific concrete initiatives. Several multi-stakeholder projects were announced at the WSSD, including Capacity 2015 of the UNDP, the Water and Energy Partnership of the EU, the UNESCO *Encyclopaedia of Life Support Systems*, and the International Youth Dialogue on Sustainable Development of the Global Youth Network.

Partnership arrangements recognise the importance of including a wide range of economic and social actors in the promotion of sustainable development. They help secure participation through their focus on concrete development initiatives. However, critics have pointed to the danger that partnership arrangements, especially those involving the private, business sector, may result in the commodification of the environment, turning more and more environmental assets into resources that are to be managed, albeit more responsibly. There is also concern that state agencies may lose authority over their sustainable development policies and programmes, particularly in selected areas, such as forests, by allowing control to shift to powerful business or industry interests that do not necessarily prioritise the promotion of sustainable development. Political corruption and a lack of democratic transparency and accountability can also enhance these dangers, as discussed in Chapter 6. The conflict over Type II partnerships can be seen in the different opinions expressed over one such initiative, the Congo Basin Forest Partnership.

Box 5.6

The Congo Basin Forest Partnership

The Congo Basin contains a quarter of the world's tropical forest. It is a region of extraordinary biological richness. However, deforestation in the Congo Basin has accelerated in recent years.

The partnership is an initiative by the United States, the six governments of the Congo Basin, other partner governments, conservation and business groups, and organisations representing civil society. It was launched by the US secretary of state, Colin Powell, at the WSSD in Johannesburg, 2002. Action is focused on eleven key landscapes in Cameroon, Central African Republic, Democratic Republic of the Congo, Equatorial Guinea, Gabon, and Republic of the Congo.

Aim

The aim of the Congo Basin Forest Partnership is to manage the Congo Basin forest in a sustainable fashion. The partnership provides support for:

- a network of national parks and protected areas;
- the issuing of forestry concessions to logging companies;
- the creation of economic opportunities for communities who depend upon the conservation of the outstanding forest and wildlife resources of the Congo Basin.

The US pledged to invest in the partnership through an expansion of USAID's 'Central African Regional Program for the Environment' (CARPE). CARPE is USAID's largest land management programme in Africa.

Partners

Governments: the United States, Cameroon, Central African Republic, Democratic Republic of the Congo, Equatorial Guinea, Gabon, Republic of Congo, United Kingdom, Japan, Germany, France, Canada, South Africa, and the European Commission.

International Organisations: World Bank, International Tropical Timber Organisation and World Conservation Union.

Civil Society: Jane Goodall Institute, Conservation International, Wildlife Conservation Society, World Wildlife Fund, World Resources Institute, Forest Trends, Society of American Foresters, American Forest & Paper Association, Association Technique Internationale des Bois Tropicaux, and the Center for International Forestry Research.

Conflicting interpretations

The US contributions to the Congo Basin Forest Partnership will promote economic development, alleviate poverty, and improve local governance, through natural resource conservation programs.

(Bureau of Oceans and International Environmental and Scientific Affairs, 2002)

The Congo Basin Initiative is a key example of the Bush Administration's flawed partnership approach. While supposedly benefiting forest protection and management in the highly biodiverse Congo Basin, the initiative will actually put more money into flawed programmes that have not reduced illegal logging, empowered local communities or enabled sustainable forest management. The US has also dismissed concerns of local environmental groups about corruption in these countries and the close collusion between government officials and timber barons.

(FoE UK, 2002b)

(*Sources*: Congo Basin Forest Partnership, 2013; Megevand et al., 2013)

Concerns have also been raised that the Type II partnerships could replace or weaken Type I agreements. Research since WSSD has found that partnerships have failed to engage smaller, poorer communities and that developing countries usually do not take a leadership role. They often serve to confirm rather than change existing power imbalances (Glasbergen, 2011). Evidence their lack of transparency has also raised concerns, giving rise to calls that Type I arrangements must continue to remain a priority, allowing substantive multilateral commitments by governments to act as the primary vehicle for promoting sustainable development (Hale, 2003).

In addition to negative views on Type II partnerships, there was also widespread criticism about the role played by the US during the WSSD. Colin Powell, then US secretary of state, was heckled at the WSSD as the US tried to block agreement on substantive timetables and targets across several key issue areas. Ever since 1992 at Rio, the US has rejected targets and timetables, an action that has considerably weakened summit outcomes. Friends of the Earth have argued that the US position is particularly egregious given the disproportionate share of global resources it consumes and environmental damage it does (FoE UK, 2002b). In addition, there were disputes with the US about the relationship between MEAs and free trade agreements, as the US pushed hard for the WTO agreements to take precedence over environmental agreements, against the wishes of the EU and G77. This points to the marked tension between the international trade and environmental regimes.

However, a more positive interpretation of the WSSD stresses the new level of dialogue in Johannesburg between governments, civil society and the private sector, as reflected in the Type II partnerships.

> The Summit was anything but a complete failure ... It was an opportunity ... to exchange ideas and information on achieving sustainable development, and to strengthen networks.
>
> (Ken Ruffin, head of the OECD Environment
> Directive [Ruffing, 2003])

> The Summit will be remembered not for the treaties, the commitments, or the declarations it produced, but for the first stirring of a new way of governing the global commons – the beginning of a shift from the stiff formal waltz of traditional diplomacy to the jazzier dance of improvisational solution-orientated partnerships that may include non-governmental organisations, willing governments and other stakeholders.
>
> (Jonathan Lash, president, World Resource
> Institute [Lash, 2002])

This interpretation is closely linked to the view that the WSSD was more than anything else an act of international environmental diplomacy.

The 2012 Rio+20 Conference

The Rio+20 Conference took place in Brazil on 20–22 June 2012 to mark the twentieth anniversary of the 1992 Earth Summit and the tenth anniversary of the WSSD. It was envisaged that the Conference would take place at the highest possible level, including attendance by heads of state and government. It was organised around two key themes: promoting a green economy in the context of poverty eradication; and enhancing the institutional framework for implementing sustainable development commitments.

The Conference was severely hampered by a lack of credibility, caused not least by the failure of many leaders of powerful, developed states to attend. Heads of state from important developed countries, including the US, Germany, Japan, the United Kingdom and Russia, stayed away. In the end, the Rio+20 Conference was attended by fifty heads of state and government and 190 environment ministers, less than half the number attending Rio 1992. Even the president of the host country, Dilma Rousseff, was reported to have been distinctly unenthusiastic about the Summit. While over 45,000 participants from governments, the private sector, NGOs and major groups attended, one of the striking features was that many governments were represented by their environment or development ministers, while their counterparts from finance, planning and business were absent (IIED, 2012). This showed how sustainable development is still seen as an environmental issue, reflecting limited efforts to mainstream sustainable development considerations into other policy fields, especially at the sectoral level. The event also did not garner much media or public attention. Failing to gather momentum as a whole, a day before the arrival of the heads of state, only 28 per cent of the Conference outcome declaration had been finalised, resulting in the Brazilian government rushing off a 'take it or leave it' communiqué which stripped out any strong commitments to produce a declaration that could be endorsed by all. In contrast, the presence of the heads of corporations such as Unilever and Puma, alongside large Brazilian companies, was very evident at Rio.

The Rio+20 document, 'The Future We Want'

The Conference outcome document, 'The Future We Want', introduced a grand vision for addressing global challenges within the framework of sustainable development (Clémençon, 2012). But the document was disappointing, serving only to reiterate promises made elsewhere (UN, 2012). It failed to lay out a coherent roadmap for implementing prior commitments, much less to define binding targets with specific deadlines (IIED, 2012). However, despite its disappointing failure to translate declaratory intent into committed action, the document does reflect a change in international negotiations, as developing countries played a much more assertive role than in previous summits. This allowed poverty eradication to be pushed forward as the overarching priority, with the opening section of the 'Future We Want' stating: 'Eradicating poverty is the greatest global challenge facing the world today and an indispensable requirement for sustainable development' (UN, 2012: 1). In this sense, the Rio+20 document is seen as a victory for developing countries in that it reaffirms poverty eradication as the prime sustainable development challenge (Clémençon, 2012). The critical importance of poverty eradication is also emphasised in the document's statements about the need to achieve the Millennium Development Goals by the target date of 2015.

'The Future We Want' also reaffirmed Rio's Principle 7 of common but differentiated responsibilities. The US was at first opposed to singling out one Rio principle over others. However, developing countries would not budge, regarding Principle 7 as the bedrock of multilateral environmental agreements. But attempts by developed countries to commit countries to embed poverty eradication within the framework of a transition to a green economy, and link this to concrete sustainable development goals, failed. The green economy was one of the two main focal areas of the conference but turned out to be one of the most controversial issues on the agenda.

African countries expressed strong interest in the notion of 'green economy', seeing it as a potential means to plan their future development trajectories and as having the potential to lift people out of poverty through a better use of natural resources, without acting as a brake on development (IIED, 2012). Developing countries also insisted on seeing a green economy as just one possible pathway to sustainable development. As a result of these difficulties, 'The Future We Want' merely states that 'We affirm that there are different approaches, visions, models and tools available to each country … to achieve sustainable development …' (UN, 2012: 10). The stress on difference reflects the disagreement between countries on the role a green economy should play in promoting sustainable development (Clémençon, 2012).

More than anything else, the outcome document from Rio+20 was shaped by deep, ideological divisions over the role that the private and public sectors and the free market economy should play. This, in turn, shaped attitudes to the green economy agenda. The free market approach was strongly advocated by the US and supported by what is known as the Washington Consensus. This consensus advocates market-led growth for poverty alleviation and for the elimination of inequality (United Nations Department of Economic and Social Affairs, 2010). At Rio+20, there were marked differences of opinion among countries as to the balance between public and private investments and thus of markets. The US found consensus with developing and especially newly emerging countries with respect to its reluctance to commit to government interference in global free market forces, a stance that was opposed by the European Union (Clémençon, 2012). In this, and through many other examples, Rio+20 was marked by the emergence of new and fluid patterns of coalition building, where opposing positions were not always oriented around the traditional lines of developed versus developing world.

The Rio+20 Conference was also noticeable in the way many member states stressed the importance of economic growth. In emerging countries in particular, there is a strong belief in a 'grow first, clean-up later' approach to development, as discussed in Chapter 12. Stressing the importance of private investment and market forces, Rio+20 saw the UN Global Compact convene more than 2,500 participants for the Rio+20 Corporate Sustainability Forum and in over 100 sessions discuss how business can help promote sustainable development. The UN Global Compact is a strategic policy initiative for businesses that are committed to aligning their operations and strategies with ten principles in the areas of human rights, labour, environment and anti-corruption (UN, 2013). More than 200 concrete commitments for sustainable development were made by companies at Rio20 (Haertle, 2012). This development has been referred to as the growth of private governance, a development that does not suggest a straightforward power shift away from states and towards

RIO+20
United Nations
Conference on
Sustainable
Development

Figure 5.1 *The Rio+20 Conference was marked by deep ideological divisions over the role of economic growth and the free market economy in the promotion of sustainable development.*

Logo courtesy of the UN.

firms, but a more complex interdependence between private and public actors (Falkner, 2003), a matter discussed in Chapter 3. While these events provide evidence of the increasingly important role that non-governmental stakeholders play in implementing sustainable development, the problem with Rio+20 is that it left limited room 'for reining in or redirecting markets to better serve long-term sustainable development objectives' (Clémençon, 2012: 333–334).

The issue of funding plagued the Conference, and was closely related to the controversy over the role of markets and of private investment. Developing countries made traditional demands for more financial compensation from rich countries for investments into sustainable development activities, especially if they do not immediately benefit their development needs (Clémençon, 2012). However, the US made clear that there would be no additional financial commitments forthcoming, placing emphasis instead on encouraging more private sector investments. Developing countries maintained that the failure to provide adequate and predictable funding violated commitments rich countries had made on previous occasions (Clémençon, 2012). Similarly, attempts to get concrete commitments by rich countries on targets for reducing unsustainable luxury consumption and production patterns also failed. The US was opposed to singling out rich countries' consumption and production patterns for special treatment. US president George Bush senior had famously proclaimed at the original Rio 1992 Summit that 'the American way of life is non-negotiable'. This motto continued to set the tone for US negotiators at Rio+20, despite the fact that developing countries, such as India, have long demanded a distinction between luxury and subsistence consumption (Clémençon, 2012). Further controversy was generated on the role of markets as many Latin American countries voiced their strong rejection of the 'commodification of nature' by the developed world (IIED, 2012) through their use of policy tools such as payments for ecosystem services and valuation systems, which put a monetary value on environmental resources, a matter discussed in greater depth in Chapter 7. 'The Future We Want' made no mention of the intrinsic value of nature. However, care should be taken lest it be assumed that developing countries presented a stronger sustainable development agenda at Rio+20. Thus, for example, while the outcome document acknowledged the importance of corporate sustainability reporting, developing countries opposed more specific commitments to request companies to start reporting on how sustainable their operations are. However, the UK government announced that it would require all publicly listed companies in the UK to report on their carbon footprint as of the following year. Similarly, reference in 'The Future We Want' to the right of women to have access to reproductive health services, advocated by the US and European countries, was eventually dropped because of opposition by developing countries.

Rio+20 brought no new policy progress on climate change, forests or biodiversity conservation, or on the interrelated issue of desertification, land degradation and drought. Nor was there much progress on establishing sustainable development goals, designed to come into effect in 2015 when the Millennium Development Goals expire, although subsequent discussions have begun to put flesh on these new goals. Despite warnings of dramatic planetary changes by scientific researchers, and highlighted in a string of publications, including the UNEP Global Environmental Outlook 5 (UNEP, 2012), there was no mention made of planetary boundaries nor of limits to the carrying capacity of the Earth. For many observers this reflects a disturbing disconnection between science and public policy (Clémençon, 2012).

Institutional reform

Reforming the institutional framework for sustainable development was the second theme of the Rio+20 Conference. The current architecture for managing international environmental issues was set up by (and for) nation states in the second half of the twentieth century (Haertle, 2012). Many have questioned whether it is able to keep pace with global economic and environmental change and the shifting power balance between the developed, emerging and developing world. However, discussions at Rio+20 were confined to the familiar proposals to establish a centralised global environment organisation, modelled along the lines of the WTO, or to restructure the UNEP to fulfil this role. Proposals for elevating UNEP to International Environment Agency status had already been discussed at the 1992 Rio Summit, but instead led to the establishment of the CSD, which, as mentioned above, never gained much influence. But by the time Rio+20 was mooted, the need for reform had become urgent. As Achim Steiner, Executive Director of the UNEP, argued: 'We're inventing a million times the different solutions, but unless we can create the structures and systems to promote change we will remain the laboratories and test grounds to show what could be done' (IIED, 2012).

Rio+20 resulted in two decisions regarding institutional matters: the creation of a high-level political forum to replace the CSD and a strengthening of the status of UNEP, short of turning it into a special UN agency.

The significance of both these changes, and whether they will support better governance for sustainable development, remains to be seen. However, despite these institutional developments, the end of the Conference was marked by a growing sense of unease with the governance process set in motion by the UN.

Assessments of the Summit have been largely negative. The International Institute for Environment and Development argued that Rio+20 showed that sustainable development is not a policy priority, especially for major powers. Rather, 'if anything, the Summit constituted a step back in the decades-long effort that started with the Brundtland Commission in the late 1980s, to place equity, sufficiency and international collaboration at the heart of global and national policy' (IIED, 2012). The outcomes of the Conference have been described as 'woefully thin' on commitments or agreed actions (IIED, 2012). It can also be considered a disappointment when set against the mounting scientific evidence about the severity of global environmental change. The issue of how to live within planetary boundaries, while dealing with legitimate development needs to address poverty, hunger, and lack of access to basic services, clean water, food security, job opportunities and health services, remains unresolved. Whether this means that the UN no longer has a purposeful role to play in global environmental governance is a matter discussed below.

Box 5.7

High-level Political Forum on Sustainable Development

Remit

- Provide political leadership and recommendations for sustainable development;
- Review progress in implementing commitments;
- Enhance the integration of economic, social and environmental dimensions;
- Have a focused, dynamic and action-oriented agenda;
- Consider new and emerging sustainable development challenges;
- Adopt negotiated declarations.

(*Sources*: Abbott, 2012; UN, n.d.)

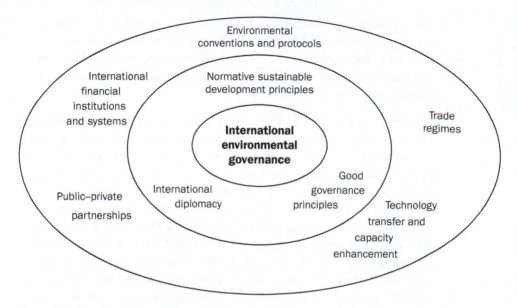

Figure 5.2 *The UN system of the promotion of sustainable development penetrates all areas of international governance.*

Global environmental governance: fit for purpose?

Environmental summits and conferences organised under the auspices of the UN have served several important functions. To see the range and significance of these functions, it is useful to draw upon the work of Hass (Hass, 2002), who has argued that the functions of conference diplomacy include:

- Agenda setting
- Popularising issues and raising consciousness
- Generating new information and new challenges for government
- Providing general alerts and warning of new threats
- Galvanising administrative reform
- Adopting new norms and doctrinal consensus
- Promoting mass involvement.

There is little doubt that the UN environmental summits and conferences have contributed to each and every one of these developments. They have established an agenda of international relations aimed at promoting sustainable development, and at a global scale. This has shifted attention from earlier preoccupations with the environment as merely a technical issue of pollution control to the current efforts to reconcile social, economic and ecological goals of development. Closely associated with this has been the articulation of a common set of normative as well as good governance principles. The system of global environmental governance that has developed since Brundtland is marked by its adherence to an increasingly sophisticated web of such principles. It has also strengthened institutional arrangements for global governance. It also helped in the formation of new environmental ministries and agencies in many countries, a development that creates order in the management of the environment at the international level.

This international process has also given voice to the developing world, allowing countries to raise concerns about equity, poverty and just distribution in access to and use of the world's resources. UN summits and conferences have also contributed to the increased participation of both economic and social actors in sustainable development initiatives. Global environmental diplomacy has been, at least in part, democratised as a result of summit practice. This has also occurred through the work of Agenda 21, which requires governments to involve major societal groups in decision making and implementation strategies in pursuit of sustainable development and to give access to information on environmental problems and policies, as discussed in Chapter 8. In addition, the UN has also officially accredited thousands of transnational NGOs and given them a voice at summit meetings.

Furthermore, conferences and summits have also helped develop 'soft law' norms that become, over time, acceptable to a large group of nations and evolve into 'hard law'. They have helped to build a body of MEAs, including environmental conventions, such as the UN FCCC, which resulted in the Kyoto Protocol and the UNCBD. Support has also been provided for developing a scientific evidence base, which has in turn raised the profile of environmental issues on national and international agendas. The reports prepared for the PreCom meetings, in advance of Rio, UNGASS, the WSSD and Rio+20, have helped identify areas for priority action. Over time, it is hoped that similar attention can be paid to implementation, as the CoPs reach agreement on rules, obligations and reporting requirements, and agree on resource transfer to aid implementation, particularly in the developing world.

Despite the positive functions of the UN summits, conferences and their follow-up activities, these events are nevertheless subject to very valid criticisms. These include concern about the exclusion of certain issues from centre stage on the agenda of global sustainable development policy, and on the unwillingness of powerful states to move beyond narrow self-interest to address our common future. The ability of the most powerful global player, the USA, to hold the UN negotiations to ransom by refusing funding transfers and, when all else fails, opting out of its conventions, is a case in point.

Much of the criticism has been directed at the inadequacies of UN governance. Concern has been expressed that the agenda of international sustainable development politics that it fosters is not sufficiently inclusive, that legislation is not sufficiently strict and that funding, particularly from the industrialised world, has not been forthcoming. Accepting the validity of the system, the desire is to make that system more efficient, effective and just. These criticisms recognise that it has been no mean feat to have established and sustained a process of high-level meetings of state leaders to address the urgent problems of environment and development (Falkner and Lee, 2012). Underpinning this sense of frustration is a broader crisis of international environmental diplomacy. The years since the Rio Summit in 1992 have seen the system of global environmental governance come under increasing strain, with global divisions blocking progress on a number of fronts, including with respect to climate change negotiations, as discussed in Chapter 6. Many of these divisions are caused by states prioritising their own national interests. There is also a greater variety of voices and positions now presented on the international stage, together with more fluid and unstable patterns of coalition-building (Falkner and Lee, 2012). There is also rising tension concerning funding, where the willingness of developing countries to participate in global governance regimes depends heavily on international assistance from developed countries. However, such funding arrangements are increasingly difficult to sell to constituencies in industrialised economies at a time of fiscal austerity and growing economic competition from emerging economies (Falkner and Lee, 2012).

However, there is also another, more fundamental critique, one that stems from within deep green theory, and that rejects the system of global environmental governance entirely. Here, the UN process is seen as part of a general trend in modern society towards the increased rationalisation of everyday life. This is a process whereby science, technology and also the policy process becomes dominated by concerns about means and procedures. Thus, new institutions are needed, which in turn requires negotiated agreements and technical tools to measure progress and monitor compliance. This means that the sustainable development agenda becomes crowded by debates about the means of achieving given ends, rather than debates about the ends themselves (Death, 2010: 55–58). This instrumental rationality does not call for fundamental changes in values or behaviour, such as in existing patterns of production or consumption. Rather, it shifts attention to devising better managerial strategies and more effective and efficient institutional control (Sachs, 1999). The Rio+20 Summit provides an important example, as discussions about consumption levels and productions patterns were sidelined, while attention was focused on negotiating institutional reforms (Death, 2014). For deep greens, this approach tames the agenda of environmental politics, and it has led some to reject the whole notion of sustainable development as flawed managerialism (Blühdorn and Welsh, 2008).

An equally damming critique has formed around the argument that, in the hands of the UN, the sustainable development agenda has become part of a legitimising project for the spread of neoliberalism. Neoliberalism places emphasis on maintaining high levels of economic growth and promotes and justifies the use of markets to manage the environment. It is closely linked with weak models of sustainable development. This perspective was very evident at Rio+20, with its emphasis on the role of markets in the promotion of sustainable development. It is also found in the growing use of market instruments and pricing mechanisms, including the pricing of ecosystem services, a matter discussed in detail in Chapters 6, 7 and 9. The emphasis on markets means that little is done to develop an approach that deals with the sustainable development as an issue of common purpose (Barnes, 1995; Baker, 2012; Dryzek, 1990). This failure has led to a growing sense of frustration regarding the ability of the international community to implement lasting solutions to environmental problems in a collective manner (Clémençon, 2012). Rather, summits and conference activity form one of the means by which late modernity manages to sustain what is known to be unsustainable (Blühdorn, 2007).

In short, the UN environmental summits can be interpreted in two conflicting ways. First, they can be seen as acts of global diplomacy, in which case they have served several important functions. Alternatively, they can be judged from a strong sustainable development stance, as ways in which capitalism seeks to address the environmental crisis through technical management and the commodification of nature. From this perspective, they are to be viewed with deep scepticism and mistrust.

Box 5.8

Summarising the role of the UN summits and conferences

Positive aspects	Negative assessment
Facilitate international negotiations	Lack substantive output
Help increase co-operation	Promote rhetoric of co-operation
Encourage participation	Allow only limited or elite participation
Strengthen global focus	Particularistic in focus and ignore the local
Develop mutual understandings	Lowest common dominator agreements; national interests remain dominant
Lead to treaties and binding conventions	Overly cautious, inadequate targets
Establish institutional of global governance	Ineffective, lack competence and political clout, poorly funded

Conclusion

Much of the discussion in this chapter has focused on exploring the institutions of governance through the UN system. It has looked at how international governance has developed through the holding of regular meetings, the formation of new administrative procedures and specific institutions, the negotiation of new laws and establishment of monitoring and reporting systems. These institutional arrangements were shown to have become, over time, more open, democratic and more principled in their approach. While governance structures have been opened up by the UN engagement, governance processes continue to operate within traditional hierarchies shaped by asymmetrical distributions of power. Here the differential exercise of political and economic power, be it at the international or the local level, limits the ability of actors effectively to engage in the promotion of sustainable development. Recognition of this difference lies at the heart of the Brundtland call for good governance practice. An important guiding principle for the reform of global governance is the fair and equitable distribution of bargaining power, so as to ensure that the influence and voice of the world's poor is heard and reflected in international decisions and outcomes that seek to promote sustainable development. Nevertheless, it is possible to embrace the agenda of sustainable development, while remaining critical of its expression in and through international governance regimes.

Summary points

- The UN has played a key role in shaping the international response to the environmental crisis and in structuring that response around the norm of sustainable development.
- As a result of the development of an international governance system, the promotion of sustainable development is not seen as coterminous with a delimited political territory, nor is it seen as the exclusive business of governments.
- While states remain key actors in the UN system, the UN has helped to open up the system of international governance to a wide range of groups from within civil society and the economy. This has facilitated the rise of private governance.
- The role of the UN is subject to two conflicting interpretations. On the one hand, it is seen as making a positive contributing to our collective future.
- In contrast, the UN can be criticised as a management agent, preoccupied with means and not ends. This displaces a more fundamental critique of the flaws of the traditional, growth orientated model of development, with its emphasis on the use of market for realising individual preferences, to the detriment of the common good.
- It is helpful to return to the Brundtland formulation and make a distinction between the formation of new institutions of governance and the processes of governance, the latter remaining constrained by wider relationships of political and economic power. Embedded in the Brundtland plea for good governance is recognition that promoting sustainable development requires changes at a more fundamental level in the processes of international governance.

Further reading

Dauvergne, P. (ed.) (2012), *Handbook of Global Environmental Politics* (Cheltenham: Edward Elgar).

Death, C. (2010), *Governing Sustainable Development: Partnerships, Protest and Power at the World Summit* (London: Routledge).

DeSombre, E. R. (2014), 'Global Environmental Governance', in T. G. Weiss and R. Wilkinson (eds), *International Organization and Global Governance* (London: Routledge), 580–592.

International Council for Science (ICSU) (2012), 'The Five Rio+20 Policy Briefs Released by the GEC Programmes', available online at www.icsu.org/news-centre/news/rio-20-policy-briefs-released-by-the-gec-programmes

Kelemen, R. D. and Vogel, D. (2010), 'Trading Places: The Role of the United States and the European Union in International Environmental Politics', *Comparative Political Studies*, 43 (4): 427–456.

Schreurs, M. A. (2012), 'Twentieth Anniversary of the Rio Summit: Taking a Look Back and at the Road Ahead', *GAIA: Ecological Perspectives for Science and Society*, 2 (1): 13–16, available online at www.ieg.earthsystem-governance.org/sites/default/files/files/publications/Schreurs_GAIA.pdf

Useful websites

Earth Negotiations Bulletin, an internet publication covering international environmental and sustainable development governance, published by the International Institute for Sustainable Development, www.iisd.ca/enbvol/
UN official website: www.un.org/geninfo/bp/enviro.html

References

Abbott, K. (2012), 'Engaging the Public and the Private in Global Sustainability Governance', *International Affairs*, 88 (3): 543–564.

Baker, S. (2012), 'Climate Change, the Common Good and the Promotion of Sustainable Development', in J. Meadowcroft, O. Langhelle and A. Ruudin (eds), *Governance, Democracy and Sustainable Development: Moving Beyond the Impasse* (Cheltenham: Edward Elgar), 249–271.

Barnes, I. (1995), 'Environment, Democracy, Community', *Environmental Politics*, 4 (4): 101–133.

Blühdorn, I. (2007), 'Sustaining the Unsustainable: Symbolic Politics and the Politics of Simulation', *Environmental Politics*, 16 (2): 251–275.

Blühdorn, I. and Welsh, I. (eds) (2008), *The Politics of Unsustainability: Eco-Politics in the Post-Ecologist Era* (London: Routledge).

Breitmeier, H. (1997), 'International Organisations and the Creation of Environmental Regimes', in O. R. Young (ed.), *Global Governance: Drawing Insights from the Environmental Experience* (Cambridge, MA: MIT Press), 87–114.

Bureau of Oceans and International Environmental and Scientific Affairs (2002), *Fact Sheet: The Congo Basin Forest Partnership* (Washington, DC: US Department of State, Bureau of Oceans and International Environmental and Scientific Affairs, December), available online at http://2001-2009.state.gov/g/oes/rls/fs/2002/15617.htm

Clémençon, R. (2012), 'Welcome to the Anthropocene: Rio+20 and the Meaning of Sustainable Development', *Journal of Environment and Development*, 21 (3): 311–338.

Congo Basin Forest Partnership (2013), 'About the Partnership', available online at http://pfbc-cbfp.org/

Death, C. (2010), *Governing Sustainable Development: Partnerships, Protest and Power at the World Summit* (London: Routledge).

Death, C. (2014), 'Summits', in C. Death (ed.), *Critical Environmental Politics* (Abingdon: Routledge), 247–256.

Dryzek, J. (1990), *Discursive Democracy* (Cambridge: Cambridge University Press).

Elliott, L. (2002), 'Global Environmental Governance', in R. Wilkinson and S. Hughes (eds), *Global Governance: Critical Perspectives* (London: Routledge), 57–74.

Falkner, R. (2003), 'Private Environmental Governance and International Relations: Exploring the Links', *Global Environmental Politics*, 3 (2): 72–87.

Falkner, R. and Lee, B. (2012), 'Introduction, Special Issue on Rio+20 and the Global Environment: Reflections on Theory and Practice', *International Affairs*, 88 (3): 457–462.

FoE UK (2002a), 'US Wrecks Earth Summit', Press release, 4 September, available online at www.foe.co.uk/resource/press_releases/0904powl

FoE UK (2002b), 'Ricardo Navarro, Chair of FoE International Earth Summit, End of Term Report', Press release, 2 September 2002, available online at www.foe.co.uk/resource/press_releases/0902scor

Glasbergen, P. (2011), 'Understanding Partnerships for Sustainable Development Analytically: The Ladder of Partnership Activity as a Methodological Tool', *Environmental Policy and Governance*, 21: 1–13.

Haertle, J. (2012), 'Why Rio+20 Was Still a Success: The Contribution of Business and Academic Institutions in Support of Sustainable Development and the Rio+20 Process' (UN: UN Global Compact), available online at http://unglobalcompact.wordpress.com/2012/07/05/why-rio20-was-still-a-success-the-contribution-of-business-and-academic-institutions-in-support-of-sustainable-development-and-the-rio20-process/

Hale, T. (2003), *Managing the Disaggregation of Development: How the Johannesburg 'Type II' Partnerships Can Be Made Effective* (Princeton, NJ: Princeton University, Woodrow Wilson School of Public and International Affairs), available online at www.princeton.edu/~mauzeral/wws402f_s03/JP.Thomas.Hale.pdf

Hass, P. M. (2002), 'UN Conferences and Constructivist Governance of the Environment', *Global Governance*, 8 (1): 73–91.

Hens, L. and Nath, B. (2003), 'The Johannesburg Conference', *Environment, Development and Sustainability*, 5: 7–39.

IIED (International Institute for Environment and Development) (2012), 'Five Things We've Learnt from Rio+20', available online at www.iied.org/five-things-we-ve-learnt-rio20

International Institute for Sustainable Development (1997), 'Summary of the Nineteenth United National General Assembly Special Session to Review Implementation of Agenda 21', *Earth Negotiations Bulletin*, 5 (88), 30 June, available online at www.iisd.ca/csd/enb0588e.html.

Kohler-Koch, B. and Rittberger, B. (2006), 'The "Governance Turn" in EU Studies', *Journal of Common Market Studies*, 44 (Supplement s1): 27–49.

Lash, J. (President, World Resource Institute) (2002), 'The Johannesburg Summit Test: What Will Change?' *What's New: News from the Official Site of the Johannesburg Summit 2002*, feature story of 25 September, available online at www.johannesburgsummit.org/html/whats_new/feature_story41.html

Megevand, C., with Aline Mosnier, Joël Hourticq, Klas Sanders, Nina Doetinchem and Charlotte Streck (2013), *Deforestation Trends in the Congo Basin Deforestation: Trends in the Congo Basin Reconciling Economic Growth and Forest Protection* (Washington, DC: International Bank for Reconstruction and Development/The World Bank), available online at https://www.forestcarbonpartnership.org/sites/fcp/files/2013/Deforestation%20in%20Congo%20Basin_full%20report_feb13.pdf

OECD (2002), *Governance for Sustainable Development: Five OECD Case Studies* (Paris: OECD), available online at www.ulb.ac.be/ceese/nouveau%20site%20ceese/documents/oecd%20governance%20for%20sustainable%20development%205%20case%20studies.pdf

Osborn, D. and Bigg, T. (1998), *Earth Summit 11: Outcomes and Analysis* (London: Earthscan).

Pepper, D. (1996), *Modern Environmentalism: An Introduction* (London: Routledge).

Ruffing, K. (Head, OECD Environment Directorate) (2003), 'Johannesburg Summit: Success or Failure?' *OECD Observer*, 6 March, available online at www.oecdobserver.org/news/archivestory.php/aid/902/Johannesburg_summit.html

Sachs, I. (1999), 'Social Sustainability and Whole Development: Exploring the Dimensions of Sustainable Development', in E. Becker and T. Jahn (eds), *Sustainability and the Social Sciences: A Cross-Disciplinary Approach to Integrating Environmental Considerations into Theoretical Reorientation* (London: Zed Books), 25–36.

Santiso, C. (2001), 'Good Governance and Aid Effectiveness: The World Bank and Conditionality', *The Georgetown Public Policy Review*, 7 (1): 1–22, available online at www.sti.ch/fileadmin/user_upload/Pdfs/swap/swap108.pdf

Ecologist (1993), *Whose Common Future? Reclaiming the Commons* (London: Earthscan).

UN (n.d.), 'The United Nations High-level Political Forum on Sustainable Development Provides Leadership and Reviews Progress on Sustainable Development', available online at http://sustainabledevelopment.un.org/index.php?menu=1556

UN (2002a), 'Plan of Implementation of the World Summit on Sustainable Development', available online at www.un.org/esa/sustdev/documents/WSSD_POI_PD/English/WSSD_PlanImpl.pdf

UN (2002b), *National Implementation of Agenda 21: A Summary* (New York: UN, August), available online at https://www.un.org/esa/agenda21/natlinfo/wssd/summarypublication.pdf; National reports are available at www.un.org/esa/agenda21/natlinfo/

UN (2007), 'Non-legally Binding Instrument on all Types of Forests' (New York: UN General Assembly, document A/C.2/62/L.5), available online at www.fordaq.com/www/news/2007/UN_Instrument%20on%20all%20 types%20of%20forests.pdf

UN (2012), 'The Future We Want' (New York: Resolution adopted by the General Assembly on 27 July 2012, A/66/L.56, 66/288), available online at www.un.org/ga/search/view_doc.asp?symbol=A/RES/66/288&Lang=E

UN (2013), 'Overview of the UN Global Compact', available online at www. unglobalcompact.org/AboutTheGC/index.html

UNDP (1997), 'Governance for Human Development: A UNDP Policy Document', available online at http://mirror.undp.org/magnet/policy/

UNEP (2003), 'The Voice of the Environment' (Nairobi, Kenya: UNEP), available online at www.unep.org/about/

UNEP (2012), 'Global Environmental Outlook 5: Environment for the Future We Want' (Nairobi, Kenya: UNEP), available online at www.unep.org/geo/ pdfs/geo5/GEO5_report_full_en.pdf

United Nations Department of Economic and Social Affairs (2010), 'Growth, Poverty and Inequality: From Washington Consensus to Inclusive Growth', Alfredo Saad-Filho (New York: UN DESA Working paper no. 100 ST/ ESA/2010/DWP/100), available online at www.un.org/esa/desa/papers/2010/ wp100_2010.pdf

Vogel, D. (2010), 'The Private Regulation of Global Corporate Conduct: Achievements and Limitations', *Business Society*, 49 (1): 68–87.

WCED (World Commission on Environment and Development) (1987), *Our Common Future* (Oxford: Oxford University Press).

World Bank (2002), 'Reforming Public Institutions and Strengthening Governance: A World Bank Strategy – Implementation Update' (Washington, DC: World Bank), available online at http://www1.worldbank. org/publicsector/StrategyUpdate1.pdf

Yamin, F. (1998/99), 'The CSD Reporting Process: A Quiet Step Forward for Sustainable Development', *Yearbook of International Co-Operation on Environment and Development* (Lysaker, Norway: Fridtjof Nansen Institute), available online at www.fni.no/ybiced/98_05_yamin.pdf

Young, O. R. (1997), 'Rights, Rules and Resources in World Affairs', in O. R. Young (ed.), *Global Governance: Drawing Insights from the Environmental Experience* (Cambridge, MA: MIT Press), 87–114.

Climate change and sustainable futures

Key issues

- Climate impacts and climate justice
- Common but differentiated responsibility
- Small Island Developing States
- Intergovernmental Panel on Climate Change
- UN Framework Convention on Climate Change; Kyoto Protocol
- USA climate action
- Mitigation and adaptation.

Introduction

The 1992 Rio Earth Summit led to two multinational environmental agreements, the UN Framework Convention on Climate Change (UN FCCC) and the Convention on Biological Diversity (CBD). Framework conventions contain loose obligations, with details subsequently worked out through meetings of the conference of the parties (known as CoPs), protocol negotiations, Secretariat work and national implementation. This chapter looks at climate change, while the following chapter explores the relationship between biodiversity and sustainable development.

Climate change is a critical global problem that affords an opportunity to put sustainable development commitments into practice, while at the same time risks furthering existing unsustainable patterns of development and related inequalities. Exploring climate change means touching upon issues at the core of the sustainable development agenda, including the consequences of traditional models of economic development and issues of global equity. Policy responses also have to cut across several sectors, including transport and energy, and across several behavioural areas, including consumption habits and lifestyle patterns. Furthermore, a complex and highly controversial global environmental management regime has developed at the UN level that has raised issues of North–South relations, as well specific concerns about the function, operation and funding of UN conventions and their related governance regimes.

> Climate policy intersects with other societal goals creating the possibility of co-benefits or adverse side effects ... Mitigation and adaptation can positively or negatively influence the achievement of other societal goals, such as those related to human health, food security, biodiversity, local environmental quality, energy access, livelihoods, and equitable sustainable development; and vice versa, policies toward other societal goals can influence the achievement of mitigation and adaptation objectives.
>
> (IPCC, 2014a: 5)

This chapter begins with a brief overview of climate change, its causes and impacts. Attention is then turned to the links with sustainable development, including as they relate to matters of justice. International efforts to address climate change through the UN FCCC, as well as mitigation and adaptation, are explored, before some concluding remarks are made.

Understanding climate change

The main cause of climate change is the enhancement of the natural greenhouse effect as a result of human activities. Certain gases in the atmosphere, known as greenhouse gases (GHGs), block heat from escaping into space. These GHGs are building up in the atmosphere beyond their natural levels, mainly from the burning of fossil fuel and the release of carbon dioxide (CO_2), one of the main GHGs, from industrial production, agriculture, transport and energy generation. In addition, deforestation causes approximately 15 per cent of current GHG emissions. These emissions have greatly intensified the natural greenhouse effect, and are seen as 'forcing' climate change, that is, changes to long-term weather patterns, such as temperature, rainfall or snowfall. This build-up not only affects the global climate system, but is raising ocean acidity, causing changes to forests and other ecosystems, and influencing agricultural crop yields and affecting human health. These changes have the potential to seriously disrupt many of the environmental processes, economic activities and built structures upon which human society depends.

Climate change and global environmental change

Climate change refers to changes attributed to human activity that have altered the composition of the atmosphere, and thereby the functioning of the Earth's climate system. 'Climate variability' refers to natural alterations in the Earth's climate.

Global environmental change encompasses broader changes to all aspects of our world, including the availability of water resources, sea-level rise, ocean acidification, and loss of biodiversity.

Climate change contributes to global environmental change, as does population growth and high consumption.

(*Sources*: NASA, n.d.a; Parliament of Australia, 2008)

As a result of GHG emissions, some degree of climate change is now irreversible, known as 'committed warming', so the Earth will continue to warm even if GHG levels in the atmosphere stopped growing immediately (NASA, n.d.b). The extent of any further change depends on what is done now to reduce GHG emissions (EPA USA, 2014a). Increased use is made of climate models, particularly by the Intergovernmental Panel on Climate Change (IPCC), to project different aspects of future climate, including temperature, precipitation, snow and ice, ocean level, and ocean acidity. These have allowed different estimates to be made for different future GHG emission scenarios (EPA USA, 2014b).

Climate change first attracted the attention of the international policy community in the 1980s. As a result, in 1988 the UNEP and the World Meteorological Organisation established the IPCC.

The Intergovernmental Panel on Climate Change assesses scientific, technical and socio-economic information relevant for the understanding of climate change, its potential impacts and options for adaptation and mitigation.

The IPCC published its *First Assessment Overview* in 1990 (IPCC, 1990). So worrying were its findings that formal international negotiations to deal with climate change were opened at the UN level (Meadowcroft, 2002). Subsequent IPCC reports not only confirmed the presence of climate change, but have also attributed this to human activities. The 2014 IPCC *Fifth Assessment Report* evidenced the continuous increase in GHG emissions:

> Total anthropogenic GHG emissions have continued to increase over 1970 to 2010 ... Despite a growing number of climate change mitigation policies, annual GHG emissions grew on average by ... 2.2% per year from 2000 to 2010 compared to 1.3% per year from 1970 to 2000. Total anthropogenic GHG emissions were the highest in human history from 2000 to 2010.
>
> (IPCC, 2014a: 6)

The 2014 Report also warned that, without additional efforts to reduce emissions, average global temperatures are projected to increase worldwide. As a result, climate change is likely to intensify and the rate of increase is very likely to be unprecedented within the past 10,000 years or more of human history (IPCC, 2007a).

In 2006 the British government presented a comprehensive study, the *Stern Review on the Economics of Climate Change* (HM Treasury, 2006). The Review argued that the scientific evidence is now overwhelming: climate change presents very serious global risks, and it demands an urgent global response as it threatens the basic elements of life, including access to water, food production, health, and the use of land and the environment. Seeing climate change as the greatest market failure the world has ever seen, the Stern Review is concerned about how this failure interacts with other market imperfections, acknowledging that the impacts are not evenly distributed, with the poorest countries and people suffering earliest and most. The Review was stark, estimating that if we don't act, the overall costs will be equivalent to losing at least 5 per cent of global GDP each year, now and forever. If a wider range of risks and impacts is taken into account, the estimated damage could rise to 20 per cent of GDP or more. The risks involved in not acting are on a scale similar to those associated with the great wars and the economic depression of the first half of the twentieth century. Nevertheless, the Review is optimistic, believing that there is still time to act to deal with the threat.

Causes of climate change

The most problematic aspect of GHG emissions is that the burning of fossil fuels is an activity that has powered economic development since the Industrial Revolution (Meadowcroft, 2002). As such, dealing with

Countries contribute different amounts of GHG emissions, with developed countries and major emerging economies leading in total CO_2 emissions (Matthews et al., 2014). Developed nations typically have high CO_2 emissions per capita, and have made a strong, historical contribution to global warming, while some BRICS countries, discussed in Chapter 12, lead in the *growth rate* of emissions. The figures are stark: industrialised countries have been responsible for 80 per cent of emissions during the twentieth century. The US is the clear leader in historical contributions, as measured by their cumulative emissions from 1850 to 2007. China, however, is now the largest global emitter of GHGs, with Russia, Brazil and India seeing rising emission rates. Brazil and India are interesting cases as most of their CO_2 emissions originate from deforestation. This is different from the other top GHG emitting countries, whose emissions are related to the burning of fossil fuels (Matthew and Hammill, 2009).

Comparing nations can, however, be misleading, given their varied sizes and populations. To get a more meaningful picture, per capita emissions are also measured. Measured this way, the emission list is topped by small countries with energy-intensive industries, such as Qatar and Bahrain, whereas the large emerging economies, such as India and China, produce significantly less emissions (PBL, 2012). Irrespective, however, of how emissions are measured and that the numbers have shifted in the direction of emerging economies, levels are much higher in the developed world than in poorer nations, with levels very low in Africa (PBL, 2012). These uneven contributions are at the core of the challenges facing the world community in its efforts to find effective and equitable solutions to the problem of climate change (Union of Concerned Scientists, 2010). Furthermore, since GHG emissions can stay for centuries in the atmosphere, taking account of historical emissions is also important. The tricky question of historical responsibility is another key tension in the process of negotiating a global response, discussed further below.

Impact of climate change

While some regions are expected to derive some benefit from climate change, its overall global impact will be negative. Changing weather patterns are likely to bring increased risk of droughts and flooding, more extreme weather events and sea level rises.

The ecological consequences will include increased risk of species extinction, especially in fragile environments such as the Artic and Antarctica. The combined impacts of ocean acidification, warmer sea temperatures and other human-induced stresses make tropical coral reef ecosystems vulnerable to collapse, with catastrophic loss of biodiversity and ecosystem functioning, threatening the livelihoods and food security of hundreds of millions of people. In addition, the geographical distribution of species and vegetation types is projected to shift radically. Changes in the abundance and distribution of species pose potentially serious consequences for human societies and for future well-being. Migration of marine species to cooler waters could make tropical oceans less diverse, while both boreal and temperate forests face widespread dieback, with impacts on fisheries, wood harvests, recreation opportunities and other services. These impacts will not, however, be evenly felt and it is likely that climate change will show regional and local variations (IPCC, 2014c).

So severe are the impacts of human activity on the Earth system that a growing number of scientists say we are living in a new geological epoch, the Anthropocene. This term dramatises the sheer scale of human activity, wherein human behaviour is changing Earth's life support system (Steffen et al., 2007: 614). An IPCC report in November 2014 warned that climate changes are having 'widespread impacts on human and natural systems' (IPCC, 2014b).

There is also growing concern about the possibility of *abrupt climate change* and/or abrupt changes in the Earth system, with potentially catastrophic consequences. Candidates for such 'tipping points' include the West Antarctic Ice Sheet, whose disintegration could raise sea levels by 4–6 metres over several centuries. Similarly a shutdown of the North Atlantic Thermohaline Circulation, sometimes called the ocean conveyor belt, could bring far-reaching, adverse ecological and agricultural consequences (IPCC, 2007c). The Greenland ice sheet may be nearing a tipping point. There are also concerns that if anthropogenically forced lengthening of the dry season in the Amazon rainforest continues, and droughts increase in frequency or severity, the rainforest could reach a tipping point, resulting in dieback of up to 80 per cent of the forest, and its replacement by savannah. Similarly, India's summer monsoon is considered at risk of tipping point and is probably already being disrupted (Lenton, 2010). Biodiversity-related tipping points are discussed in Chapter 7.

Box 6.2

Observed impacts of climate change

Impact on natural systems

- In many regions, changing precipitation or melting snow and ice are altering hydrological systems, affecting quantity and quality of water resources.

- Glaciers continue to shrink worldwide, affecting runoff and water resources downstream.

- Permafrost thawing is occurring in high-latitude and high-elevation regions.

- Many terrestrial, freshwater, and marine species have shifted their geographic ranges, seasonal activities, migration patterns, abundances, and species interactions.

- Negative impacts on crop yields, including wheat and maize, are found for many regions and in the global aggregate.

Impact on society

- At present the impact on human health is relatively limited. However, there has been increased heat-related mortality and decreased cold-related mortality in some regions.

- Local changes in temperature and rainfall have altered the distribution of some water-borne illnesses and disease vectors.

- Non-climatic factors and multidimensional inequalities also shape differential risks: people who are marginalised are especially vulnerable.

- Climate-related extremes, such as heat waves, droughts, floods, cyclones, and wild-fires, reveal significant vulnerability and exposure of some ecosystems and many human systems. Impacts include alteration of ecosystems, disruption of food production and water supply, damage to infrastructure and settlements, morbidity and mortality, and consequences for mental health and human well-being.

- For countries at all levels of development, there is a significant lack of preparedness in some sectors.

- Climate-related hazards exacerbate other stressors, often with negative outcomes for livelihoods, especially for people living in poverty.

- Violent conflict increases vulnerability and harms assets that facilitate adaptation, including infrastructure, institutions, natural resources, social capital, and livelihood opportunities.

(*Source*: IPCC, 2014c)

In short, it is clear that climate change will bring major ecological and social consequences. These include loss of ecosystems and biodiversity, disruptions to agriculture, erosion of food security, spread of disease, and threats to low-level settlements in many countries. Furthermore, the ability of countries to take mitigation measures and to adapt varies greatly. It is expected that the impacts will be most acute in developing countries because they are already more vulnerable and have less capacity to adapt (Meadowcroft, 2002). This raises issues of climate justice, which are discussed in the next section.

Linking sustainable development and climate change

Addressing climate change is necessary to achieve sustainable development, while it is also clear from the above discussion on impacts that climate change will amplify some of the challenges that have slowed the pace of sustainable development (Matthew and Hammill, 2009: 1127). Furthermore, existing mitigation and adaptation strategies could also undermine efforts to promote sustainable development and its related MDGs and SD goals, such as poverty eradication and equity. This means that climate change is inevitably bound up with issues of distributive justice and the legitimate development aspirations of many poor countries (Meadowcroft, 2002).

The links between climate change and sustainable development can be viewed by:

1 Identifying the links between climate change and social and ecological impacts;
2 Exploring how policy responses reflect and impact upon inequalities.

These connections are increasingly articulated through the lens of climate justice, including at the international level within the UN. The climate justice perspectives put justice and equity at the heart of understanding the impacts of, and responses to, climate change.

Climate justice

There are already strong links made in the literature between sustainable development and social justice, including in the seminal work of Andrew Dobson, *Justice and the Environment* (1998). Dobson assesses

Box 6.3

Links between climate change and justice

1 Responsibility for climate change is not equally distributed. Some groups and societies emit more GHG than others, an idea known as 'ecological debt'.
2 Impacts will not affect all groups and societies equally; some are more vulnerable than others.
3 The developed world has greater capacity to protect itself against climate change (adaptation).
4 Vulnerability is determined by political-economic processes that benefit some and disadvantage others, with the disadvantaged frequently being the most vulnerable.
5 Climate change will compound under-development, given the factors above.
6 Policy responses may themselves create unfair outcomes by exacerbating, maintaining, or ignoring existing and future inequalities.
7 The developed world has more financial and technical ability to undertake mitigation efforts.

(*Source*: modified from Barnett, 2006: 115)

the relationship between environmental sustainability and social justice, including the equitable distribution of material resources and the need to take account of future generations. Similarly, there are several links that can be made between climate change and matters in relation to equity and justice.

Dealing with the direct impacts, we have already discussed the asymmetrical impacts on the poor, marginalised and vulnerable. Climate change contributes to injustice, as those who will be hit first and hardest by climate impacts have contributed least to the problem. The most vulnerable communities are those who already suffer from deprivation, exclusion and inequality, with the impacts of climate change constituting what is known as a 'compound injustice', in other words, it will exacerbate their poverty. The right to food also highlights the links between human rights and climate change.

These links have been pointed out in several of the IPCC reports, including those directed at policy makers:

> Issues of equity, justice, and fairness arise with respect to mitigation and adaptation. Countries' past and future contributions to the accumulation of GHGs in the atmosphere are different, and countries also

face varying challenges and circumstances, and have different
capacities to address mitigation and adaptation.

(IPCC, 2014a: 5)

Climate change also raises issues of gender inequality (MRF CJ, 2013;
MRF CJ, 2014a). Men and women are affected differently given their
socially constructed roles and responsibilities. For example, women are
often the primary food producers and providers of water and cooking fuel
for their families and have greater responsibility for family and community welfare. In these circumstances, climate change places a greater burden on women as they try to meet family needs. Furthermore, women
may be constrained by social and cultural structures that place them in
inferior social positions, limiting their access to income, education, public voice, and survival mechanisms. The 1991 cyclone in Bangladesh
illustrates many of these issues. More than 90 per cent of the estimated
140,000 fatalities were women; their limited mobility, skills, and social
status exacerbated their vulnerability to this extreme weather event (Cameron et al., 2013). Yet, applying the discussion on the gender dimensions
of sustainable development in Chapter 11, women's economic contribution can also be seen as central to tackling climate change. Women play a
vital role as agents of change within their communities, such as through
adapting their agricultural practices, steps that can bring co-benefits,
including improving household nutrition and generating income.

Another way to look at the issue of equity and climate justice is from
the perspective of spatial impact. This perspective recognises that the
distribution of impacts is likely to lean towards regions with the least
capacity to adapt. Studies suggest that Africa will be hit hardest with
climate damages, while in Asia about 1 billion people face risks from
reduced agricultural yields and water supplies, and increases in extreme
weather events (Cameron et al., 2013). Small Island Developing States
(SIDS) are often cited as the most vulnerable countries and the first
nations on Earth to face critical climate change thresholds. The Rio
Earth Summit recognised that promoting sustainable development
would present special problems for SIDS given their ecological and
economic fragility. Most are archipelagos, with small and dispersed
land areas, possessing a rich diversity but relatively few natural
resources, and are geographically isolated. The small size of their economies and the narrow range of products they can produce often makes
then highly vulnerable to international trading conditions. Agenda 21
pointed to the need for special planning for sustainable development in
these circumstances.

Box 6.4

Impact of climate change on Small Island Developing States

Background

The UN recognises forty-one SIDS, including the Maldives, Papua New Guinea, Solomon Islands, Malta, the Bahamas, Haiti, and Trinidad and Tobago, thirty-eight having ratified the UN FCCC.

SIDS and climate change

SIDS are particularly vulnerable to climate change. As their population, agricultural land and infrastructure tend to be concentrated in the coastal zone, sea-level rise will have significant effects on their economies and living conditions, as will any increased frequency and intensity of storm events. These may affect vegetation, lead to saline intrusion and reduced freshwater resources, and damage their coral reefs, affecting both subsistence and commercial fisheries production. The very survival of some low-lying SIDS is threatened. In 2007, the IPCC estimated that by 2100 global warming will lead to a sea-level rise of 180–590 mm, while more recent research suggests that these figures are likely to be at least twice as large, up to about 2 metres. Nations such as Kiribati, the Maldives, Marshall Islands and Tuvalu will become uninhabitable in this later scenario.

Policy response

In 1994 a UN Global Conference on the Sustainable Development of SIDS was held in Barbados. This resulted in the Barbados Programme of Action, which provided a framework for planning and implementing policies aimed at promoting sustainable development. Fourteen priority areas were identified in the Programme. The WSSD also identified several areas for priority action, including technology transfer and capacity building to improve sustainable management of coastal areas, marine and coastal biodiversity protection, sustainable fisheries management, and dealing with problems of waste, pollution, water management and tourism. A high level meeting held in Mauritius in 2005 served as the culmination of a ten-year comprehensive review of the Barbados Programme of Action. At the 2010 UN Climate Change Conference held in Cancún, Mexico, a new framework of adaptation was agreed, including for the SIDS, known as the Cancún Adaptation Framework.

However, for many SIDS, adaptation has been constrained by limited financial resources, including an overall decline in Official Development Assistance. A review in 2014 of the Mauritius Strategy of Implementation showed that some SIDS, especially low-lying atolls, are already losing significant amounts of territory due to rising sea levels. They are also facing a problem of food security as land and fresh water resources are threatened by the encroaching salt water. In extreme cases, SIDS are exploring options for resettling populations in other countries.

(*Sources*: Ghina, 2003; UN, 2014)

Exploring SIDS gives us insight into the interconnections that exist between environmental resources, economic viability and social capacity to adapt, also known as resilience. Ironically, while they have contributed least to GHG emissions, such states will be subjected to some of the most significant adverse effects.

Although climate impacts have received the bulk of attention, there is a growing body of work stressing the potential implications of policy responses. Some of this draws upon a rights based approach, for example, the right to food and shelter, to link human rights and sustainable development (MRF CJ, 2014b). An important example is provided by biofuel policies, which aim at reducing GHG emissions through the provision of more carbon-neutral fuel, but have been shown to threaten the right to food by reducing the land available for food production and thereby increasing food prices. Similarly, the design of carbon taxes can also undermine rights if not designed to protect low income families from increased costs of basic necessities, such as domestic heating and energy supply (Cameron et al., 2013). A rights framework also helps to frame a series of governance requirements that can support morally appropriate responses to climate change, while being rooted in the sustainable development principles of equality and justice.

Several NGO groups wish to see justice and rights based approaches integrated in the UN FCCC, for example through ensuring that policies

---

Box 6.5

Governance principles for climate justice

- Respect and protect human rights;
- Support the right to development;
- Share benefits and burdens equitably;
- Ensure that decisions are participatory, transparent and accountable;
- Highlight gender equality and equity;
- Harness the transformative power of education for climate stewardship;
- Use effective partnerships.

(*Source*: MRF CJ, 2014a)

balance equity in terms of sharing both the burden and benefits of climate change mitigation and adaptation policies.

International climate change governance

As mentioned above, the UN FCCC is the main international agreement dealing with climate change. By 2014, the Convention had been signed by 196 parties and is the parent treaty of the 1997 Kyoto Protocol, discussed below. Article 2 of the UN FCCC calls upon member states to:

> Stabilize greenhouse gas concentrations in the atmosphere at a level that would not disrupt the Earth's climate system.
>
> (Article 2, UN FCCC)

The objective of the UN FCCC is to achieve stabilisation of GHG concentrations in the atmosphere at a level that would prevent dangerous anthropogenic interference with the climate system. The Convention stipulates that such a level should be achieved within a time frame that allows ecosystems to adapt naturally, to ensure that food production is not threatened and to enable economic development to proceed in a

---

Box 6.6

The UN FCCC principles

Overriding principle

● Framing policy within the commitment to promote sustainable development.

Normative principles

● Promoting equity, including the obligation of industrialised countries to take the lead in mitigation efforts.

Governance principles

● Accepting of the principle of common but differentiated responsibilities;
● Making use of the precautionary principle;
● Adhering to the principle of cost effectiveness.

---

sustainable manner. To constrain impacts within limits that society will reasonably be able to tolerate and to which ecosystems can adapt, in 2010 governments agreed that global average temperatures need to be stabilised within 2° Celsius. The UN FCCC has also laid down several principles to guide the governance of climate change, all of which are framed within a commitment to promote sustainable development.

To put these principles into effect the Convention makes a distinction between states according to their level of industrialisation. The industrialised countries that were members of the OECD in 1992 plus countries with economies in transition (the EIT parties), including the Russian Federation, the Baltic States, and several Central and Eastern European States, are classified as Annex I countries. Annex I countries are held responsible for the largest share of emissions and are obligated to take the lead in mitigation efforts. Annex II parties consist of the OECD members of Annex I, but not the EIT parties. Annex II countries are the most prosperous of the industrialised states and have the additional obligation of providing financial resources and technology to enable developing countries to undertake emissions reduction activities and to help them adapt. In addition, there are also non-Annex I parties, consisting mostly of developing countries. Certain groups of developing countries are recognised by the Convention as being especially vulnerable, including countries with low-lying coastal areas and those prone to desertification and drought. Forty-nine least developed countries (LDCs) are given special consideration under the Convention on account of their limited capacity to adapt. The fact that China is regarded as a non-Annex party was the source of considerable controversy in the USA, as discussed below.

Kyoto Protocol

By 1995, several countries had come to realise that emission reductions provisions of the UN FCCC were inadequate, and two years later they adopted the Kyoto Protocol, which legally binds developed countries to quantified emission reduction targets. Using the principle of differentiated obligation, targets were set for countries according to their level of economic development. The Protocol's first commitment period started in 2008 and ended in 2012, with a second commitment period beginning in 2013. In the first period, Annex I countries were committed to reduce emissions of six GHGs by an average of 5 per cent, relative to 1990 levels. Different targets were set within the Annex I group. The EU, for example, had a target of 8 per cent reduction, but so-called 'burden sharing' allowed these targets to be redistributed among the member states, so long as they sum to the overall EU allowance. There were no reduction targets for developing countries for the first period.

However, despite the use of differentiated obligation, the developing states, the G77 + China, approached climate change policy within the context of their broader disputes with the industrialised world (Meadowcroft, 2002). They argued for increased financial flows, technological assistance to shift to cleaner energy sources and help to develop adaptation. From the onset, the Alliance of SIDS urged vigorous action, while the OPEC nations, fearing decreases in oil prices, tried to slow proceedings (Meadowcroft, 2002). Another major problem arose because the Kyoto Protocol assigned emission targets based on countries' 1990 emission levels. By basing future caps on past levels, these targets were seen as rewarding historically high emitters and penalising low emitters (Baer et al., 2000). Furthermore, in the wake of the massive deindustrialisation that followed the fall of communism, Eastern European countries found themselves sitting on a huge stockpile of unused pollution credits. This 'hot air', as it is known, could be sold or hoarded for later, thus enabling them or others to meet their targets, while avoiding the undertaking of any actual emission reductions.

Kyoto Mechanisms

Under the Protocol, countries must meet their targets primarily through national measures. However, the Protocol allows countries to use three

---

Box 6.7

The Kyoto Mechanisms

1 **International Emission Trading**: this allows countries that can achieve low cost abatement to sell emission entitlements to other countries that are having difficulty meeting their targets.
2 **Joint Implementation**: this allows states to gain credit for emission reductions they helped to achieve in another country, within the Annex I group. For example, they could undertake a particular project, such as helping to replace an outmoded power plant with a more carbon efficient alternative, provided this is not a nuclear plant.
3 **Clean Development Mechanism**: this allows developed countries to implement offset projects in developing countries to reduce their cost of compliance.

market-based mechanisms as additional ways to meet their targets. These are known as the Kyoto Mechanisms.

China is the dominant player on the CDM market, hosting about half of all projects. A CDM project might involve, for example, a rural electrification project using solar panels or the installation of more energy-efficient boilers. The mechanism is seen as giving industrialised countries some flexibility in how they meet their Kyoto targets. Joint implementation allows a host country to benefit from foreign investment and technology transfer. Emission trading allows countries that have emission units to spare to sell this excess to countries that are over their targets. This has allowed a new commodity to be created – carbon – that is now traded on the 'carbon market'. Forests have a separate place in the Kyoto agreement, included under so-called 'land use, land-use change and forestry' (LULUCF) activities.

> There are two different ways to deal with forests in the context of climate change:
>
> 1 Increasing their capacity as carbon sinks: reforestation or afforestation.
> 2 Reducing emissions due to deforestation (REDD).

The Protocol focused only on the first LULUCF option, but excluded the second option. Reforestation, as an ecological restoration measure, is discussed in Chapter 7. Despite its exclusion from the Kyoto Mechanisms, however, Reducing Emissions from Deforestation and Forest Degradation (REDD) has subsequently gone on to play a key role in climate change actions. REDD provides incentive for developing countries to protect their forest resources so as to help address climate change. Deforestation and forest degradation, which result from agricultural expansion and infrastructure development, account for nearly 20 per cent of global GHG emissions. These emissions amount to more than the entire global transportation sector and are second only to that of the energy sector.

Despite their advantages, however, the Kyoto Mechanisms are surrounded with problems. The lack of compliance mechanisms makes it difficult to enforce targets. Countries have also tried to exploit numerous loopholes in international agreements. Canada, for example, wanted to use the sequestration of carbon in its forests to excuse it from making any substantial reductions in emissions (Young, 2003). The mechanisms

Box 6.8

UN Programme on Reducing Emissions from Deforestation and Forest Degradation in Developing Countries (REDD and REDD+)

REDD aims to make forests more valuable standing than they would be if cut down, by creating a financial value for the carbon stored in trees. Once this carbon is assessed and quantified, REDD involves *developed* countries paying *developing* countries carbon offsets for their standing forests.

REDD had national programmes in several countries, but tended to focus more on achieving 'sustainable' logging. However, a number of parties to the UNFCCC, led by the Coalition of Rainforest Nations, argued that there was a need to move beyond this limited focus, to protect the vast carbon stocks locked up in forests.

REDD+ is similar to REDD, but encourages developing countries to contribute to mitigation actions by conserving forest carbon stocks, sustainably managing forests and enhancing forest carbon stocks. Countries that successfully implement these measures are financially compensated for their efforts. The REDD Programme's strategy for 2011–2015 focuses on providing capacity support for technical needs in a limited number of work areas.

As of July 2012, REDD had led to transfers of approximately US$117.6 million to developing countries, and such transfers are predicted to extend to US$30 billion a year. This is a significant North–South financial flow that has the potential to support sustainable development, help conserve biodiversity and secure vital ecosystem services.

Critique

This initiative has helped policy makers recognise the economic value of the carbon stored in forests. It also creates economic incentives for the protection of forested lands and the investment in low-carbon economic development paths for developing countries. While the initiative has improved over time, it still faces several challenges, including how to build technical capacity so that countries can measure their forest loss and degradation accurately, and how to ensure that the rights of local and indigenous communities are respected in policy implementation. REDD is also just a partial approach in that it only addresses incentives for conservation in tropical forests.

Illicit deforestation continues, and the programme has to work within the corruption that has long beset the forestry sector. Abuses in the implementation of REDD+ have included the falsification or exaggeration of carbon credits from projects and favouritism in the allocation of projects and permits. Journalists have exposed the abuses by rogue businessmen, known as 'carbon cowboys', who coerce and bribe local villagers into handing over the rights to the carbon in their forests.

To make an effective contribution to climate change, REDD+ activities in developing countries must complement, not substitute for, deep cuts in developed countries' emissions.

(*Sources*: Oakes et al., 2012; Parker et al., 2009; Transparency International, 2014; UN REDD Organisation, 2011; 2013; UN REDD Programme, 2009; UNEP, 2013; REDD-net 2010)

are also seen as allowing industrialised countries to avoid domestic emission reduction, while enabling them to buy up cheap emission reduction from developing counties. This is seen as a complex way of distributing pollution entitlements (Agarwal, 2002). Furthermore, African governments have argued that, because their carbon emissions and energy consumption levels are so low, there are few options for implementing CDM projects on the continent. In addition, the focus on emissions reductions is out of kilter with the sustainable development needs of many of Africa's poor, such as their need for food and energy security. In addition, both offset schemes and emission trading have been mired in corruption scandals.

Several of these problems can be seen more clearly by looking more closely at the carbon market and the emissions trading that has developed under the Kyoto Protocol. We focus on the EU GHG Emission Trading System (EU ETS), launched in 2005, as it is the largest mandatory cap-and-trade scheme operating to date

The sharpest and most public differences of opinion on the Kyoto Protocol have arisen between the EU and the USA. To understand this difference, a short overview of US climate policy is now provided.

US climate policy

From the onset, the EU has always worked for a stringent, legally binding approach, while negotiations leading up to the Kyoto Protocol saw the US hold out against legally binding measures amidst concerns about compliance costs and possible negative economic impacts. In addition the US wanted binding reduction targets for developing countries so as to avoid giving them unfair advantages in global markets (Bryner, 2000). Then-president George W. Bush was also sceptical

Box 6.9

The EU Emission Trading System

The EU ETS is a cornerstone of the EU's climate change policy. The scheme is based on the belief that trading facilitates greater cost-effectiveness and allows greater scope for technological innovation, compared to traditional regulation. The ETS includes around 11,000 installations, accounting for about 45 per cent of EU CO_2 emissions and operates in thirty countries. Coverage was expanded in 2013. Airlines were scheduled to join in 2012, but this gave rise to disputes, including with China, which argued that the scheme placed too high a financial burden on its airlines.

'Cap and trade'

The scheme works on a 'cap and trade' principle: a limit is placed on the total amount of certain GHGs that can be emitted by the factories, power plants and other installations in the scheme. Within this cap, companies receive emission allowances that they can sell to or buy from one another. The limit on the total number of allowances available ensures that they have a market value.

At the end of each year a company must surrender enough allowances to cover all its emissions. If a company reduces its emissions, it can keep the spare allowances to cover future needs or else sell them to another company that is short of allowances. The number of allowances is reduced over time so that, in principle, total emissions fall.

The EU hopes to link its ETS with other systems around the world to form a global carbon market. In 2012 Australia approved an emissions trading market, expected to cover approximately 60 per cent of the country's annual GHG emissions by 2015. Similar initiatives exist in California, the Canadian province of Québec, Mexico, and the Republic of Korea. In 2012 the first GHG trade took place in China, in Guangdong Province. China plans to open six further regional emissions trading schemes and to integrate these and link them to global markets.

Critique

The ETS has not operated exactly as predicted. During the first phase of the program (2005–2007), emission permits were over-allocated, which resulted in a 2.1 per cent increase in emissions from levels existing before the scheme began. The scheme has also suffered from corruption, including the resale and misreporting of used carbon offsets, sophisticated computer hacking schemes leading to theft from national carbon emission registries, and value-added-tax frauds. In 2010, the discovery of VAT fraud led to the deflation of the European carbon market by approximately 90 per cent and forced the momentary suspensions of trading. In 2011, the theft of carbon credits was reported in Austria, the Czech Republic, Germany, Greece, Italy, and Romania.

Together, these problems raised concerns about not only the fate of the EU ETS, but also the increasing number of domestic cap-and-trade systems being implemented around the world. The Rio+20 Conference explicitly recognised that corruption is an impediment to effective environmental stewardship.

ETS has been heavily criticised by environmental NGOs for giving industry 'a right to pollute'. For many environmentalists, emission trading breaches the principle of environmental integrity as it involves allocating rights to pollute and turning pollution into a tradable commodity.

(*Sources*: *BBC News*, 2012; Bomberg, 2007; Commission of the European Communities, 2014; The Carbon Trust, 2008; UNEP, 2013)

about scientific claims about climate change. Thus in 2001 it came as no surprise when the US announced that it would not implement the Kyoto Protocol.

The US decision to abandon the Kyoto Protocol caused a storm of opposition, most noticeably from the EU. There was also opposition from within the US. In June 2001, the US National Academy of Sciences produced a report which argued that a climate warming trend was evident and that human activity was largely responsible (National Academy of Sciences, 2001). In addition, the US EPA released a report in 2002 describing in graphic detail the negative impacts of climate change on sensitive ecosystems in America (United States Department of State, 2002). There was also strong support outside government, including among the business community, especially the insurance sector. Yet the Bush administration continued to hold to the position that any international deal for limiting GHG emissions must be (1) inclusive of both industrialised and developing countries; (2) predicated upon science, and (3) economically benign.

Geopolitical rivalry also played a big role in America's rejection of the Kyoto Protocol. The failure to include China in the list of Annex I countries proved particularly contentious, as the US believed that this gave China a comparative economic advantage (Cohen and Egelston, 2003: 320). Furthermore, US rejection of the Kyoto Protocol should not be seen in isolation, as it has rejected several other international initiatives, including a treaty to ban landmines.

The USA was not alone, however, in its failure to engage in international efforts to address climate change. Canada, Australia, Norway and Japan initially supported the US stance on Kyoto. The Umbrella Group

Box 6.10

Counter-move: USA Cities for Climate Protection Campaign

Disappointed with the US government's decision to not ratify the Kyoto Protocol, 141 American cities came together under the leadership of Seattle mayor Greg Nickels and pledged to meet or exceed the 7 per cent GHG reduction target by 2012 from 1990 levels foreseen for the US under the Kyoto Protocol. The mayors also called for the establishment of bipartisan national GHG emissions reduction legislation, including a national emissions trading system. The US Conference of Mayors Climate Protection Agreement was launched on 16 February 2005 – the day when the Kyoto Protocol entered into force and became law for the 141 countries that have ratified it.

Since 2005, the Agreement has been signed by over a thousand mayors, including those of Los Angeles, Washington, Chicago, Portland, New York and many other major US cities. This work is supported by ICLEI's Cities for Climate Protection Campaign, which has been providing tools, such as a national protocol on local GHG emissions, to help implementation.

(*Source*: ICLEI, 2012)

of countries, as it is known, continues to insist that developing countries should also be required to meet quantified emission reduction targets. The Umbrella Group is a loose coalition of non-EU developed countries formed following the adoption of the Kyoto Protocol. Although there is no formal membership list, the group usually includes Australia, Canada, Iceland, Japan, New Zealand, Norway, the Russian Federation, Ukraine, and the United States. In 2011 Canada, having initially signed the Kyoto Protocol, dropped out, becoming the first signatory to announce its withdrawal from Kyoto. Australia is the fifteenth largest emitter of GHG and on a per capita basis it emits more than the Europeans or Chinese, but it too remains a laggard in its responses (Flannery et al., 2014). Australia has, however, joined the second commitment period of Kyoto, partly because this enables it to take advantage of the growing global carbon market and also ensure Australian businesses have access to international credits under the CDM.

Despite these difficulties, the election of Barack Obama as America's president has brought major advances in the USA's engagement in climate change policy. Prior to his election, Mr Obama declared that, because climate change poses a serious and imminent threat, if elected he would pursue an active mitigation agenda. Recent reports from the

Environmental Protection Agency confirmed these threats (EPA USA, 2010; 2014c). Since taking office in 2009, President Obama has reaffirmed his administration's commitment and in 2013 outlined the US Climate Action Plan. This plan sees unprecedented efforts by the US to reduce carbon emissions, promote clean energy and help prepare the US for impacts. The plan also makes a commitment to lead international efforts to address global climate change.

Box 6.11

The Obama administration's Climate Action Plan

Actions under the Climate Action Plan fall under several headings and within each heading there are several related initiatives.

1. The path toward a clean energy economy

Reducing carbon pollution from power plants: US power plants account for roughly one third of all domestic GHG emissions. In 2014, the Environmental Protection Agency proposed a Clean Power Plan, setting new carbon pollution standards for power plants and designed to cut emissions by 30 per cent from 2005 levels.

Accelerating clean energy leadership: this makes use of public–private partnerships and the streamlining of the federal permitting process to increase renewable energy use in federally subsidised housing, install renewable projects on public lands and military installations, and double wind and solar electricity generation by 2025.

Expanding and modernising the electric grid: this is designed to improve access to remote sources of solar and wind energy and reduce power outages, while also creating construction and operations jobs.

Staying on the cutting edge: an Advanced Research Project Agency-Energy began in 2009 to advance energy projects that have the potential to transform the way energy is generated, stored and used.

Building a twenty-first-century clean energy infrastructure: this tightens fuel economy standards for passenger vehicles and sets standards for commercial vehicles.

Cutting energy waste: this includes energy conservation standards for appliances and equipment, making energy efficiency accessible to rural America, completing home efficiency upgrades, and advancing energy efficiency through the Better Buildings Challenge.

Reducing other GHGs: this takes action to reduce the emissions of hydrofluorocarbons (HFCs) and shift to safer and more sustainable alternatives within the federal administration. In 2014 a Strategy to Reduce Methane Emissions took steps to cut methane emissions from landfills, coal mining, agriculture, and oil and gas systems.

Leading by example: the federal government, as the largest energy consumer in the US, plans to reduce its GHG emissions from direct sources, such as building energy use, by 28 per cent by 2020 and emissions from indirect sources, such as from employee commuting, by 13 per cent by 2020. The 2012 Operational Energy Strategy Implementation Plan serves as a roadmap to transform the way the Department of Defense, a large energy consumer, uses energy in military operations.

2. Preparing for the impacts

This aims at building a more climate-resilient America through several initiatives.

Assessing the impacts of climate change: in 2014, the administration released the Third US National Climate Assessment, which detailed the impacts across all US regions and key economic sectors.

Supporting climate-resilient investments: a Task Force on Climate Preparedness and Resilience has been established. Members include state, local, and tribal leaders who have experienced building climate preparedness and resilience in their communities.

Rebuilding and learning from Superstorm Sandy: the 2012 Superstorm Sandy highlighted the US's vulnerability to extreme weather events. Rebuilding is now guided by the need to make communities more resilient to emerging challenges, such as rising sea levels, extreme heat, and more frequent and intense storms. This rebuilding strategy is now serving as a model for other communities recovering from disasters. The need for such lessons was also highlighted by Hurricane Katrina, which also raised other public policy issues including emergency management, environmental policy, and the links between poverty and vulnerability to environmental risks.

Promoting resilience in the health sector: the Department of Health and Human Services is developing a set of resources and tools to ensure that the medical system is resistant to climate impacts.

Maintaining agriculture sustainability: the Agriculture Department created seven new Regional Climate Hubs to support farmers, ranchers and forest landowners to adapt to climate change and weather variability.

Providing tools for climate resilience: in 2014, the Climate Data Initiative was launched to provide the information and tools needed to plan for climate impacts. The first phase of the Initiative focuses on coastal resilience and will then expand to include food resilience.

Reducing risk of droughts and wildfires: in 2013 the National Drought Resilience Partnership was launched, an alliance of federal agencies designed to help prepare for droughts and reduce their impact. In 2014, the final phase of the National Wildfire Cohesive Strategy was launched to prepare for a drier future.

3. International re-engagement

The Obama administration has also re-engaged in international climate negotiations. It has played a role in launching several new international partnerships, including the Global Alliance for Climate-Smart Agriculture, the Oil and Gas Methane Partnership, the Pilot Auction Facility for Methane and Climate Change Mitigation, and the Cities Climate Finance Leadership Alliance.

Leading public-sector financing toward cleaner energy: public financing for new coal-fired power plants overseas has ended, except in rare circumstances. Other nations, including the United Kingdom, the Netherlands, and the Nordic countries, have joined the initiative, and the World Bank has adopted a similar policy.

Bilateral co-operation: climate change is now an issue in bilateral relations, including with China and India.

Expanding clean energy use and cutting waste: the Energy Department is leading the Clean Energy Ministerial, a high-level global forum to improve access to energy efficiency and clean energy supply.

Cutting short-lived climate pollutants: in 2013 both the US and China expressed their support for using the institutions of the Montreal Protocol to phase out HFCs. The US is also involved in the Climate and Clean Air Coalition to reduce emissions of methane, HFCs, and black carbon.

Supporting forests: Norway, the United Kingdom and the US have launched a public–private partnership, the Initiative for Sustainable Forest Landscapes, to reduce GHG emissions from deforestation and promote sustainable agriculture and land use planning and policies.

Promoting free trade in environmental goods: in 2014, a US-led coalition of countries launched talks to eliminate tariffs on a range of environmental goods under the WTO.

Mobilising climate finance: in 2014, the US, UK and Germany launched the Global Innovation Lab for Climate Finance, a public–private platform to spur private-sector investment in low-carbon, climate-resilient infrastructure in developing countries.

(*Sources*: EPA US, 2013; Román and Carson, 2009; Goulder and Stavins, 2011; United States Department of State, 2010; The White House, 2011; The White House, Office of the Press Secretary, 2014; The White House, President Barack Obama, 2014)

The shift in climate politics under the Obama presidency has been accompanied by changes at the regional, state and local levels, and in public opinion. However, even though climate change ranks high on the Obama administration's priority list, there remain numerous domestic, structural, institutional and political hurdles blocking effective action. Given the foundering economy, failing health care system and the costs of two expensive and complicated wars, climate change shares its space on a crowded policy agenda (Román and Carson, 2009). In addition, Republican opposition has repeatedly blocked Obama's efforts, including efforts to pass a law on climate change, although a 2007 Supreme Court ruling allows the 1970 Clean Air Act to be used to regulate CO_2 emissions. Furthermore, climate policy is increasingly seen as a means to achieve other policy objectives, as reflected in the use of energy infrastructure investment to create jobs and invigorate the economy and support important geopolitical goals, such as achieving energy security. Thus US policy is better understood as part of a broader set of strategic goals that include national economic recovery and long-term economic growth, national energy security, and the desire to maintain international leadership. In this context, the promotion of sustainable development remains of marginal concern.

The next steps for the Kyoto Protocol

The 2011 UN Climate Change Conference in Durban hoped to establish a new treaty to promote action beyond the first commitment period of the Kyoto Protocol. It was also expected to finalise at least some of the Cancún Agreement's objectives (see Box 6.13 below), such as the promised transfer of funds from rich countries to poorer countries to help them protect forests, adapt to climate impacts, and green their economies. After two weeks of negotiations a deal was reached only on the last day, following a 60-hour marathon negotiation session. However, with no treaty deal, the Conference instead agreed to establish a legally binding pact by 2015, to take effect in 2020. This promises to be the first international agreement to require action by all major emitters, including the US, Japan, India and China. There was also progress made on the creation of a Green Climate Fund to help poor countries adapt to climate impacts. A decision was also made to conduct a Global Review to explore whether the commitment to keep climate change to within a 2 °C rise is appropriate, or whether an even lower 1.5 °C rise is required.

Many were disappointed by the Durban outcomes. Scientists and environmental groups warned that agreements reached were not sufficient to avoid global warming beyond 2 °C and more urgent action is needed. Greenpeace, for example, issued a statement calling on conference participants to ensure a peak in global emissions by 2015, deliver better climate finance, and set up a framework for protecting forests in developing countries (Greenpeace, 2011). Similarly, Nnimmo Bassey, chair of Friends of the Earth International, argued:

> delaying real action until 2020 is a crime of global proportions …
> An increase in global temperatures of 4 degrees Celsius, permitted
> under this plan, is a death sentence for Africa, Small Island States,
> and the poor and vulnerable worldwide. This summit has amplified
> climate apartheid, whereby the richest 1% of the world has decided
> that it is acceptable to sacrifice the 99%.
>
> (Climate Justice Now, 2011)

A year later, the World Bank published a report demanding that moral responsibility be taken on behalf of future generations, arguing that a 4 °C warmer world must be avoided and that warming needs to be held *below* 2 °C (World Bank, 2012). The World Bank, alongside the International Energy Agency, also warned that the world is heading for unprecedented warming of between 4 °C and 6 °C if trends are not reversed. This would result in droughts, floods, heat waves and fiercer storms, reduce agricultural productivity, bring plant and animal extinctions, and widespread population movements.

In the midst of these growing concerns, the 2012 CoP meeting was held in Doha, Qatar. However, in the lead-up to the conference the US, Japan, Russia, New Zealand and Canada indicated they would not sign up to a second Kyoto commitment period. The Conference nonetheless saw confirmation of the extension of the Kyoto Protocol to 2020, as well as the date of 2015 for the development of a successor document, to be implemented from 2020. Wording adopted by the conference incorporated for the first time the concept of 'loss and damage', an agreement in principle that richer nations could be financially responsible to other nations for their failure to reduce carbon emissions. The outcome of the Doha talks received mixed responses, with SIDS critical of the overall package. This is not least because the Kyoto second commitment period applies to only about 15 per cent of annual global emissions of GHGs. Similar failures surrounded CoP19, held in Warsaw, Poland in 2013. Difficulties arose because of the US insistence that China and India agree to binding reduction commitments, whereas Chinese and Indian delegates argue that funding from industrialised countries is needed before such emissions cuts can be executed without impacting their growth rates. The G77 + China bloc walked out during talks about 'loss and damage' compensation for the consequences of global warming. Furthermore, Australia was accused of not taking the talks seriously, especially as the country did not send high-ranking officials to the meeting. On the last day of the conference, WWF, Oxfam, ActionAid, the International Trade Union Confederation, Friends of the Earth and Greenpeace walked out. Organisers from the Ministry of the Economy in Poland were also strongly criticised for co-hosting an event with the World Coal Association alongside the UN FCCC talks, seen as a provocation, given Poland's large coal reserves and its heavy reliance on coal. The dismissal of the Conference President, Marcin Korolec, from his cabinet position as Minister of Environment during the negotiations was also seen by delegates as a further sign of Poland's lack of commitment to global action.

In November 2014, mindful of the need to ensure that negotiations move beyond their current impasse, and committed to taking a leadership role in international climate governance, the US secretly negotiated an agreement with China to lower GHG emissions. Under the agreement, China commits for the first time to cap its output of CO_2 emissions by 2030 and to increase its use of zero-emission energy sources, such as wind and solar power, to 20 per cent by 2030. In return, the US agreed to double the pace of the cuts in its emissions, reducing them to between

---

Box 6.12

The problem of China

Because of the principle of 'common but differentiated responsibility' China was not originally subject to quantified emission reduction targets under the Kyoto Protocol. However, the map of global GHG emissions has changed dramatically since the Protocol was negotiated. Since 2007, China has overtaken the US to become the world's largest GHG emitter. Furthermore, China's fossil fuel consumption, and hence GHG emissions, are expected to continue to grow. Without China's meaningful participation, any climate treaty will fail to limit emissions and prevent temperatures from rising to dangerous levels. As a result, pressure has been mounting on China to engage with the international community to address climate change.

A major concern is that China is still at an early stage of industrialisation and urbanisation. The country still has a large population living in poverty, that has a right to develop out of that poverty. In addition, were China to focus too heavily on global pollutants, this could reduce investment available to deal with its local and regional environmental problems, including air and water pollution, problems that have direct and immediate health consequences and are prioritised over what is seen as the longer term issue of climate change.

(*Source*: Zhang, 2012)

26 per cent and 28 per cent below 2005 levels by 2025. The agreement removes a key plank from domestic interests blocking Obama's efforts to cut carbon pollution, namely the claim that China is unwilling to undertake similar cuts. Obama nonetheless faces opposition at home, especially given the Republican control of Congress. China also faces technical challenges in reaching its targets. Nevertheless, the agreement is significant and can help in the lead-up to international climate negotiations in Paris in 2015, and may even give these much needed momentum. Although significant, however, the targets are weak compared to that of the EU, which has already endorsed a binding 40 per cent GHG reduction target by 2030 (*The Guardian*, 12 November 2014).

Limitations of the current governance regime

As we see from the above outline of global efforts to address climate change, there are both weaknesses and strengths in the current approach, especially as embedded in the Kyoto Protocol. Although critical, some support the Protocol because it is 'the only game in town', providing the architecture for future, more stringent, international agreements (Aldy and Barrett, 2003). As Meadowcroft has argued, the Kyoto Protocol was never intended as a comprehensive solution, but to be

> understood as part of a long-term process to create global institutions to stabilise atmospheric concentrations of greenhouse gasses. It is a first step that commits Annex I countries to modest initial reductions, and can serve as a bridge towards more substantial future cuts, and the extension of the pool of participating countries, in subsequent commitment periods.
>
> (Meadowcroft, 2002: 15)

Others criticised existing commitments as being too weak (Grubb, 2000; Gupta et al., 2007). The World Bank, for example, has argued that despite the UN FCCC and its Protocol, CO_2 emissions continue to grow (World Bank, 2010). Others fear that under the current governance regime climate policy has become a victim of market environmentalism, to the detriment of both the climate and local communities (Liverman, 2008). Furthermore, the gap between current commitments to reduce emissions and the actual reductions required to keep average global temperatures from rising no more than 2 °C above their pre-industrial level presents a major problem. In addition, even if fully implemented, which is doubtful, the Kyoto Protocol will not halt the rise in GHG emissions, let alone address climate change. Rapidly rising emissions from developing countries will more than neutralise whatever abatement is

Box 6.13

International climate change governance: schedule of activities

2014   UN Climate Summit, New York, to raise political momentum for a meaningful climate agreement in Paris in 2015 and to galvanise action.

2014   IPCC Fifth Assessment reports.

2013   CoP19 Warsaw includes decisions on further advancing the Durban Platform, the Green Climate Fund and Long-term Finance, the Warsaw Framework for REDD + and the Warsaw International Mechanism for Loss and Damage.

2012   The Doha Amendment to the Kyoto Protocol adopted.

2011   The Durban Platform for Enhanced Action accepted at CoP17.

2010   Cancún Agreements to provide finance, technology and capacity building support to developing countries to help them adapt and to adopt sustainable paths to low emission economies.

2009   Copenhagen Accord at CoP15. Countries later submitted emissions reductions pledges or mitigation action pledges, all non-binding.

2007   PCC Fourth Assessment Report. Climate science entered into popular consciousness. At CoP13, parties agreed on the Bali Road Map, which charted the way for developments after 2012. IPCC shares the Noble Peace Prize.

2005   Entry into force of the Kyoto Protocol. The Nairobi Work Programme on Adaptation agreed.

2001   IPCC Third Assessment Report. Bonn Agreements adopted, based on the Buenos Aires Plan of Action of 1998. Marrakech Accords adopted at CoP7, detailing rules for implementation of Kyoto Protocol, setting up new funding and planning instruments for adaptation, and establishing a technology transfer framework.

1997   Kyoto Protocol formally adopted at CoP3.

1996   UN FCCC Secretariat set up to support action under the Convention.

1995   The first Conference of the Parties (CoP1) takes place in Berlin.

1994   UN FCCC enters into force.

1992   UN FCCC is opened for signature.

1991   First meeting of the Intergovernmental Negotiating Committee.

1990   IPCC First Assessment Report. IPCC and second World Climate Conference call for a global treaty. UN General Assembly negotiations on a framework convention begin.

1988   The Intergovernmental Panel on Climate Change set up.

1979   The first World Climate Conference takes place.

(*Source*: Modified and updated from UN FCCC Secretariat, 2014)

reached, even if proportionate reductions are agreed in the next commitment period. Emissions from countries that have not signed up will also
mean GHG levels in the atmosphere will keep rising.

Conclusion

In this chapter, the causes and impact of climate change were explored,
before attention was turned to international efforts to address the problem. By exploring the range of issues that have come to be associated
with the global governance of climate change, further insight has been
gained into the complex nature of the cross-cutting tasks involved in
promoting sustainable development.

Its location within the UN system frames the global governance of climate change within a broader sustainable development agenda. However, it also means that very complex negotiation processes and
agreements, centred on the UN FCCC, have come into play. This has
meant that Third World countries continue to approach climate change
within the context of their broader disputes with the industrialised
world, while embroiling the Convention in the controversial relationship
between certain states, such as China, Australia, Canada and the US,
and the UN. Transition to a post-carbon future remains uncertain in the
face of this global political wrangling.

Conventions set out broad areas of agreement, making it necessary for
CoPs to flesh out the details, agree targets and timetables, and establish
reporting and monitoring mechanisms. However, CoP meetings have
presented an ideal opportunity for the articulation of narrow self-interest,
aimed at achieving short-term, often commercial, gains. Corruption has
also marred efforts. This is very different from the democratic, participatory processes envisaged by Brundtland. Despite its limitations, however,
efforts at global governance have helped redefine climate change in
social, economic and political terms. There is now clear understanding
that the promotion of sustainable development requires efforts to address
climate change, even if efforts to date have proved inadequate, although
vibrant subnational, local responses continue to be seen.

Summary points

- Climate change will bring major social consequences. The ability of
  countries to undertake adaptation varies greatly.

- Issues of climate justice relate the problem of climate change directly to sustainable development, but the danger remains that responses will undermine the promotion of sustainable development.
- The UN FCCC has helped to frame the problem of climate change in social, economic and political terms.
- The 1977 Kyoto Protocol is part of a long-term process to create global institutions to stabilise atmospheric concentrations of GHGs.
- Global efforts to address climate change to date have been ineffective, although a vibrant response is discernible at sub-national levels, including within cities.

Further reading

Adger, W. N. (2003), 'Social Capital, Collective Action, and Adaptation to Climate Change', *Economic Geography*, 79 (4): 387–404.

Agarwal, A. (2002), 'A Southern Perspective on Curbing Global Climate Change', in Stephen H. Schneider, Armin Rosencranz and J. O. Niles (eds), *Climate Change Policy: A Survey* (Washington, DC: Island Press), 375–391.

Bomberg, E. (2012), 'Mind the (Mobilization) Gap: Comparing Climate Activism in the United States and European Union', *Review of Policy Research*, 29 (3): 408–430.

Dobson, A. (1998), *Justice and the Environment: Conceptions of Environmental Sustainability and Dimensions of Social Justice* (Oxford: Oxford University Press).

International Bank for Reconstruction and Development/World Bank (2012), *Turning Down the Heat: Why a 4 °C Warmer World Must Be Avoided*, Report for the World Bank by the Potsdam Institute for Climate Impact Research and Climate Analytics (Washington, DC: World Bank), available online at www-wds.worldbank.org/external/default/WDSContentServer/WDSP/IB/2012/12/20/000356161_20121220072749/Rendered/PDF/NonAsciiFileName0.pdf

Matthew, R. A. and Hammill, A. (2009), 'Sustainable Development and Climate Change', *International Affairs*, 85 (6): 1117–1128.

Nordhaus, W. D. (2007), 'A Review of the Stern Review on the Economics of Climate Change', *Journal of Economic Literature*, XLV: 686–702.

Pelling, M. (2011), *Adaptation to Climate Change: From Resilience to Transformation* (London and New York: Routledge).

Shue, H. (2014), *Climate Justice: Vulnerability and Protection* (Oxford: Oxford University Press).

United Nations Task Team on Social Dimensions of Climate Change (2011), *The Social Dimensions of Climate Change: Discussion Draft*, available online at www.who.int/globalchange/mediacentre/events/2011/social-dimensions-of-climate-change.pdf

Vig, N. J. and Kraft, M. E. (2012), *Environmental Policy: New Directions for the Twenty-First Century* (Thousand Oaks, CA: Sage).

Web based resources

Encyclopaedia of the Earth: Climate Change: www.eoearth.org/topics/view/51c
   bfc78f702fc2ba8129e7b/
NGO Climate Action Network: www.climatenetwork.org
Official UN FCCC site, http://unfccc.int/

References

Agarwal, A. (2002), 'A Southern Perspective on Curbing Global Climate
   Change', in S. H. Schneider, A. Rosencranz and J. O. Niles (eds), *Climate
   Change Policy: A Survey* (Washington DC: Island Press), 375–391.
Aldy, J. E. and Barrett, S. (2003), 'Thirteen Plus One: A Comparison of Global
   Climate Policy Architectures', *Climate Policy*, 3 (4): 373–397.
Baer, P., Harte, J., Aya, B., Herzog, V., Holdren, J., Hultman, N. E., Kammen,
   D. M., Norgaard, R. B. and Raymond, L. (2000), 'Climate Change, Equity
   and Greenhouse Gas Responsibility', *Science*, 289: 2287.
Barnett, J. (2006), 'Climate Change, Insecurity and Injustice', in W. Adger,
   J. Paavola, H. Saleemul and M. J. Mace (eds), *Fairness in Adaptation to
   Climate Change* (Cambridge, MA: MIT Press), 115–129.
BBC News (2012), 'UK, China "Bans" Airline from Joining EU Carbon
   Scheme', www.bbc.co.uk/news/business-16901106, 6 February.
Bomberg. E. (2007), 'Policy Learning in an Enlarged European Union:
   Environmental NGOs and New Policy Instruments', *Journal of European
   Public Policy*, 14 (2): 248–268.
Bryner, G. (2000), 'The United States: Sorry, Not Our Problem', in W. M. Lafferty
   and J. Meadowcroft (eds), *Implementing Sustainable Development: Strategies
   and Initiatives in High Consumption Societies* (Oxford: Oxford University
   Press), 273–302.
Cameron, E., Shine, T. and Bevins, W. (2013), 'Climate Justice: Equity and
   Justice Informing a New Climate Agreement', Working paper (Washington,
   DC: World Resources Institute, and Dublin: Mary Robinson Foundation –
   Climate Justice), available online at www.mrfcj.org/media/pdf/climate_
   justice_equity_and_justice_informing_a_new_climate_agreement.pdf
The Carbon Trust (2008), *Cutting Carbon in Europe: The 2020 Plans and the
   Future of the EU ETS*, available online at www.carbontrust.com/
   media/84896/ctc734-cutting-carbon-in-europe-2020-plans.pdf
Climate Justice Now (2011), '2011 CoP17 Succumbs to Climate Apartheid',
   available online at www.climate-justice-now.org/2011-cop17-succumbs-
   to-climate-apartheid-antidote-is-cochabamba-peoples%E2%80%
   99-agreement/
Cohen, M. J. and Egelston, A. (2003), 'The Bush Administration and Climate
   Change: Prospects for an Effective Policy Response', *Journal of
   Environmental Policy and Planning*, 5 (4): 315–331.

College of Earth and Mineral Science, Pennsylvania State University (2014a), 'What Is Geoengineering?' available online at https://www.e-education.psu. edu/meteo469/node/179

College of Earth and Mineral Science, Pennsylvania State University (2014b), 'Carbon Capture and Sequestration', available online at https://www.e-education. psu.edu/meteo469/node/223

Commission of the European Communities (2014), *Emission Trading Scheme*, http://ec.europa.eu/clima/policies/ets/index_en.htm

Dobson, A. (1998), *Justice and the Environment: Conceptions of Environmental Sustainability and Dimensions of Social Justice* (Oxford: Oxford University Press).

Environmental Protection Agency (EPA USA) (2010), *Climate Change Indicators in the United States* (Washington, DC: EPA), available online at www.epa.gov/climatechange/indicators.html.

EPA USA (2013), 'EPA Proposes Carbon Pollution Standards for New Power Plants'/'Agency Takes Important Step to Reduce Carbon Pollution From Power Plants as Part of President Obama's Climate Action Plan', News releases, 20 September, available online at http://yosemite.epa.gov/opa/ admpress.nsf/0/da9640577ceacd9f85257beb006cb2b6!

EPA USA (2014a), 'Future Climate Change', available online at www.epa.gov/ climatechange/science/future.html

EPA USA (2014b), 'Climate Change Science Overview', available online at www.epa.gov/climatechange/science/overview.html

EPA USA (2014c), 'Third National Climate Assessment', available online at http://nca2014.globalchange.gov/report

Flannery, T., Hueston, G. and Stock, A. (2014), *Lagging Behind: Australia and the Global Response to Climate Change* (Potts Point, NSW: Climate Council of Australia), available online at www.climatecouncil.org.au/uploads/ 211ea746451b3038edfb70b49aee9b6f.pdf

Ghina, F. (2003), 'Sustainable Development in Small Island Developing States: The Case of the Maldives', *Environment, Development and Sustainability*, 5: 39–165.

Goulder, L. H. and Stavins, R. N. (2011), 'Challenges from State–Federal Interactions in US Climate Change Policy', *American Economic Review*, 101 (3): 253–257.

Greenpeace (2011), 'Polluticians Occupy the Climate', Press release, 23 November, available online at www.greenpeace.org/international/en/press/ releases/Polluticians-occupy-the-climate-/

Grubb, M. (2000), 'The Kyoto Protocol: An Economic Appraisal', FEEM working paper no. 30, *Social Science Research Network*, doi:10.2139/ ssrn.229280. SRN 229280.

The Guardian (2014), 'US–China Climate Deal Boosts Global Talks but Republicans Vow to Resist', Wednesday 12 November, available online at www.theguardian.com/environment/2014/nov/12/us-china-climate-deal-boosts-global-talks-but-republicans-vow-to-resist

Gupta, S., Tirpak, D. A., Burger, N., Gupta, J., Höhne, N., Boncheva, A. I., Kanoan, G.M., Kolstad, C., Kruger, J. A., Michaelowa, A., Murase, S., Pershing, J., Saijo, T. and Sari, A. (2007), 'Policies, Instruments and Co-operative Arrangements', in B. Metz, O. R. Davidson, P. R. Bosch, R. Dave and L. A. Meyer (eds), *Climate Change 2007: Mitigation*, Contribution of Working Group III to the Fourth Assessment Report of the Intergovernmental Panel on Climate Change (Cambridge and New York: Cambridge University Press), available online at www.ipcc.ch/publications_and_data/ar4/wg3/en/ch13html

HM Treasury (2006), *The Economics of Climate Change: The Stern Review* (London: HM Treasury), available online at http://webarchive.nationalarchives.gov.uk/20100407172811/www.hm-treasury.gov.uk/stern_review_report.htm

ICLEI (2012), *Local Sustainability 2012: Taking Stock and Moving Forward – Global Review* (Bonn, Germany: Local Governments for Sustainability World Secretariat), available online at http://local2012.iclei.org/fileadmin/files/LS2012_global_review_www.pdf)

IPCC (1990), *First Assessment Overview and Policymaker Summaries* (Geneva, Switzerland: IPCC).

IPCC (2007a), *Climate Change 2007, Synthesis Report: Summary for Policy Makers*, available online at www.ipcc.ch/pdf/assessment-report/ar4/syr/ar4_syr_spm.pdf

IPCC (2007b), *Fourth Assessment Report Synthesis Report, Glossary of Terms*, IPCC, 2007, available online at www.ipcc.ch/pdf/assessment-report/ar4/syr/ar4_syr_appendix.pdf

IPCC (2007c), *Fourth Assessment Report: Climate Change 2007 Climate Change 2007: Working Group III: Mitigation of Climate Change*, available online at www.ipcc.ch/publications_and_data/ar4/wg3/en/ch2s2-2-4.html

IPCC (2014a), 'Summary for Policymakers', in O. Edenhofer, R. Pichs-Madruga, Y. Sokona, E. Farahani, S. Kadner, K. Seyboth, A. Adler, I. Baum, S. Brunner, P. Eickemeier, B. Kriemann, J. Savolainen, S. Schlömer, C. von Stechow, T. Zwickel and J. C. Minx (eds), *Climate Change 2014, Mitigation of Climate Change: Contribution of Working Group III to the Fifth Assessment Report of the Intergovernmental Panel on Climate Change*, (Cambridge: Cambridge University Press), available online at http://report.mitigation2014.org/spm/ipcc_wg3_ar5_summary-for-policymakers_approved.pdf

IPCC (2014b), *IPCC Fifth Assessment Report: Climate Change 2014: Synthesis* www.ipcc.ch/pdf/assessment-report/ar5/syr/SYR_AR5_SPM.pdf

IPCC (2014c), *Fifth Assessment Report, AR5, Climate Change 2014: Impacts, Adaptation and Vulnerability: Summary for Policy Makers*, http://ipcc-wg2.gov/AR5/images/uploads/WG2AR5_SPM_FINAL.pdf

Lenton, T. M. (2010), *Earth System Tipping Points* (USA EPA), available online at http://yosemite.epa.gov/ee/epa/eerm.nsf/vwAN/EE-0564-112.pdf/$file/EE-0564-112.pdf

Liverman, D. M. (2008), 'Conventions of Climate Change: Constructions of Danger and the Dispossession of the Atmosphere', *Journal of Historical Geography*, 35 (2): 279–296. doi:10.1016/j.jhg.2008.08.008.

Mary Robinson Foundation – Climate Justice (MRF CJ) (2013), *Climate Justice: An Intergenerational Approach* (Dublin: MRF CJ), available online at www.mrfcj.org/media/pdf/Intergenerational-Equity-Position-Paper-2013-11-16.pdf

Mary Robinson Foundation – Climate Justice (2014a), *Principles of Climate Justice* (Dublin: MRF CJ), available online at www.mrfcj.org/about/principles.html.

Mary Robinson Foundation – Climate Justice (2014b), *Position Paper: Human Rights and Climate Justice* (Dublin: MRF CJ), www.mrfcj.org/media/pdf/PositionPaperHumanRightsandClimateChange.pdf

Matthew, R. A. and Hammill, A. (2009), 'Sustainable Development and Climate Change', *International Affairs*, 85 (6): 1117–1128.

Matthews, H. D., Graham, T. L., Keverian, S., Lamontagne, C., Seto, D. and Smith, T. J. (2014), 'National Contributions to Observed Global Warming', *Environmental Research Letters*, 9, 014010. doi: 10.1088/1748-9326/9/1/014010.

Meadowcroft, J. (2002), *The Next Step: A Climate Change Briefing for European Decision-Makers* (Florence, Italy: European University Institute, Robert Schuman Centre for Advanced Studies, Policy Paper, 02/13).

NASA (n.d.a), 'Global Climate Change, What's the Difference Between Climate Change and Global Warming?' http://climate.nasa.gov/climatechangeFAQ/#Q2

NASA (n.d.b), 'Global Climate Change, What Is the Greenhouse Effect?' http://climate.nasa.gov/climatechangeFAQ/#Q8

[US] National Academy of Sciences (2001), *Science of Climate Change* (Washington, DC: National Academy Press).

Oakes, N., Leggett, M., Cranford, M. and Vickers, H. (2012), *The Little Forest Finance Book* (Oxford: Global Canopy Programme).

Parker, C., Mitchell, A., Trivedi, M. and Mardas, N. (2009), *The Little REDD+ Book* (2nd edn), available online at www.theredddesk.org/redd_book/framework

Parliament of Australia (2008), 'Climate Change and Global Warming: What's the Difference?' available online at www.aph.gov.au/About_Parliament/Parliamentary_Departments/Parliamentary_Library/Browse_by_Topic/ClimateChange/theBasic/climate

PBL, Planbureauvoor de Leefomgeving (Netherlands Environmental Assessment Agency) (2012), *Trends in Global CO2 Emissions: 2012 Report* (The Hague: PBL), available online at www.pbl.nl/sites/default/files/cms/publicaties/PBL_2012_Trends_in_global_CO2_emissions_500114022.pdf

REDD-net (2010), *Catalysing REDD+ at the National Level: Summary of Experience so far* (London: REDD-net ODI).

Román, M. and Carson, M. (2009), *Sea Change: US Climate Policy Prospects under the Obama Administration* (Stockholm: Commission on Sustainable Development, Regeringskansliet), available online at www.sei-international.org/mediamanager/documents/Publications/Climate/seachange_roman.pdf

Steffen, W., Crutzen, P. J. and McNeill, J. R. (2007), 'The Anthropocene: Are Humans Now Overwhelming the Great Forces of Nature?' *Ambio*, 36 (8): 614.

Transparency International (2014), *Protecting Climate Finance: An Anti-Corruption Assessment of the UN-REDD Programme* (Berlin: Transparency International), available online at www.transparency.org/whatwedo/publication/protecting_climate_finance_un_redd_programme

The Union of Concerned Scientists (2010), *Each Country's Share of CO$_2$ Emissions*, available online at http://webcache.googleusercontent.com/search?q=cache:zNPneuLpmy8J:www.ucsusa.org/global_warming/science_and_impacts/science/each-countrys-share-of-co2.html+&cd=1&hl=en&ct=clnk&gl=uk#.VCP9Pe49_ng

UN (2014), *Trends in Sustainable Development: Small Island Developing States (SIDS)* (New York: United Nations Department of Economic and Social Affairs Division for Sustainable Development).

UNEP (United Nations Environment Programme) (2013), *The Impact of Corruption on Climate Change: Threatening Emissions Trading Mechanisms?* available online at http://na.unep.net/geas/archive/pdfs/GEAS_Mar2013_EnvCorruption.pdf

UN FCCC Secretariat (2014), 'Background on the UNFCCC: The International Response to Climate Change', available online from: http://unfccc.int/essential_background/items/6031.php

United States Department of State (2002), *United States Climate Action Report* (Washington, DC: U.S. Government Printing Office), available online at www.epa.gov/globalwarming/publications/car

United States Department of State (2010), *US Climate Action Report 2010* (Washington, DC: Global Publishing Services, June), available online at www.state.gov/g/oes/rls/rpts/car/index.htm

UN REDD Organisation (2011), *The UN-REDD Programme Strategy, 2011–2015*, available online at www.un-redd.org/PublicationsResources/tabid/587/Default.aspx#foundation_docs

UN REDD Organisation (2013), *Legal Analysis of Cross-Cutting Issues for REDD+ Implementation: Lessons Learned from Mexico, Viet Nam and Zambia* (Geneva, Switzerland: UN-REDD Programme).

UN REDD Programme (2009), 'About the UN-REDD Programme', available online at www.un-redd.org/AboutUN-REDDProgramme/tabid/102613/Default.aspx

The White House, Office of the Press Secretary (2014), 'Fact Sheet: President Obama Announces New Actions to Strengthen Global Resilience to Climate Change and Launches Partnerships to Cut Carbon Pollution' (Washington, DC: 23 September), available online at www.whitehouse.gov/the-press-office/2014/09/23/fact-sheet-president-obama-announces-new-actions-strengthen-global-resil

The White House, President Barack Obama (2014), 'The President Has Taken Unprecedented Action to Build the Foundation for a Clean Energy Economy,

Tackle the Issue of Climate Change, and Protect Our Environment', available online at www.whitehouse.gov/energy/climate-change

The White House, Washington (2011), *Blueprint for a Secure Energy Future* (Washington, DC: 30 March), available online at www.whitehouse.gov/blog/2011/03/30/obama-administration-s-blueprint-secure-energy-future

World Bank (2010), 'Integrating Development into a Global Climate Regime', in World Bank, *World Development Report 2010: Development and Climate Change* (Washington, DC: World Bank), ch. 5, available online at http://siteresources.worldbank.org/INTWDRS/Resources/477365-1327504426766/8389626-1327510418796/Chapter-5.pdf

World Bank (2012), *Turning Down the Heat: Why a 4 °C Warmer World Must Be Avoided*, A Report for the World Bank by the Potsdam Institute for Climate Impact Research and Climate Analytics (Washington, DC: World Bank), available online at http://www-wds.worldbank.org/external/default/WDSContentServer/WDSP/IB/2012/12/20/000356161_20121220072749/Rendered/PDF/NonAsciiFileName0.pdf

Young, O. R. (2003), 'Environmental Governance: The Role of Institutions in Causing and Confronting Environmental Problems', *International Environmental Agreements: Politics, Law and Economics*, 3: 337–393.

Zhang, J. (2012), *Delivering Environmentally Sustainable Economic Growth: The Case of China* (San Diego: School of International Relations and Pacific Studies, University of California), available online at http://asiasociety.org/policy/environmentally-sustainable-economic-growth-possible-china

7 Addressing biodiversity loss

Key issues

- Biodiversity loss
- Tipping point
- UN Convention on Biological Diversity
- Biosafety
- Aichi targets
- Ecological restoration
- Ecosystem services.

Introduction

This chapter examines the links between biodiversity and sustainable development. It explores the consequences of biodiversity loss for both ecosystem resilience as well as the maintenance of societal well-being. Attention is also paid to the global governance of biodiversity and the role played by the UN, including through the Convention on Biological Diversity (CBD). This includes exploring the strategies and action plans devised under the remit of the CBD, alongside the use of ecological restoration as a tool for addressing biodiversity loss. The recent emphasis on ecosystem services, and the valuation of such services in monetary terms, is explained and its significance explored.

Understanding biodiversity and its loss

Biodiversity

Biodiversity can be defined as 'the variability among living organisms from all sources including terrestrial, marine and other aquatic ecosystems and the ecological complexes of which they are part; this includes diversity within species, between species and of ecosystems' (UNEP, 1992). Biodiversity is a crucial indicator of planetary health (Iles, 2003). It also contributes to ecosystem resilience, broadly understood to mean the ability of ecosystems to continue to function under changing environmental conditions, such as climate change. Biodiversity establishes the conditions for well-being and survival, including that of human society, and a healthy ecosystem is central to the aspirations of humankind. However, biodiversity can also been seen to have value in its own right, over and above any use value to human beings.

> Biological diversity or biodiversity is the variety of life on Earth, within species, between species and across ecosystems.

Loss of biodiversity

Recent decades have witnessed a substantial and largely irreversible loss in the diversity of life on Earth, caused by anthropogenic pressures. In addition, these pressures are also disrupting ecosystem processes and the services they provide. Biodiversity loss can be understood as the 'long-term or permanent qualitative or quantitative reduction in components of biodiversity and their potential to provide goods and services, to be measured at global, regional, and national levels' (UNEP, 2004).

There are no shortages of reports giving warnings about the rapid decline in biodiversity. The Global Biodiversity Outlook (GBO) reports, which were initiated by the UN to provide information on the status of biodiversity, have identified multiple indications of continuing decline in biodiversity in all three of its main components – genes, species and ecosystems. The third GBO, the GBO-3, also showed that the ecological footprint of humanity exceeds the biological capacity of the Earth by a wider margin (UNEP, 2010a).

According to the IUCN *Red List of Threatened Species*, an estimated 24 per cent of all mammals, 12 per cent of birds, 25 per cent of reptiles, 20 per cent of amphibians, 30 per cent of fish and 16 per cent of conifers are currently threatened with extinction (IUCN, 2013). There is also evidence of severe genetic erosion of cultivated plants and animals. The UN Food and Agricultural Organisation adds to this concern by showing that some of the leading 'provider countries' for crop plants, such as wheat and maize, have lost more than 80 per cent of their plant varieties

---

Box 7.1

Overview of current status of biodiversity

Species assessed for extinction risk are on average moving closer to extinction. Amphibians face the greatest risk and coral species are deteriorating most. Nearly a quarter of plant species are estimated to be threatened with extinction.

● The abundance of **vertebrate** species, based on assessed populations, fell by nearly a third on average between 1970 and 2006, and continues to fall globally, with especially severe declines in the tropics and among freshwater species.

● **Natural habitats** in most parts of the world continue to decline in extent and integrity, although there has been significant progress in slowing the rate of loss in some regions. Freshwater, wetlands, sea ice habitats, salt marshes, coral reefs, seagrass beds and shellfish reefs are all showing serious decline.

● Extensive **fragmentation** and degradation of forests, rivers and other ecosystems has also led to loss of biodiversity and ecosystem services.

● **Crop and livestock** genetic diversity continues to decline in agricultural systems.

(Source: UNEP, 2010a)

(FAO, 2009). Agricultural systems based on industrial monoculture are highly susceptible to plant diseases, climate change and ecological shifts, and thus lack resilience. This can potentially have severe consequences for the poor and for food security and thus for efforts to promote sustainable futures (UNEP, 2002).

A regional perspective also shows similar patterns. In Europe, it has been shown that a majority of species do not have healthy populations and that the overall status of grassland, wetland and coastal habitat types is particularly poor (EEA 2010a, 2010b, 2010c). Most future scenarios project continuing high levels of extinctions and loss of habitats throughout this century, with associated decline of some ecosystem services important to human well-being (UNEP, 2010a).

There is also a high risk of dramatic biodiversity loss and accompanying degradation of a broad range of ecosystem services if ecosystems are pushed beyond certain thresholds or tipping points, discussed also in Chapter 6.

> A *tipping point* is a situation in which an ecosystem experiences a shift to a
> new state, with significant changes to biodiversity and the services to people
> it underpins, at a regional or global scale.
>
> (*Source*: UNEP, 2010a)

Tipping points have several characteristics, including that change
becomes self-perpetuating through so-called positive feedbacks, for
example deforestation reduces regional rainfall, which increases fire
risk, which causes forest dieback and further dying. Such changes are
long lasting and hard to reverse. Furthermore, there is a significant
time lag between the pressures driving the change and the appearance
of impacts. Candidates for tipping point changes include dieback of
the Amazon forest, due to the interaction of deforestation, fire and climate change, with parts of the forest moving into a self-perpetuating
cycle of more frequent fires and intense droughts leading to a shift to
savannah-like vegetation. The Arctic marine ecosystem is also seen as
likely to experience an ecological tipping point. Tipping points pose
major challenges for efforts to promote sustainable development
because their occurrence is unpredictable given current knowledge,
and it is very difficult for society to adapt to such abrupt and irreversible changes.

Drivers of biodiversity loss

The causes of biodiversity loss are complex and multidimensional, but
all relate ultimately to the adoption of our growth-orientated, economic
development model. Biodiversity loss is caused by the interplay
between particular economic sectors (such as agriculture, energy and
transport) and individual ecosystems. Biodiversity is also under threat
from global problems, such as GHG emissions, which affect all ecosystems. Discussions on biodiversity loss tend to divide the causes
into so-called 'direct drivers' and 'underlying drivers'. Most attention
is paid in policy documents to the need to address the direct drivers of
biodiversity loss.

Habitat loss is the largest direct driver of the current global species
extinction event (Fahrig, 2001). Agriculture also exerts high pressure
on the natural environment, despite several decades of agricultural
mitigation measures in many developed countries. There are other

---

Box 7.2

Direct drivers of biodiversity loss

1. Habitat loss and degradation
2. Climate change
3. Excessive nutrient load and other forms of pollution
4. Over-exploitation and unsustainable use
5. Invasive alien species.

(*Source*: UNEP, 2010a)

---

drivers, including over-exploitation of forests and marine fisheries, while urban sprawl and abandonment of agricultural land are putting pressure on natural and semi-natural areas. Moreover, invasive alien species, that is animals and plants that are introduced accidently or deliberately into a natural environment where they are not normally found, bring serious negative consequences for their new environment, and continue to present a threat, increasingly in marine systems. The five principal pressures directly driving biodiversity loss are either constant or increasing in intensity.

Attention has also to be paid to the underlying causes of loss. These are related to human population growth, the expansion of human activity into formerly natural areas, and the intensification of human activities in existing areas. Over-population and changes in land-use are causing problems with respect to nitrogen deposition and increasing nutrient loading in ecosystems, desertification and land degradation, and over-fishing. Rising human production and consumption activities are also an underlying factor. Europe, for example, is unable to meet its consumption demands from within its own borders, where demand for goods and services exceeds its total capacity for biological production and absorption of waste (WWF, 2005). Moreover, this gap between demand and biocapacity has been growing progressively since 1960.

Biocapacity refers to the capacity of a given biologically productive area to generate an ongoing supply of renewable resources and to absorb its spill-over wastes. Unsustainability occurs if the area's ecological footprint exceeds its biocapacity.

(*Source*: Greenfacts, 2014)

We can thus see that biodiversity loss is ultimately driven by unsustainable models of consumption and production, a loss that is pushed further by population growth, although the relationship between population growth and consumption is not straightforward, as discussed in Chapter 4. Addressing biodiversity loss heightens the importance of placing limits to growth on the sustainable development agenda.

Policies to protect biodiversity and use its components sustainably have the potential to reap rich rewards for society, through underpinning better health, greater food security and community well-being, while enabling society to adapt to climate change. However, preventing further human-induced loss is extremely challenging. This is not least because the issue has only recently emerged as a visible item on the policy agenda and we still lack a long-term vision about how biodiversity is to be maintained. Nevertheless, biodiversity is increasingly subject to a regime of international governance.

The international governance of biodiversity

The Rio Earth Summit of 1992 saw the adoption of the UN Convention on Biodiversity (CBD). It has three main objectives.

Box 7.3

Main objectives of the CBD

1 The conservation of biological diversity;
2 The sustainable use of its resources;
3 The equitable sharing of the benefits arising from the use of genetic resources.

The CBD addresses a wide range of concerns. These range from ecosystems protection to the exploitation of genetic resources; from conservation to issues of environmental and social justice; from commerce to scientific knowledge, and from the allocation of rights to the imposition of responsibility (Le Prestre, 2002).

Links to the wider sustainable development agenda

Like its sister convention, the UN FCCC, the CBD forms an integral part of, and reflects the UNCED understanding of sustainable development.

It promotes the sustainable and equitable use of biodiversity resources. It affirms the primacy of social and economic development, while aiming to couple that development with biodiversity protection. The CBD also explicitly draws upon the precautionary principle, stating that:

> Where there are threats of serious or irreversible damage, lack of full scientific certainty shall not be used as a reason for postponing cost-effective measures to prevent environmental harm.

(CBD, Article 15)

The links between the CBD and the wider sustainable development agenda are also reflected in the Convention's adoption of what is called 'the ecosystem approach'.

Ecosystem means a dynamic complex of plant, animal and micro-organism communities and their non-living environment interacting as a functional unit.

(*Source*: UNEP, n.d.a)

The ecosystem approach sees humans as an integral component of ecosystems. This differs from earlier approaches to conservation that sought to remove humans from 'protected' ecosystems, such as wild life parks, to better conserve their species. In contrast, the ecosystem approach seeks to integrate human activities with biodiversity protection, including through devising integrated land, water and living resource management plans.

The objectives of the CBD also go well beyond narrowly defined conservation measures because it is based on the belief that 'addressing the threats to biodiversity requires immediate and long-term fundamental changes in the ways that resources are used and benefits are distributed' (UNEP, 2002). The Convention makes it clear that there is need to address the *causes* of biodiversity loss. These include addressing issues such as property rights, trade patterns, inequitable social relations and unsustainable patterns of economic development and resource consumption.

The CBD also aims to reinforce other UN policy priorities, including improving the well-being of poor and vulnerable people. Biodiversity is a key economic, financial, cultural, and strategic asset for developing countries, and is critical for economic and social development as well as poverty reduction. The majority of the world's biodiversity resides in the Global South and its loss undermines the potential to promote sustainable development not only within developing countries, but globally. Making such links within the CBD not only helps increase the perceived legitimacy of biodiversity policy but can also lead to new, synergistic policy developments. This link is evident from the incorporation of CBD objectives into the Millennium Development Goals (MDGs) in 2002 (Sachs et al., 2009) and in the promotion of 'co-benefits'. The Johannesburg Summit also highlighted the essential role of biodiversity in meeting the MDGs, as discussed in Chapter 11.

However, the trade-offs between achieving the MDG targets for 2015 and reducing the rate of biodiversity loss have yet to be addressed in practice. Often aid and other efforts to reduce extreme poverty and improve health bring relatively high rates of habitat loss and associated

---

Box 7.4

South–South co-operation for biodiversity protection

What is South–South and triangular co-operation?

South–South co-operation is a term used to describe the exchange of resources, technology and knowledge between developing countries.

South–South co-operation can help enhance technical, financial, scientific and technological exchanges and innovations for development.

Why is it important?

Developing countries have been accumulating knowledge, experience and expertise on biodiversity. With this increased capacity, South–South co-operation can complement North–South exchanges.

Plan of action

A Multi-Year Plan of Action for South–South Co-operation on Biodiversity for Development was adopted by the group of developing countries, G77 and China at the South–South Co-operation Forum in 2010 in Nagoya, Japan. This is seen as having the potential to make an important contribution to the implementation of the 2011–2020 Strategic Plan of the CBD, discussed below.

(*Source*: UNEP, n.d.b)

loss of species. For example, improving rural road networks, a common feature of hunger reduction strategies, often accelerates rates of biodiversity loss directly through habitat fragmentation, and indirectly by facilitating unsustainable harvests, for example, of bush meat (World Resources Institute, 2005). One way in which efforts are being made to address these trade-offs at the UN level is through facilitating South–South co-operation through the CBD.

Efforts to promote South–South co-operation serve to highlight the material or other benefits that biodiversity provides for human well-being.

Because of the scope of issues it addresses and the links that it makes between the ecological, economic and social dimensions, it has been argued that the CBD is 'the first truly and for the moment the foremost sustainable development treaty' (Le Prestre, 2002: 270). However, these very characteristics, of wide scope and deep ambition, are also a source of weaknesses, leading to major tensions among the signatories to the Convention. The group of developing countries (G77) support the CBD precisely because of the links it makes between its three basic goals of conservation, sustainable use and benefit-sharing. However, this is set against a background of profound suspicion on the part of many developing countries that much of the environmental business of the UN is aimed at limiting development. Suspicion that the CBD regime prioritises not so much the preservation of biodiversity, but ensuring commercial access to plant and animal genetic resources for Western biotechnology and pharmaceutical companies, continues to abound (Baker, 2003). Others, particularly the US, would prefer to see the three objectives of the Convention decoupled, because these links move the Convention too deeply into the political arena and thus into the muddy waters of North–South relations.

While some complain that the Convention is too wide, there are concerns that a number of important issues have not been addressed, including ensuring participation of indigenous groups and respect for different cultural values and types of knowledge as they relate to the use of plant and animal genetic resources. There are also calls to take account of different understandings of rights, as some traditions see these as vested in the individual, while others understand rights only in relation to the community, tribe or group. Over time, we can expect that the CBD will face even further issues, such as the need to maintain the biodiversity of the high seas, as well as within cities and the urban landscape (IUCN, 2010).

As a wide ranging, ambitious and deeply political Convention (Le Pres-tre, 2002), it is not surprising to find that, from its birth, the CBD has been plagued with problems. The US refused to add its signature, as it could not accept the clause dealing with intellectual property rights. Japan, as well as two EU member states (the UK and France) feared competitive disadvantages for their growing biotechnology sector (Baker, 2003). Later, at the New York Summit in 1997, the ongoing dis-pute over the regulation of biotechnology and the protection of intellec-tual property rights led to virtual stalemate.

Getting to work: the biosafety protocol

As is typical of UN procedures, regular conferences of the parties to the Convention (CoP) meetings take place that aim to establish programmes of work and set priorities for implementation. Given the wide scope of the CBD and the highly contentious issues it raises, it is not surprising to find that negotiations among the CoPs have been difficult (Herken-rath, 2002).

Arguably, the most successful work to emerge from the CBD is the agreement reached in 2000, the Cartagena Protocol on Biosafety, although it too has not escaped its share of controversy.

The issue of liability and redress for damage resulting from the trans-boundary movements of LMOs was one of the themes on the agenda during the negotiation of the Cartagena Biosafety Protocol (see Box 4.5). This relates to what would happen if the transboundary movement of LMOs caused environmental damage. Developing countries wanted to have substantive provisions on liability and redress included in the Pro-tocol. The negotiators were, however, unable to reach any consensus regarding liability and disputes continued to drag on over several years. Negotiations took place again in 2010 at a CBD CoP10 meeting in Nagoya, Japan.

A noticeable feature of the biosafety negotiations at Nagoya was the divide between countries concerned about the risks and adverse impacts of LMOs and those that downplayed such risks because they are, or plan to be, exporters of such products. Brazil, India, the Philippines, Para-guay and Mexico emerged at this meeting as strong deniers of the risks posed by LMOs, as did the USA. Developing countries and some devel-oped countries like Norway maintained that an international regime to deal with damage caused by LMOs was necessary given their risks and

Box 7.5

The Cartagena Protocol on Biosafety

Background

The Cartagena Protocol on Biosafety (the Biosafety Protocol) was adopted in 2000 as a supplementary agreement to the CBD. It entered into force in 2003. The Protocol provides legally binding measures to promote and monitor transfer, handling and use of 'living modified organisms' (LMOs), which are the result of genetic engineering, and that may have adverse effects on biological diversity. In everyday usage LMOs are usually considered to be the same as GMOs (Genetically Modified Organisms), but definitions and interpretations of the term GMO vary widely.

Aim

The Protocol is designed to protect a nation's domestic environment from the accidental release and spread of LMOs. It is based on two principles: the precautionary principle and the principle of prior informed consent of receiving countries.

Its objectives are to contribute to the 'safe transfer, handing, and use of living modified organisms resulting from modern biotechnology that may have adverse effects on the conservation and sustainable use of biological diversity, taking into account risks to human health, and specifically focusing on transboundary movements' (Protocol, Article 1, 'Objectives').

Trade

The Protocol is premised on the belief that trade and environment agreements should be mutually supportive. It represents a good example of an MEA with trade implications.

Under the Bush administration, the USA criticised the Protocol, arguing that it could be used to control international trade, particularly in agricultural products. Others argue that it will do the opposite: by developing a management and regulatory regime for LMOs, the Protocol could encourage trade in biotechnological products despite the fact that little is known about their impact on biodiversity and despite growing public disquiet.

Significance

The adoption of the Protocol gave rise to a 'feel good factor', and helped channel increased resources into the Convention. On the other hand, there is a danger that the management of the Biosafety Protocol will become a central activity, diverting attention and resources away from other dimensions of the CBD.

(*Sources*: Anderson, 2002; UNEP, 2011)

the need to ensure that those responsible would be held liable. The Nagoya meeting eventually led to an international agreement, known as the 'Nagoya–Kuala Lumpur Supplementary Protocol on Liability and Redress to the Cartagena Protocol on Biosafety'. The Supplementary Protocol left it to countries to determine and implement 'response measures' at the national level, a weak position that reflected the lack of consensus between countries involved in the negotiations. Many developing countries lamented the lack of support for implementing more robust biosafety measures. They called for the establishment of a special biosafety fund within the Global Environmental Facility (GEF). However, the EU refused to include such a request. The final compromise language urged parties to give priority to biosafety when applying for GEF funding (Lin and Stabinsky, 2010).

Aside from concerns about biosafety, there is also anxiety about access to genetic resources, about who has 'ownership' of such resources, that is, the legitimate right to their use, or to negotiate access to them. This in turn is related to unease about the international system of intellectual property rights. The growing commercial significance of biodiversity resources is also raising concerns about 'biopiracy' among developing countries, as biotechnological and pharmaceutical firms, particularly from the industrialised world, seek out the biological resources of developing countries, in the hope of identifying new candidates for commercial exploitation. This has the potential to threaten biodiversity through commercial over-exploitation and the disruption of traditional patterns of biological resource use (Iles, 2003).

In an effort to address some of these issues, a new Protocol to the CBD was agreed in 2010, the Nagoya Protocol on Access to Genetic Resources and the Fair and Equitable Sharing of Benefits Arising from Their Utilization. This aims at sharing the benefits arising from the utilisation of genetic resources in a fair and equitable way, including by appropriate access to genetic resources and by appropriate transfer of relevant technologies. The fair and equitable sharing is one of the three objectives of the CBD (UNEP, n.d.c).

Despite these advances, there are still several controversial issues on the agenda, including the use of 'Genetic Use Restriction Technologies', especially so-called 'terminator technology', a technology that induces sterility in the second generation of crops. Their use can lock farmers in developing countries into structures of dependency as they are forced to buy seeds from biotechnology companies at the start of each sowing season.

Plans, targets and actions to 2020

The CBD is a very difficult convention to implement. There are several reasons for this, including, as discussed above, its complexity and scope and its political ramifications (Le Prestre, 2002). The issue of biodiversity loss has also a relative lack of public visibility, especially when compared to climate change, although this is, at least in part, counteracted by growing policy engagement with ensuring the delivery of ecosystem services, dicussed below.

Strategic plan and targets

It was not until 2002 that the parties to the CBD were able to agree a Strategic Plan, which committed them to 'achieve by 2010 a significant reduction of the current rate of biodiversity loss at the global, regional, and national level as a contribution to poverty alleviation and to the benefit of all life on earth' (UNEP, 2002). The 2010 Target, as it is called, was adopted at the 2005 WSSD in Johannesburg.

The year 2012 marked the twentieth anniversary of the signing of the CBD and the period leading up to the anniversary witnessed a flurry of activity at the international level, including new declarations, strategies and research, including into the economic significance of biodiversity. The UN also launched a global scientific body, the Intergovernmental Platform on Biodiversity and Ecosystem Services (IPBES). The IPBES provides independent advice and scientific evidence on the state and trends of biodiversity for policy makers worldwide. The period is also marked by the recognition that the anniversary occured at a time when the international community was struggling to find ways to address the continuing decline in biodiversity, all the more marked by the dismal failure to meet the 2010 Target. Attention was also increasingly drawn to the need to devise ways to link biodiversity policy to the maintenance of ecosystem functionality and services. This was set against a backdrop of recognition that links between biodiversity policy and climate change adaptation and mitigation efforts were urgently required.

In preparation for the anniversary of the CBD, the Nagoya CoP meeting in 2010 agreed a revised and updated *Strategic Plan for Biodiversity*, which included the so-called Aichi Biodiversity Targets for 2010–2020 (UNEP, 2010b). The new Strategic Plan aims to

> take effective and urgent action to halt the loss of biodiversity
> in order to ensure that by 2020 ecosystems are resilient and

continue to provide essential services, thereby securing the planet's variety of life, and contributing to human well-being, and poverty eradication.

The Plan aims to reduce the pressures on ecosystems, to engage in the restoration of lost ecosystems, to ensure that biological resources are sustainably used and that adequate financial resources are provided. In short, the Plan aims to ensure that biodiversity is mainstreamed within policy.

> Biodiversity mainstreaming: the integration of biodiversity considerations into other policy fields.

To build support and momentum for the Plan, and especially given that the previous 2010 targets were not met, the UN declared 2011–2020 as the UN Decade on Biodiversity

> The vision for the new plan is: 'Living in Harmony with Nature' where by 2050, biodiversity is valued, conserved, restored and wisely used, maintaining ecosystem services, sustaining a healthy planet and delivering benefits essential for all people.
>
> (UNEP, 2010b)

The Plan contains five strategic goals and twenty targets, collectively known as the Aichi Targets.

Box 7.6

Biodiversity: strategic goals 2011–2020

Strategic Goal A Address the underlying causes of biodiversity loss by mainstreaming biodiversity across government and society.

Strategic Goal B Reduce the direct pressures on biodiversity and promote sustainable use.

Strategic Goal C Improve the status of biodiversity by safeguarding ecosystems, species and genetic diversity.

Strategic Goal D Enhance the benefits to all from biodiversity and ecosystem services.

Strategic Goal E Enhance implementation through participatory planning, knowledge management and capacity building.

(*Source*: UNEP, 2010c)

The twenty headline Aichi Biodiversity Targets for 2015 or 2020 are organised under the five strategic goals. Specific targets were also set, and include, *inter alia*: that by 2020 the rate of loss of all natural habitats including forests is halved and, where feasible, brought close to zero; to establish a conservation target of 17 per cent of terrestrial and inland water areas and 10 per cent of marine and coastal areas; to restore at least 15 per cent of degraded areas through conservation and restoration activities; and to make special efforts to reduce the pressures faced by coral reefs.

The goals and targets can be seen as not only aspirations for achievement at the global level, but also as providing a framework for the

---

Box 7.7

Examples of Aichi biodiversity targets

Target 2

By 2020, at the latest, biodiversity values have been integrated into national and local development and poverty reduction strategies and planning processes and are being incorporated into national accounting, as appropriate, and reporting systems.

Target 5

By 2020, the rate of loss of all natural habitats, including forests, is at least halved and where feasible brought close to zero, and degradation and fragmentation is significantly reduced.

Target 6

By 2020 all fish and invertebrate stocks and aquatic plants are managed and harvested sustainably, legally and with the application of ecosystem-based approaches.

Target 12

By 2020 the extinction of known threatened species has been prevented and their conservation status, particularly of those most in decline, has been improved and sustained.

(*Source*: UNEP, 2010d)

establishment of national or regional targets (UNEP, 2010e). This flexibility could be positive in that it allows countries to take into account national needs and priorities. However, there is a risk that efforts at the national and sub-national levels will fail to meet the enthusiasm reflected in the ambitious plans devised at the UN level. This is not least because actions to protect biodiversity receive a tiny fraction of funding compared to activities aimed at promoting economic growth. In the EU, for example, spending on biodiversity amounts to less than 0.1 per cent of the EU budget in any one year (EEA, 2009).

There are also concerns about the emphasis on target setting (Mace and Baillie, 2007; Tear et al., 2005). The 2010 Target has been described as a 'data free' target because it was set independently of any empirical analysis or sound scientific support (Svancara et al., 2005: 989). Furthermore, even with the best efforts, it is very difficult to meet such targets, because additional species will be lost from most regions due to the extinction debt, that is, the time lag between loss and fragmentation of habitat and loss of biodiversity. A more positive way is to look at the 2020 Target as a 'stretch target', that is, as an indication of a desired achievement that is considerably beyond current performance. This can force policy beyond an incremental approach to a 'breakthrough' improvement. The EU has often adopted such an approach, for example, with respect to 'technology forcing' legislation in the car industry, stemming from an underlying belief that regulation can improve, as opposed to hamper, innovation. However, this tactic may backfire: stretch targets that are set without leadership and resources can be particularly misguided and create scepticism about the usefulness of targets as policy tools (Walsh, 2000: 185).

Using ecological restoration

One approach in the Nagoya Strategic Plan that has grabbed attention is the idea of using ecological restoration as a tool for the protection of biodiversity. Ecological restoration is growing in policy importance and several international actors, including the UNEP and the EU, have made a declaratory commitment to engage in ecological restoration as a means of addressing global environmental change. Several countries are also using restoration as a tool for the provision of ecosystem services, including China, which provides an exemplar of best practice, as discussed in Chapter 12.

> Ecological restoration can be understood as 'the process of assisting the
> recovery of an ecosystem that has been degraded, damaged, or destroyed'.
>
> (*Source*: SER, 2004)

Aside from stressing the importance of restoration for achieving the
2020 Biodiversity Targets, climate change mitigation and adaptation
policy is increasingly relying upon restoration, for example, through
reforestation for carbon sequestration or restoring wetlands for flood
protection, as discussed in Chapter 6. Furthermore, it is in increased use
as a compensation tool in planning decisions and as a tool for addressing environmental damage from industrialisation, mining and quarrying,
as seen, for example, in Germany where an ambitious programme of
ecological restoration and remediation was undertaken in the new
Länder following reunification.

Box 7.8

Ecological restoration as a tool for biodiversity

The full range

Restoration projects can target many different ecological systems or landscapes and be
conducted in both urban and rural areas. In addition, restoration activities can occur
across a variety of scales, from limited and highly localised experiments, to remediation
of industrial, quarrying or mining sites, to what are best described as 'mega projects',
such as the Kissimmee River restoration in central Florida, restoration of the prairies in
the USA or contemporary water management in the Netherlands.

Restoration can include returning specific types of damaged ecosystems to a more 'natural' state or even creating ecosystems *de novo*. It can also involve the deliberate reintroduction of apex or highly interactive species that have been lost at the local level due
to changes in land use and other development pressures. Wolf reintroduction provides a
typical example and has led to controversies between ecologists and farmers in the eastern part of Germany, Scandinavia and in North America. River restoration is another
focus of project attention, involving the removal of dams, river re-meandering and
re-bouldering, 'daylighting' of culvert rivers, or remediation of urban riverbanks.

Restoration has also attracted the attention of policy makers as a solution to flooding
arising from more extreme weather events. It is also indicative of a more positive
relationship with our natural surroundings, heralding a move from earlier 'hard'
engineering, for example, the burial of rivers within subterranean pipes, to a newer,
more environmentally sensitive approach that uses softer engineering styles and materials to integrate rivers with their flood plains.

Critique

A different view is presented by those who see ecological restoration as a deeply anthropocentric and morally questionable endeavour. For Katz in particular the belief that restoration activities can replace natural value by the creation of functionally equivalent natural systems is an expression of human *hubris* regarding technical power and human capacity to master the natural world. This places restoration on the lower rungs of the ladder of sustainable development. Restored sites have been criticised by Elliot for their lack of authenticity, interruption of historical continuity and change of origin. Compensatory restoration, such as the creation of *de novo* wetlands for compensatory planning purposes, denies the place base and place connectivity of a particular site. Furthermore, such activities are part of the increased *humanisation* of the natural world, as they require continuing management interventions. From these perspectives, ecological restoration is not seen as an appropriate route to the promotion of sustainable development.

(*Sources*: Baker et al., 2013; Elliot, 1997; Katz, 1997)

Biodiversity action plans

While the CBD seeks to promote a sense of shared responsibility for the protection of biodiversity at the global level, it nevertheless reaffirms the role of the state as the key actor in biodiversity protection. In keeping with UN traditions, the CBD upholds the principle of subsidiarity, laying down that 'the authority to determine access to genetic resources rests with the national governments and is subject to national legislation'. Furthermore, implementation of the CBD crucially depends on the creation of national strategies and plans.

In many countries, there is a lack of basic knowledge about biodiversity and a lack of capacity to undertake research in the area. There are also problems with funding. At the international level, biodiversity protection is financed through the GEF and is their largest portfolio. However, at least 40 per cent of biodiversity finance from the GEF is in the form of borrowing. This increases the risk that developing countries may escalate their debt load in order to fund biodiversity protection. The irony is that the need to finance debt repayments may lead, in turn, to pressure for unsustainable exploitation of the genetic and biological resources that the CBD is designed to protect (Iles, 2003). Furthermore, the GEF funds are not intended to meet the total costs of achieving the CBD's objectives, but only the so-called 'incremental costs', that is, that part of the total costs that yields global benefits and would not be incurred by a country in the course of its 'normal' development. This makes it very difficult for the GEF to deal with the underlying causes of biodiversity loss. There are also difficulties of a more political nature that have soured the relationship between the parties to the Convention and the GEF. Because of its close relationship with the World Bank, the GEF is viewed with suspicion by both developing countries and by environmental NGOs. The Japan Biodiversity Fund, established in support of the implementation of the Nagoya Biodiversity Outcomes, may help with this problem. One of its key objectives is to support, at regional and sub-regional levels, the translation of the new Strategic Plan for Biodiversity 2011–2020 into national priorities.

Further problems with implementation arise because national action plans often call for the creation of protected sites and parks, but such sites are often established in sparsely populated areas, to minimise conflict over land use, but not necessarily in areas with the highest biodiversity or that most urgently need protection. In addition, the protected areas do not necessarily cover a representative range of ecosystems, habitats, species and genetic diversity (Baker, 2003). In some cases, sites have been established without the introduction of regulations to allow indigenous and local communities to continue their traditional use of the resources within these sites. This can result in a lack of support or even open opposition by the inhabitants of an area to the designation of their lands as nature parks (Swiderska et al., 2008). It can also lead to accusations that biodiversity protection is another example of the closure of the commons, undermining the social pillar of sustainable development.

> In the guise of conserving biodiversity, parks can be created that deprive people of their land and livelihoods, and are open to foreign corporations for bio prospecting without sharing the benefits.
>
> (Iles, 2003: 232)

Furthermore, such cases reveal a gap between the rhetoric of adopting an ecosystem approach that integrates humans into, on the one hand, biodiversity protection policy, and practices in biodiversity management on the other.

In summary, the goal of either slowing or halting the loss of biodiversity is the subject of several key international agreements under the remit of the UN and has been approved in a variety of forums and under various formulations at international, regional and state level. These agreements recognise the rapid degradation of ecosystems and habitats, the increasing threat to many species, and indeed the ecosystem itself, and the urgent need to take action. However, in most cases action to implement the CBD has not been taken on a sufficient scale to address the pressures on biodiversity. The 2010 Target has not been met and prospects for meeting the 2020 Targets are poor. There has been insufficient integration of biodiversity issues into broader policies, strategies and programmes, and the underlying drivers of biodiversity loss have not been addressed. However, one recent development has proved very significant for highlighting the problem of biodiversity loss and attracting government attention. We refer here to growing use of the concept of ecosystem services.

Emphasising ecosystem services

Biodiversity underpins the functioning of ecosystems that provide a wide range of services to human societies. These links between nature and the economy are increasingly described using the concept of 'ecosystem services'. Ecosystem services refer to the flows of value to human societies arising from the state and quantity of natural capital (TEEB, 2010).

> Ecosystem services are the direct and indirect contributions of ecosystems to human well-being, such as carbon dioxide absorption, clean water, plant pollination by insects and nitrogen fixation.

A forest at the source of a river, for example, will provide more than fruits or timber. It will also play a role in water quality protection (filtering the water as it flows through roots and soil), flood control (reducing runoff and erosion), carbon storage and sequestration (in the form of additional biomass), biodiversity conservation (providing habitat for plants or animals living) and landscape aesthetics (MEA, 2005a).

In order to help our understanding of the services that ecosystems provide to human beings, ecosystem services have been divided into four categories: provisioning, regulating, cultural and supporting services (MEA, 2005a).

The Millennium Ecosystem Assessment, conducted under the auspices of the UN, provided an international assessment of the consequences of the loss of ecosystem services to human well-being (MEA, 2005a). Carried out between 2001 and 2005, the MEA found that over the past fifty years, humans have changed ecosystems more rapidly and extensively than in any comparable period of time in human history, largely to meet rapidly growing demands for food, fresh water, timber, fibre, and fuel. The MEA found that over 60 per cent of the total services that human beings depend on were in decline, while only four

---

Box 7.9

Classification of ecosystem services

Provisioning services

Ecosystem services that describe the material or energy outputs from ecosystems, including food, water and other resources.

Examples: food, raw materials, fresh water, medicinal resources.

Regulating services

Services that ecosystems provide by acting as regulators, such as regulating the quality of air and soil or by providing flood and disease control.

Examples: Local climate and air quality; carbon sequestration and storage; moderation of extreme weather events; waste-water treatment; erosion prevention and maintenance of soil fertility; pollination; biological control.

Supporting services

Habitat or supporting services underpin almost all other services. Ecosystems provide living spaces for plants or animals; they also maintain a diversity of different breeds of plants and animals.

Examples: habitats for species; maintenance of genetic diversity.

Cultural services

The non-material benefits people obtain from contact with ecosystems. They include aesthetic, spiritual and psychological benefits.

Examples: recreation and mental and physical health; tourism; aesthetic appreciation and inspiration for culture, art and design; spiritual experience and sense of place.

(*Sources*: TEEB, 2010; MEA, 2005a)

were improving (three of these four improving services are related to an increase in food production). This decline affects 70 per cent of the regulating services, such as climate regulation and water and air quality regulation, on which food production relies heavily and upon which human life depends. The report has shown that modern agriculture has been very successful at providing the ecosystem services for which markets exist – crops, livestock, fish, fibre and wood – in ever greater quantities. However, producing these provisioning services has often come at a cost to the other regulating, supporting and cultural services that are not directly covered by markets (MEA, 2005a).

If ecosystems are disrupted, this reduces the richness and variety of the natural world, which is important in and of itself. It also means that the provision of food, fibre, medicines and fresh water, pollination of crops, filtration of pollutants, and protection from natural disasters are potentially threatened. Human societies may also encounter newly emerging diseases. Opportunities for knowledge and education, as well as recreational and aesthetic enrichment, are also threatened. At the heart of the MEA assessment is a stark warning: human activity is putting such strain on the natural functions of the Earth that the ability of the planet's ecosystems to sustain future generations can no longer be taken for granted.

With nearly two thirds of the services provided by nature to humankind found to be in decline worldwide, this means that the benefits reaped from our growth-orientated development model have been achieved by running down natural resources. The MEA has referred to this as 'spending the capital' (MEA, 2005b). We are depleting assets at the expense of our grandchildren. As will be recalled from Chapter 2, this breaches the inter-generational equity principle of sustainable development. It also breaches the principles of intra-generational equity, as starkly illustrated by the MEA:

> The cost is already being felt, but often by people far away from those enjoying the benefits of natural services. Shrimp on the dinner plates of Europeans may well have started life in a South Asian pond built in place of mangrove swamps – weakening a natural barrier to the sea and making coastal communities more vulnerable. ... We place in jeopardy the dreams of citizens everywhere to rid the world of hunger, extreme poverty, and avoidable disease.

> (MEA, 2005b: 4)

The emphasis on ecosystem services has heightened awareness of the environmental, economic, social, and cultural value of nature. It has also grasped the imagination of policy makers, in ways that the more abstract notion of biodiversity has failed to do. The ecosystem services approach can be seen, for example, in the Malawi Principles for biodiversity management under the CBD (Rauschmayer et al., 2009). In addition, it has also resulted in a shift of attention from an earlier 'protectionist' approach to nature conservation, which focused on the protection of species and habitat by establishing conservation areas, to a new and wider emphasis on the need to maintain ecosystem functions and services. The example of protection of ecosystem services in China, including through ecological restoration and the use of payments for ecosystem services (PES), is discussed in Chapter 12. The focus on ecosystem services has also heightened concerns that the poor would face the earliest and most severe impacts of such changes, but ultimately all societies and communities would suffer.

> We can no longer see the continued loss of and changes to biodiversity as an issue separate from the core concerns of society: to tackle poverty, to improve the health, prosperity and security of our populations, and to deal with climate change.
>
> (UNEP, 2010a: section 6673)

It has also supported a new emphasis on the development of ecological infrastructure, that is, strategically planned and managed networks of natural lands, working landscapes and other open spaces that conserve ecosystem values and functions and provide associated benefits to society (Stephens, 2011).

However, there is also a danger in this approach and its deeply anthropocentric values. It may prioritise the need to ensure that ecosystems are functioning adequately in terms of their provision of services to humankind. This may marginalise the stronger models of sustainable development, and their requirement to value nature over and above any utilitarian considerations. This problem can be explored in more depth by looking at a key report, the Economics of Ecosystems and Biodiversity Report, which sharpened awareness of the value of biodiversity and ecosystems and made a compelling economic case for the conservation of ecosystems and biodiversity (TEEB, 2010).

The economics of ecosystems and biodiversity

The Economics of Ecosystems and Biodiversity report (TEEB) evaluated the economic costs associated with the loss of biodiversity and the associated decline in ecosystem services worldwide (TEEB, 2010). The authors compared this with the costs of effective conservation and sustainable use.

The TEEB report argued that the failure to account for the full economic values of ecosystems and biodiversity has been a significant factor in their continuing loss and degradation. The invisibility of many of nature's services to the economy results in widespread neglect of natural capital, leading to decisions that degrade ecosystem services and biodiversity. The TEEB report was extremely important in counteracting this trend: it helped make nature's values visible.

The report also drew links between poverty alleviation and ecosystem services, while recognising that poverty is a complex phenomenon and the relationship between poverty and biodiversity is not always clear-cut. In many countries, poor households rely on natural capital for a disproportionately large fraction of their income, for example, in agriculture, forestry and fisheries. Moreover, these households have few means to cope with losses of critical ecosystem services, such as drinking water purification or protection from natural hazards. Sustainable

---

Box 7.10

The economics of ecosystem services: some numbers

Conserving forests avoids GHGs worth US$3.7 trillion.

Some 30 million people in coastal and island communities are totally reliant on reef-based resources as their primary means of food production, income and livelihood.

The *total economic value* of insect pollination worldwide is estimated at $153 billion, representing 9.5 per cent of world agricultural output in 2005.

Local authorities in Canberra have planted 400,000 trees to regulate microclimate, reduce pollution and thereby improve urban air quality, reduce energy costs for air conditioning, and store and sequester carbon. These benefits are expected to amount to some US$20–67 million over the period 2008–2012.

(*Sources*: Gallai et al., 2008; TEEB, 2010)

management of natural capital was thus shown to be a key element in achieving poverty reduction objectives as reflected in the MDGs.

The TEEB study made a strong case for significant changes in the way we manage nature, based on economic concepts and tools. It has led to wider recognition of nature's contribution to human livelihoods, health, security, and culture. The application of economic value to the use of biodiversity and ecosystem services has also helped clarify why prosperity and poverty reduction depend on maintaining the flow of benefits from ecosystems. The report also points out how the management of ecosystem services can be seen as an economic opportunity rather than as a constraint on development (Kumar, 2010).

Despite the benefits of framing biodiversity in economic terms, there remain the twin dangers of seeing biodiversity in purely utilitarian and highly anthropocentric terms, and in over-emphasising ecosystem services at the expense of biodiversity. The total amount of biodiversity that would be conserved based strictly on utilitarian considerations may well be less than the amount present today. In addition, steps to increase the production of specific ecosystem services may well require the simplification of natural systems, for example, in agriculture. Moreover, managing ecosystem services may not necessarily require the conservation of biodiversity. A forested watershed, for example, could provide clean water and timber, whether it was covered by a diverse native forest or a single-species plantation, but a single-species plantation may not provide significant levels of biodiversity. Even if utilitarian benefits were taken fully into account, decline in biodiversity may continue as other economic priorities take precedence. Thus, we would argue, the level of biodiversity that survives on Earth would need to be determined by ethical concerns rather than merely utilitarian considerations.

Conclusion

There are clear links between the maintenance of biodiversity and meeting sustainable development objectives. However, attempts to address biodiversity loss are struggling. The main tool of global governance, the CBD, is faced with an enormous task of reconciling biodiversity protection with ongoing anthropogenic pressures. Several controversial interfaces have developed between biodiversity and the spheres of politics and commerce that add to the difficulties. There is also the urgent need to integrate biodiversity policy into climate change adaptation and mitigation strategies, and to address any conflicts that may emerge between targets for biodiversity conservation and those for climate mitigation.

Insofar as the CDB can stimulate structural change, it has the potential to contribute to the promotion of strong forms of sustainable development. Actions to date would suggest, however, that the CBD is more likely to promote a weaker form of sustainable development, dominated by utilitarian, particularly commercial views, of nature and its biodiversity resources. While the focus on ecosystem services and their economic worth has helped win support for biodiversity preservation from across a wide section of societal actors, including economic interests, this approach makes biodiversity protection vulnerable to shifting cultural, political and economic perceptions and values. Stronger forms of sustainable development would promote a different relationship with nature, one that is based upon recognition of the intrinsic value of biodiversity, rather than merely upon the utilitarian argument that biodiversity should be protected insofar as it is of use to us.

It is also necessary to address potential trade-offs between the goals of protecting ecosystems, species and biodiversity and the promotion of the social pillar of sustainable development. Biodiversity policy can sometimes have negative welfare consequences, where the costs of implementing biodiversity policies can accrue locally, for example, to those whose access to biodiversity resources are restricted by conservation efforts. At the same time, those from outside the locality may reap the benefits. Similarly, while the links between the maintenance of biodiversity and the enhancement of societal well-being are clear, existing strategies to address poverty can often undermine biodiversity.

Despite these problems, however, this chapter has shown that the span of nearly two decades since the CBD was signed has brought recognition of the complexity and seriousness of the problem of biodiversity loss. The issue has gained political saliency at global, regional and national levels. It is now clear that the promotion of sustainable development is dependent upon the maintenance of biodiversity, especially at a global level. Similarly, the continued loss of biodiversity poses a direct challenge to efforts to promote sustainable development.

Summary points

- The main significance of the CBD is that it has helped to make a link between biodiversity protection and the promotion of sustainable development. Biodiversity preservation is now an integral part of the construction of our sustainable future.
- Biodiversity is now redefined in social and economic terms, and no longer seen only as a technical, scientific issue.
- Despite impressive accomplishments, including the development of an international biodiversity governance regime; international, regional and national declarations of intent; publication of action plans and strategies; and the rapid development of monitoring capacity, biodiversity is under accelerating threat of loss.
- The 2010 Target has not been met and the 2020 Target is likely to meet the same fate.
- Most of the direct drivers of biodiversity loss are projected to remain constant or to increase in the near future. The objective of slowing, let alone halting, biodiversity loss remains a major challenge.
- Recognition of its economic value has significantly enhanced recognition among the policy community of the severe consequences of biodiversity loss.
- The underlying drivers of biodiversity loss arising from demographic, economic, technological, socio-political and cultural pressures remain. To address such underlying drivers, an economics of limits is needed.

Further reading

Greenpeace (2010), *Biodiversity Conservation Beyond 2010: A Global Plan for Life on Earth*, Briefing, available online at www.greenpeace.org/international/global/internationa/publications/oceans/2010/CBDplan.pdf

Helm, D. and Hepburn, C. (eds) (2014), *Nature in the Balance: The Economics of Biodiversity* (Oxford: Oxford University Press).

Jeffries, M. J. (2006), *Biodiversity and Conservation* (London: Routledge, 2nd edn).

Swanson, T. (2013), *Global Action for Biodiversity: An International Framework for Implementing the Convention on Biological Diversity* (London: Routledge).

References

Anderson, S. (2002), *Identifying Important Plant Areas* (London: Plantlife International).

Baker, S. (2003), 'The Dynamics of European Union Biodiversity Policy: Interactive Functional and Institutional Logics', *Environmental Politics*, 12 (3): 24–41.

Baker, S., Eckerberg, K. and Zachrisson, A. (2013), 'Political Science and Ecological Restoration', *Environmental Politics*, doi: 10.1080/09644016.2013.835201.

EEA (European Environment Agency) (2009), *Progress towards the European 2010 Biodiversity Target* (Copenhagen: EEA, Technical report no. 4/2009).

EEA (2010a), *Assessing Biodiversity in Europe: The 2010 Report* (Copenhagen: EEA, Technical report no. 5/2010).

EEA (2010b), *EU 2010 Biodiversity Baseline: Post-2010 EU Biodiversity Policy* (Copenhagen: EEA), available online at www.eea.europa.eu/publications/eu-2010-biodiversity-baseline

EEA (2010c), *EU 2010 Biodiversity Baseline* (Copenhagen: EEA, Technical report no. 12/2010).

Elliot, R. (1997), *Faking Nature: The Ethics of Environmental Restoration* (London: Routledge).

Fahrig, L. (2001), 'How Much Habitat is Enough?' *Biological Conservation*, 100: 65–74.

FAO (2009), *How to Feed the World in 2050* (Rome: FAO), available online at www.fao.org/fileadmin/templates/wsfs/docs/expert_paper/How_to_Feed_the_World_in_2050.pdf

Gallai, N., Salles, J-M, Settele, J. and Vaissière, B. E. (2008), 'Economic Valuation of the Vulnerability of World Agriculture Confronted with Pollinator Decline', *Ecological Economics*, doi: 10.1016/j.ecolecon.2008.06.014.

Greenfacts (2014), 'Biocapacity', available online at www.greenfacts.org/glossary/abc/biocapacity.htm

Herkenrath, P. (2002), 'The Implementation of the Convention on Biological Diversity: A Non-Government Perspective Ten Years On', *Review of European Community and International Environmental Law*, 11 (1): 29–37.

Iles, A. (2003), 'Rethinking Differential Obligations: Equity under the Biodiversity Convention', *Leiden Journal of International Law*, 16: 217–251.

IUCN (International Union for the Conservation of Nature and Natural Resources) (2010), *A New Vision for Biodiversity Conservation: Strategic Plan for the Convention on Biological Diversity* (Gland, Switzerland: IUCN), available online at http://cmsdata.iucn.org/downloads/cbd_cop10_position_paper_strategic_plan_2011_2020_1.pdf

IUCN (2013), *Redlist of Threatened Species* (Gland, Switzerland: IUCN), available online at www.iucnredlist.org/

Katz, E. (1997), *Nature as Subject: Human Obligation and Natural Community* (Oxford: Rowman and Littlefield).

Kumar, P. (ed.) (2010), *The Economics of Ecosystems and Biodiversity: Ecological and Economic Foundations*. An output of TEEB: The Economics of Ecosystems and Biodiversity (London: Earthscan).

Le Prestre, P. G. (2002), 'The CBD at Ten: The Long Road to Effectiveness', *Journal of International Wildlife Law and Policy*, 5: 269–285.

Lin, L. L. and Stabinsky, D. (2010), 'The Rift at Nagoya on GMO Safety and Socioeconomic Impacts' (Geneva: Third World Network, Third World Resurgence no. 242/243, October-November), 37–39, available online at www.twnside.org.sg/title2/resurgence/2010/242-243/cover06.htm

Mace, G. M. and Baillie, J. E. M. (2007), 'The 2010 Biodiversity Indicators: Challenges for Science and Policy', *Conservation Biology*, 21 (6): 1406–1413.

MEA (2005a), *Millennium Ecosystem Assessment, Ecosystems and Human Well-being: Synthesis* (Washington, DC: Island Press).

MEA (2005b), *Living Beyond Our Means: Natural Assets and Human Well-Being, Statement from the Board* (Rome: MEA Secretariat), available online at www.maweb.org/documents/document.429.aspx.pdf

Rauschmayer, F., van den Hove, S. and Koetz, T. (2009), Participation in EU Biodiversity Governance: How Far Beyond Rhetoric?' *Environment and Planning C: Government and Policy*, 27: 42–58.

Sachs, J. D., Baillie, J. E. M., Sutherland, W. J., Armsworth, P. R., Ash, N., Beddington, J., Blackburn, T. M., Collen, B., Gardiner, B., Gaston, K. J., Godfray, H. C. J., Green, R. E., Harvey, P. H., House, B., Knapp, S., Kümpel, N. F., Macdonald, D. W., Mace, G. M., Mallet, J., Matthews, A., May, R. M., Petchey, O., Purvis, A., Roe, D., Turner, K. S. K., Walpole, M., Watson, R. and Jones, R. E. (2009), 'Biodiversity Conservation and the Millennium Development Goals', *Science*, 325 (5947) (18 September): 1502–1503.

SER (2004), 'Society for Ecological Restoration: SER International Primer on Ecological Restoration', available online at http://www.ser.org/resources/resources-detail-view/ser-international-primer-on-ecological-restoration.

Stephens, A. (2011), 'Harnessing Ecological Infrastructure and Adapting to Risk', Talk presented to the 'Agri-Food Chain: Vulnerability and Adaptation' NBI side event, CoP17, 1 December, available online at www.nbi.org.za/Lists/Events/Attachments/48/Stephens_Harnessing_Ecological_Infrastructure_And_Adapting_To_Risk.pdf

Svancara, L. K., Brannon, R., Scott. J. M., Groves, C. R., Noos, R. F. and Pressey, R. L. (2005), 'Policy-Driven versus Evidence-based Conservation: A Review of Political Targets and Biological Needs', *BioScience*, 55 (11): 989–995.

Swiderska, K., with Roe, D., Siegele, L. and Grieg-Gran, M. (2008), *The Governance of Nature and the Nature of Governance: Policy That Works for Biodiversity and Livelihoods* (London: International Institute for Environment and Development).

Tear, T. H., Karciva, P., Angermeier, P. L., Comer, P., Czech, B., Kautz, R., Landon, L., Mehlman, D., Murphy, K., Ruckelshaus, M., Scott, J. M. and Wilhere, G. (2005), 'How Much Is Enough? The Recurrent Problem of Setting Measurable Objectives in Conservation', *BioScience*, 55 (10): 835–849.

TEEB (2010), *Mainstreaming the Economics of Nature: A Synthesis of the Approach, Conclusions and Recommendations of TEEB* (Geneva: UNEP, TEEB), available online at www.teebweb.org/Portals/25/TEEB%20Synthesis/TEEB_SynthReport_09_2010_online.pdf

UNEP (n.d.a), 'Article 2: Use of Terms' (Montreal: Secretariat of the Convention on Biological Diversity), available online at www.cbd.int/convention/articles/?a=cbd-02

UNEP (n.d.b), 'South–South Cooperation' (Montreal: Secretariat of the Convention on Biological Diversity), available online at www.cbd.int/ssc/

UNEP (n.d.c), 'The Nagoya Protocol on Access and Benefit-sharing' (Montreal: Secretariat of the Convention on Biological Diversity), available online at https://www.cbd.int/abs/doc/protocol/nagoya-protocol-en.pdf

UNEP (2004), 'CBD Decision VII/30' (Montreal: Secretariat of the Convention on Biological Diversity), available online at www.cbd.int/decision/cop/?id=7767

UNEP (1992), Convention on Biological Diversity (Montreal: Secretariat of the Convention on Biological Diversity), available online at https://www.cbd.int/doc/legal/cbd-en.pdf

UNEP (2002), COP6 Decision VI/26, Strategic Plan for the Convention on Biological Diversity. CBD Secretariat, Sixth Ordinary Meeting of the Conference of the Parties to the Convention on Biological Diversity, 7–19 April, The Hague, Netherlands, available online at www.cbd.int/decision/cop/default.shtml?id=7200

UNEP (2010a), *2010 Global Biodiversity Outlook* 3 (Montreal: Secretariat of the Convention on Biological Diversity), available online at www.cbd.int/gbo/gbo3/images/GBO3-Table2-TrendsByIndicator.pdf

UNEP (2010b), *Strategic Plan for Biodiversity 2011–2020, including Aichi Biodiversity Targets* (Montreal: Secretariat of the Convention on Biological Diversity), available online at www.cbd.int/sp/

UNEP (2010c), 'Strategic Goals and Targets for 2020: Taking Action for Biodiversity' (Montreal: Secretariat of the Convention on Biological Diversity), available online at www.cbd.int/2011-2020/goals/

UNEP (2010d), 'Aichi Biodiversity Targets' (Montreal: Secretariat of the Convention on Biological Diversity), available online at www.cbd.int/sp/targets

UNEP (2010e), *Strategic Plan for Biodiversity 2011–2020, Provisional Technical Rationale, Possible Indicators and Suggested Milestones for the Achi Biodiversity Targets* (Montreal: Secretariat of the Convention on Biological Diversity, UNEP/CBD/COP/10/27/Add.1), available online at www.cbd.int/doc/meetings/cop/cop-10/official/cop-10-27-add1-en.pdf

UNEP (2011), Nagoya-Kuala Lumpur Supplementary Protocol on Liability and Redress to the Cartagena Protocol on Biosafety (Montreal: Secretariat of the Convention on Biological Diversity), available online at http://bch.cbd.int/protocol/text/article.shtml?a=cpb-0

Walsh, P. (2000), 'Targets and How to Assess Performance against Them', *Benchmarking*, 7 (3): 183–199.

WWF (2005), *Europe 2005: The Ecological Footprint* (Brussels: European Policy Office Belgium, World Wide Fund for Nature).

World Resources Institute (2005). *The Wealth of the Poor: Managing Ecosystems to Fight Poverty* (Washington, DC: World Resources Institute).

Local actions for sustainable development

Key issues

- Agenda 21, Local Agenda 21, Local Action 21
- Local authorities and planning
- Urban sustainable development
- Education for sustainable development
- Participation and civil society.

Introduction

This chapter examines the promotion of sustainable development at the local level through Agenda 21, the action plan adopted at the 1992 Earth Summit. It focuses on Local Agenda 21 (LA21), a scale at which actions can have both immediate and direct effect. LA21 engagement differs between different regions and countries across the globe, as revealed by the case studies discussed in this chapter. Because LA21 gives a key role to local authorities, the spatial dimension of sustainable development comes into play, including in relation to urban design, land use planning and transportation, particularly within cities. Looking at LA21 also provides an opportunity to critically explore participation from civil society actors operating at the local level. While attention is focused at the local level, account is also taken of how wider processes, across the different levels of global, international, national and local governance, facilitate or impede change.

Local actions

Agenda 21

Agenda 21 sets out an action plan for working towards development that is socially, economically and environmentally sustainable.

Its forty chapters present solutions that emphasise the importance of engagement from local government and the business community and of participation from civil society. It highlights 'capacity building', that is, education and training directed at individuals, organisations and government agencies and actors, including local authorities. The 'Capacity 21' initiative by the United Nations Development Programme (UNDP) supports the implementing of Agenda 21. Emphasis on capacity building reflects the belief that the promotion of sustainable development requires new forms of social learning. This can include learning how to engage in constructive dialogue and to envisage collectively the elements needed to build sustainable futures. As will be seen below, this learning forms an important part of LA21 activity and is closely related to the use of social capital. Social capital refers to networks, shared norms, values and understandings that facilitate co-operation within and between groups (Putnam, 2001). LA21 highlights the need to foster the accumulation of social capital that facilitates collective action in pursuit of sustainable development (Evans et al., 2008). However, in keeping with the UN system of governance, national governments have overall responsibility for the implementation of Agenda 21 and its related LA21 (Dresner, 2002).

Local Agenda 21

Chapter 28 of Agenda 21 is devoted to the role of local political authorities in the introduction of comprehensive planning processes aimed at promoting sustainable development within their locality. This activity is known as LA21. Chapter 28 sets detailed targets and timetables for local authorities, singled out because they have highly relevant formal functions and responsibilities, including:

- Implementing national environmental policies and regulations;
- Establishing local environmental priorities and implementing related policies;
- Developing and maintaining local economic, social and environmental infrastructure;
- Overseeing planning and regulations;
- Managing waste and dealing with transport.

As the level of government closest to the people, local authorities also play a vital role in educating and mobilising civil society. Increasingly, they are also expected to provide green infrastructure and make better, planned use of local natural resources, and ensure the continued delivery of ecosystem services. The role of local authorities in planning shapes the form of urban development within a particular place. This determines whether a city is compact or sprawled, which in turn impacts on energy and transport use, making planning an important tool, for example, in combating climate change.

Local authorities are expected to act as the catalyst in the start-up of an LA21 initiative and subsequently as a facilitator, ensuring the participation of a wide range of local actors. There is also a strong role envisaged for national governments, including through launching national campaigns directed at encouraging local authorities to take action. It is intended that LA21 will be stimulated by, and in turn help to develop, international co-operation.

> LA21 is a participatory, multi-stakeholder process to achieve the goals of Agenda 21 at the local level through the preparation and implementation of a long-term, strategic plan that addresses priority local sustainable development concerns.
>
> (ICLEI, 2002a: 3)

Chapter 28 provides guidelines for local authorities, which have to be adopted and adapted by them to suit their specific needs through the formulation of a Local LA21 Action Plan.

Box 8.1

Key steps in the LA21 process

1 **Multi-sectoral engagement**: establishing a local stakeholder group as the co-ordination and policy body;
2 **Consultation**: consultation with community groups, NGOs, businesses, churches, government agencies, professional associations and trade unions;
3 **Devising a Community Action Plan**: creating a shared vision and identifying proposals for action;
4 **Participatory assessment**: assessing local social, environmental and economic needs;
5 **Participatory target setting**: negotiations among key stakeholders or community partners to achieve the vision and goals set out in the Community Action Plan;
6 **Monitoring and reporting**: establishing procedures and indicators to track progress and to allow participants to hold each other accountable to the Plan.

(*Source*: ICLEI, 2002a)

The WSSD in Johannesburg in 2002 recognised the progress made on LA21, but also that commitments were not sufficiently translated into practices. The Local Government Declaration to the WSSD reflects this concern:

> Fighting poverty, exclusion and environmental decay is a moral issue, but one of self-interest. Ten years after Rio, it is time for action of all spheres of government, all partners. Local action, undertaken in solidarity, can move the world.
>
> (ICLEI, 2002b)

The Johannesburg Declaration reaffirmed commitment to Agenda 21, stating that 'we undertake to strengthen and improve governance at all levels, for the effective implementation of Agenda 21', and that 'national governments agree to enhance the role and capacity of local authorities as well as stakeholders implementing Agenda 21' (UN, 2002a). The WSSD saw the launch of a new phase of LA21, Local Action 21, including with the support of UN-Habitat. The Type 2 partnership initiatives (discussed in Chapter 5) were designed, at least in part, to help strengthen the implementation of Agenda 21. Local Action 21 commits local governments to enter a decade of accelerated action to create sustainable communities and protect the common good (UN, 2002b).

Later, the Rio+20 outcome document contained unprecedented reference to the importance of local governments, including in promoting urban sustainability. However, it lacked any tangible commitment by national governments to enhance the capacity of local authorities to take effective action. This added to the disappointment that Rio+20 generated (ICLEI, 2012a).

Distinguishing LA21 from general environmental policies

Despite the guidelines, principles, timetables and targets that structure LA21 activities, it has proved difficult to distinguish these efforts from general local, environmental protection measures. In reporting activities to the UN, states often list any and all environmental activity of local authorities as LA21 activities. This makes it difficult to monitor actual compliance with the objectives of Chapter 28, identify value added, evaluate impact and to ascertain how LA21 has contributed to the promotion of sustainable development at the local and, collectively, at the global levels.

Box 8.2

Characteristics needed for LA21 to be part of efforts to promote sustainable development

1 Relating environmental effects to underlying economic and political pressures;
2 Relating local issues and decisions to global impacts;
3 Commitment to the principles of sustainable development, such as justice and equity;
4 Commitment to environmental policy integration;
5 Involvement of the local community, including local stakeholders, business and organised labour;
6 Commitment to define and work with local problems within a broader ecological and regional framework and within a longer period;
7 Identification with Rio and the related processes that Rio has spawned.

(*Source*: Lafferty and Eckerberg, 1998)

Using the normative principles of sustainable development as laid down by Brundtland, Lafferty and Eckerberg have set out a range of criteria for actions to form part of an LA21.

Institutionally, LA21 is supported by the International Council for Local Environmental Initiatives – Local Governments for Sustainability (ICLEI), an international association of local governments founded in 1990 with the help of the UNEP.

International co-operation for LA21

ICLEI influenced the writing of Chapter 28 and has since played a key role in stimulating the development of LA21. Its mission is 'to build and serve a worldwide movement of local governments to achieve tangible improvements in global sustainability, with special focus on environmental conditions through cumulative local actions' (ICLEI, 2011). It acts as a co-ordinator and facilitator, especially at the start-up stage of activities; works as a clearing house for information and serves as a conduit for the transfer of expertise and best practice, especially between North and South; and represents local authorities engaged in LA21 activities at the international level. It has played an important role in shaping LA21 priorities and actions.

After the 1992 Summit, ICLEI launched an international action programme on sustainable development planning, called the LA21 Model

Communities Programme. This four-year partnership succeeded in kick-starting LA12 activities in several places across the globe. Over time, ICLEI's work has, however, moved beyond the original LA21 remit, to both widen and deepen engagement with the sustainable development agenda initiated by Brundtland. Its campaigns and programmes now cover a broad range of themes, including: climate change adaptation; biodiversity loss, water and food; municipal planning and management; land use and development; transformation of infrastructure, including transport, building and energy; and the expansion of the green economy and social development. Attention is particularly focused on cities, a focus that both reflects and has helped to highlight the importance of urban sustainable development.

Exploring LA21 practices

General trends

LA21 processes have been expanding worldwide since the Rio Summit. Surveys conducted in 1997 in preparation for the Earth Summit +5 showed that more than 1,800 local authorities had established an LA21 planning process, rising to about 6,416 in over 113 countries by the WSSD in 2002, including eighteen countries with national campaigns. However, the majority were in Europe (82 per cent) and in the Asia Pacific Region (10 per cent), with the balance (8 per cent) distributed over Africa, Latin America, North America and the Middle East.

Developments since 2002 show further progress. The first Local Agenda 21 process was started in Korea in 1995 and by 2012 86 per cent of local governments had established LA21 councils. In the US, sustainability is high on the local agenda, with 80 per cent of the biggest cities citing it as one of their top five priorities. By 2012, Spain had 3,763 active LA21 processes, involving almost half of all Spanish municipalities. In France, where the LA21 was late to take hold, by 2012 almost 850 LA21 initiatives existed, covering 70 per cent of the *communautés urbaines* and more than half of the regions and departments (ICLEI, 2012b).

The experience with the implementation of LA21 differs under different economic, social and regional conditions (ICLEI, 2012c). To begin with, the degree of stakeholder involvement can vary significantly, ranging from simply providing input at the consultative stage to direct involvement in budgetary management. In addition, priorities differ, as would be expected, between the industrialised countries and the Third World. While energy conservation and waste reduction are often European priorities, municipalities in the Third World are more concerned with poverty reduction, the provision of basic services and education. However, water is a common priority for all municipalities, regardless of their economic situation (ICLEI, 2002a). Health, safety and social integration have also been included in LA21 activities. Australia, for example, has the Cities for Healthy and Safe Communities Campaign, dealing with alcohol abuse and violence in urban areas as part of its promotion of sustainable communities. More recently, several European countries, including the UK, Italy and France, have seen environmental, social and economic issues integrated under the umbrella of social entrepreneurship or social economy (ICLEI, 2012b). These in turn can be linked to the support of social innovation.

Irrespective of the different social, economic and political contexts in which they operate, all local authorities face problems integrating LA21 plans into their policy remits, particularly economic development planning. Furthermore, municipalities, even in the richer, Northern countries, often lack resourses to support LA21 activities, especially financial assistance from central government. Most municipalities also identify the lack of a national tax structure that rewards sustainable development practices as a key obstacle (ICLEI, 2002a).

However, many activities deal with general environmental issues and not the promotion of sustainable development. In addition, reliable data on local sustainable development processes is lacking, making it difficult to assess progress or to make accurate comparisons. Furthermore, in the years since 2002 in particular, many countries have replaced the term LA21 with other terms such as 'local sustainability strategy', or 'integrated development programmes', reflecting local conditions but also a changed thematic focus.

Shifting and expanding actions

Similar to many European countries, Japan has seen a shift from air, water and soil pollution as key local sustainable development issues in the 1980s to concerns with climate mitigation, resource efficiency and biodiversity conservation in the 1990s. Over time, tackling the increasing waste stream from growing consumption has also become a priority

Box 8.3

General trends in LA21 activities

African municipalities prioritise poverty alleviation, economic development and health issues. Local processes are characterised by strong stakeholder involvement, sometimes strengthening the role of certain groups, for example, women. In addition to the lack of financial and political support, there is limited expertise.

The **Asia-Pacific** local processes are typically driven by a national campaign, have a strong environmental focus and are well integrated into the municipal governance system, through, for example, local environmental strategies. Stakeholder participation is also important, but with repeated calls from the local level for a more favourable national policy framework and tax reforms.

Europe is where most LA21 activities take place. These are characterised by strong involvement of national governments but an important role is played by national municipal associations and the European Sustainable Cities and Towns Campaign. Some European countries began early. In Sweden, for example, all municipalities had adopted LA21 by 2002. The priority issues included energy management, transportation, land use and biodiversity. Municipalities complain of the lack of commitment from national government and limited community interest, and call for alignment with national sustainable development strategies and better integration of LA21 processes into other municipal operations.

The **Latin American** priorities include community and economic development, poverty alleviation and water management. This is the only region to identify tourism as a priority and indicate heritage and culture preservation as one of the key achievements of local processes. Municipalities boast the highest rate of stakeholder involvement, even if some groups continue to be excluded, such as ethnic minorities. Lack of power at the local level, however, remains a barrier.

North American municipalities call for a revised tax structure and the removal of subsidies placed on unsustainable products and policies. However, the term 'LA21' is generally not in use, although there are myriad local sustainability processes. Activities focus mostly on land use, transportation, water resources management, economic development and air quality.

(*Source*: ICLEI, 2012b)

for Japanese local governments. The USA provides another typical trajectory, where the interest of local actors has shifted from environment protection, through to climate mitigation and on to climate adaptation concerns.

This widening of the local agenda and the loosening of ties to the original LA21 process is also reflected in the work of ICLEI, mentioned above. Much of this shift arises from growing concern about the impact of climate change at the local level. At present, when talking about local sustainable development, many communities refer mainly to their activities targeted at climate change adaptation, including for example in relation to flood defence. Attention is also paid to mitigation, at least in relation to encouraging more energy efficient transport modes. Biodiversity loss, or more specifically loss of ecosystem services, has also begun to mobilise local action, amidst concerns about food insecurity, including in urban contexts. ICLEI's Local Action for Biodiversity, a global urban biodiversity programme running since 2006, works with cities to manage and conserve biodiversity at the local level (ICLEI, 2012b). Local governments have also forged international partnerships, including, for example, through the establishment of the Cities and Biodiversity Global Partnership designed to exchange best practices.

There are also a number of cities, particularly in Asia, that perform very well in terms of environmental, economic, and to a lesser extent social indicators, but their activities are not necessarily inspired by the original LA21 approach (ICLEI, 2012b). Examples of such cities are explored in Chapter 12, where China's experience with 'eco-towns' is examined.

The Brundtland Report stressed the need to focus on cities as critical locations for the promotion of sustainable development, especially given that by 2050 it is expected that two thirds of humanity will be living in towns and cities. Cities have become important sites in the quest for ways to promote sustainable development, drawing in issues of spatial planning, housing design, transport, land use planning and, more recently, food security. This has also led to a distinction between 'green' and 'brown' environmental agendas. The green agenda focuses on reducing the impact of urbanisation on the natural ecosystem, while the brown agenda addresses environmental threats to health from overcrowding, lack of sewage treatment, inadequate waste disposal and water pollution. The brown agenda is reflected in the prioritisation of WEHAB (Water, Energy, Health, Agriculture and Biodiversity) initiatives at the Johannesburg Summit as discussed in Chapter 5. These initiatives establish a clear connection between promoting equitable, socially viable and sustainable human settlements and the need to reduce poverty.

The development of cities has resulted in a dramatic shift in the spatial and material relationship that people have with nature. Most of the human population now live and work far from the land and the biophysical processes that support them (Rees, 1999), making it difficult for people to develop a relationship of care for the land. Cities are also drains on natural resources, requiring ever greater quantities of food, material commodities and energy to sustain their inhabitants, and thus have large ecological footprints, as discussed in Chapter 2. Furthermore, poverty, a phenomenon long associated with rural areas, has increasingly become urbanised.

> Many cities – especially in the developing world – represent today the most alarming concentrations of poverty. It is estimated that some 750 million urban dwellers live in life-threatening conditions of deprivation and environmental degradation. This number is expected to double by 2025. Thus, the global trend in urbanisation implies nothing less than the 'urbanisation of poverty and deprivation'.
>
> (Tuts, 2002)

However, the very factors that make cities weigh so heavily on the ecosystem – the concentration of population and consumption – gives them enormous economic and technical advantages in the quest for sustainable development (Rees, 1999). Their economies of scale lower the costs per capita of providing infrastructure, increase the range of options for material recycling and reuse, reduce the per capita demands for occupied land, allow for electricity co-generation and offer potential for reducing fossil fuel consumption through the provision of public transport (Rees, 1999).

While both climate adaptation and biodiversity protection, especially in urban settings, have lately become very visible on the local agenda, they are neither replacing nor running in parallel to earlier LA21 processes. On the contrary, in most cases they build upon ongoing processes, often breathing new life and funding into local actions (ICLEI, 2012b).

Implementing LA21 in Europe

As mentioned above, most LA21 activity is to be found in Europe. However, there are still differences in the timing and the extent of LA21 activities between different European countries. Some countries began their LA21 activities relatively soon after the Rio Summit and many, if not all, of their local authorities went on to engage in LA21 initiatives. In contrast, other countries began late and only a few of their local authorities have since become engaged. Some latecomers, however, have rapidly caught up with practice.

There are several reasons for these differences. First, a close correlation exists between LA21 activity and the type of local government system in operation in a country (Lafferty and Eckerberg, 1998). In the Northern European system (Norway, Sweden, Denmark, Finland), local authorities have a high degree of autonomy, have their own funds raised through local taxes and have broad powers in relation to environmental matters. With their long tradition of local government autonomy, Nordic countries were front runners in LA21. The Middle European System (Germany and Austria), a federal system where local authorities are small and have varied powers, has experienced a slow start. In the Anglo-Irish system there are relatively large local authorities, but they have few powers and a weak financial basis, and are dependent on central government, although recent devolution reforms have changed the power of local governments in the UK. This system, as well as the Napoleonic system (France, Spain, Italy and the Netherlands), where there is a high degree of central government control, have both resulted in late and fewer LA21 initiatives, although some countries have subsequently caught up.

Second, to be successful, local initiatives need the support of national policies and funding. Other influences include the presence of an active and politically mobilised population, including environmental NGOs, interested and motivated civil servants and local politicians, in addition to involvement with international networks. The support of local political and administrative decision makers is crucial. Furthermore, where participatory practices are poorly developed, late and weak responses to LA21 are typical (Lafferty and Eckerberg, 1998). LA21 processes also depend on being built into existing organisations that are themselves concerned with sustainable development issues.

In Western Europe, the work of Agenda 21 relies heavily upon the 1998 Convention on Access to Information, Public Participation in Decision-Making and Access to Justice in Environmental Matters, known as the Aarhus Convention (Bell, 2004). This elaborates Principle 10 of the Rio Declaration, which stresses the need for citizen participation and access to information on the environment held by public authorities. Drawing upon this principle, the Aarhus Convention seeks to promote sustainable development through granting procedural rights. Such rights include citizen access to information and the right to public participation. It is premised on the belief that granting procedural rights will enable citizens to participate directly in environmental decision making, thereby enhancing the quality of environmental policy.

Case study: LA21 in Sweden

Sweden is regarded as an LA21 leader state (Eckerberg, 1999; Eckerberg and Forsberg, 1998; Rowe and Fudge, 2003). All municipalities have at some point been engaged in LA21. Strong commitment to the common good, advanced environmental policy frameworks and the availability of a wide range of financial tools have all helped promote LA21 work in Sweden.

Box 8.4

Växjö, Sweden: a fossil fuel free city

Växjö is a medium-sized city in southern Sweden, often referred to as 'the greenest city in Europe'.

In 1996, ahead of the UN FCCC Kyoto climate conference and building upon environmental activities undertaken since the 1970s, local politicians decided that Växjö should become fossil fuel free. The first phase was to reduce CO_2 emissions by 50 per cent per capita by 2010 from 1993 levels. The 2030 deadline for a 100 per cent reduction was agreed later.

A number of climate and energy saving measures were implemented, including conversion to largely biomass-based heating, introduction of smart metering systems and construction of the first wooden high-rise blocks to passive energy standard.

Växjö has demonstrated the viability of decoupling local economic growth from CO_2 emissions.

(*Source*: ICLEI, 2012b)

Case study: France

Compared to other European countries, France was a latecomer to LA21, as mentioned above, only getting involved towards the end of the 1990s. However, with strong leadership from the Ministry of Environment (renamed the Ministry of Sustainable Development in 2009) and other national level actors, it now boasts one of the most active local

LA21 campaigns (ICLEI, 2012b). LA21 activity is directly supported by Comité 21, created in 1994 with the mandate to support the implementation of sustainable development in the French regions by helping LA21 initiatives.

France's active LA21 engagement can be attributed to several factors. First, LA21 is supported by national strategies. The first National Strategy for Sustainable Development (2003–2008) set a goal of establishing 500 LA21 processes by 2008. The subsequent National Strategy for Sustainable Development (2009–2013) set the more ambitious goal of 1,000 LA21 initiatives by the end of 2012. Second, the development of national legislation encouraged the growth of LA21 activity. LA21 is understood as a voluntary process, but supported by a number of legislative requirements relating to the integration of sustainable development concerns into local strategic plans. New environmental laws mandate every region and every local government with more than 50,000 inhabitants to produce a report about sustainable development of their territories, and for those who do not have an LA21 Plan, to adopt and implement a climate plan (ICLEI, 2012b). Third, a framework for sustainable regional development and LA21 was established in 2006 by the Ministry of Sustainable Development in partnership with local governments. The framework sets five objectives:

1 Fight against climate change;
2 Preservation of biodiversity, environment and resources;
3 Social cohesion and solidarity between territories and generations;
4 Development of human beings;
5 Development of sustainable production and consumption methods.

(Ministère de L'Écologie, du Développement Durable et de L'Énergie, 2014a).

Sustainable development is now defined by those five objectives in French environmental law, the so-called 'Grenelle 2' law of 2010. Implementation is supported by an online management tool that provides an overview of relevant national commitments, highlights actions needed, and provides a set of evaluation criteria and accompanying indicators.

Finally, LA21 in France is characterised by broad stakeholder participation. It also benefits from well established institutions that support local governments. Regional agencies for environment or sustainable development provide technical and financial support to the local authorities.

These agencies work together under the umbrella of RARE (Réseau des Agences Régionales pour l'Environnement). LA21 is more numerous in regions that have such regional agencies (French Ministère de L'Écologie, du Développement Durable et de L'Énergie, 2014b). Support is also provided by the National LA21 Observatory, which has collected and analysed local sustainability experience since 2006.

Case study: the UK

The UK was one of the LA21 forerunners, proving a useful tool for local authorities to raise awareness of sustainable development. The Local Government Act 2000 placed a new duty on local authorities to promote the environmental, economic and social well-being of their communities. The Localism Act 2011 further devolves responsibilities to local authorities, including in relation to planning and civil society participation. Commentators have remarked on the strong overlap between the 2000 Act and LA21, including similarities between proposed Community Plans and LA21 Plans and the emphasis on integration of economic, environmental and social policies (Lucas et al., 2001).

However, local authorities have been heavily influenced by government guidance, which means LA21 is often driven by top-down policy priorities. In addition, having an LA21 Plan is not necessarily an indication of integrated social, economic and environmental policies at the local level. Many schemes are linked to regeneration programmes targeted at areas that have suffered economic, social and physical decline, often prioritising traditional economic growth (Lucas et al., 2003). Initiatives have struggled to recruit deprived and excluded communities. In addition, officers involved in regeneration are often unaware of the experiences of their counterparts in LA21 and are thus duplicating efforts and failing to learn from past engagement (Lucas et al., 2003).

In Europe, some leading states have become involved in partnership networks for the transfer of best practice to slower-moving countries, and for capacity enhancement. A good example is the Baltic LA21 Forum, a network of experts from local authorities, NGOs and various other organisations around the Baltic Sea. Special emphasis is placed on capacity enhancement and support for transfer of best practice in the countries in the eastern part of the region. In addition, it co-operates with the Union of Baltic Cities to support the sustainable development of cities in the region (Baltic Local Agenda 21 Forum, 2009).

Box 8.5

Dorset LA21

The Dorset Agenda 21 was founded in 1995 in response to the Rio Summit. It is constituted as a charity, with a mission to encourage climate-friendly and sustainable communities in Dorset.

Following awareness-raising and consultation, in 1999 it published Dorset's LA21 strategy, *Dorset in the Twenty-first century: An Agenda for Action* and subsequently the 2001 *Dorset LA21 Strategy*. The charity works with the Dorset Energy Partnership and has a range of local projects, including:

1. **Healthy Futures**: an environmental and educational awareness project for primary schools;
2. **Community Choices for Sustainable Living** (2004–2007): a project to encourage low carbon, sustainable communities;
3. **Dorset Climate Change Coalition** (2006–present): a partnership of public and community organisations working to mitigate climate change.

It also acts as a hub for community sustainability work in the county.

(*Source*: Dorset Country Council, 2013)

Other local authorities have become involved in a highly focused initiative, in particular around education for sustainable development. The educational sector plays a key role in supporting LA21 activities, a role recognised in the UN Decade of Education for Sustainable Development (2005–2014). Education for sustainable development includes incorporating key sustainable development issues into teaching and employing learning methods that motivate people to change their behaviour and take action. It promotes competencies like critical thinking, imagining future scenarios and making decisions in a collaborative way (UNESCO, 2014).

Some slow-moving countries have found it difficult to catch up on LA21 initiatives, especially when they have lacked central government support. In their place, a host of civil society initiatives have populated local actions. Post-unification Germany is a case in point, where lack of guidance from the national level during the 1990s resulted in the emergence of local movements which brought together environmental NGOs, church groups and groups active in decentralised development in a common engagement for LA21. A key focus was on improving public

Box 8.6

Education for sustainable development in Barcelona

In 2002, Barcelona published its LA21, *The Citizen Commitment to Sustainability*. This set out ten objectives to help stimulate LA21 engagement.

Of particular note is the Barcelona's School Agenda 21 Program. Phase 1 (2002–2012) worked with schools to identify environmental problems and propose solutions. Some of the problems have been fully addressed, such as the treatment of waste water, remediation of a rubbish dump and the placing of organic waste collection containers throughout the city. The initiative also supported the purchase of Fair Trade products and reduction in consumption of clean water. Several other issues identified, such as green spaces, energy conservation, waste prevention and promoting sustainable tourism, still require attention.

The city won the Dubai International Award for Best Practices to Improve the Living Environment, administered jointly with UN-Habitat for its Agenda 21 Schools Programme.

(*Source*: Franquesa, 2012)

participation in the former East Germany. Subsequent attempts by local governments to professionalise LA21 processes often met with scepticism and even resistance from local groups. As a consequence, local governments started organising parallel processes to engage citizens in such actions as climate mitigation plans, social integration strategies and model neighbourhoods (ICLEI, 2012b).

Local authorities have joined their efforts through numerous networks aimed at identifying exemplars of best practice from which lessons can be learned and policies transferred from one city to another. Framing the promotion of sustainable development within a discrete spatial scale, while participating in transnational municipal networks engaged in promoting urban sustainability, has been dubbed 'new localism' (Marvin and Guy, 1997). The most prominent of these networks is the European Sustainable Cities and Towns Campaign.

Unfortunately, in many countries, local authorities lack the capacity to address rapid urbanisation and in ways that promote sustainable development. However, as mentioned throughout this chapter, capacity building of local authorities for the promotion of sustainable cities has been advanced at the international level.

Box 8.7

LA21 in Freiburg, Germany

Freiburg is characterised by an impressive mixture of activities that justify its image as a forerunner in sustainable development.

In the Vauban district a derelict military area was converted into a car-free pilot neighbourhood. Associations of private households have replaced corporate developers as builders of multi-storey apartment houses; the light rail system runs on renewable energy; so-called 'plus energy' houses produce more energy than they consume; a high-rise apartment block was refurbished into a 'passive house'; and newly built municipal buildings have to be passive houses.

These multiple initiatives are not the product of a well-designed plan managed by local government. Instead, they are the result of activities carried out by a multitude of civil society initiatives and organisations. The fact that many original participants in such initiatives have become local councillors, officials or decision makers in local companies has gradually turned the local government, municipal utilities and the economy into crucial contributors in this patchwork of agents for sustainable development.

It was only in 2011 that a new unit for sustainability management was established at the mayor's office, which aims to link the various actors and activities and evaluate and document their collective achievements.

(*Source*: ICLEI, 2012b)

LA21 in the Americas

The United States of America

The work of the President's Council on Sustainable Development can be seen as the United States' response to the Rio Earth Summit and thus as the country's national Agenda 21 activity. When President Clinton established the Council in 1993, he asked it 'to bring people together to meet the needs of the present without jeopardizing the future'. The Council's report, *Sustainable America: A New Consensus*, did not, however, refer to LA21 by name (PCSD, 1996). Nevertheless, the Council echoed the Earth Summit's call for local government interventions to promote sustainable development. It also called for actions on a range of issues, including community-driven strategic planning, collaborative regional planning, improved building design and less urban sprawl

Box 8.8

The European Sustainable Cities and Towns Campaign

The European Sustainable Cities and Towns Campaign was launched at the end of the First European Conference on Sustainable Cities and Towns held in Ålborg (Denmark) in May 1994. The founding document of the Campaign is the Ålborg Charter, which outlines what is understood by local sustainable development and contains a commitment to engage in LA21 processes.

The mission of the Campaign is to support the exchange of experiences between cities, collect information on local activities and serve as an interface between the European Union and the local sustainable development movements.

There are now more than 2,700 participants in the Campaign, making it the largest bottom-up movement that has emerged following the LA21 mandate from Rio.

In 2013, the European Sustainable Cities Platform (www.sustainablecities.eu) was set up as an information hub for local governments and organisations.

(*Source*: ICLEI, 2003–2011)

(Bryner, 2000). Most of the recommendations that were made in the Report have, however, still not been implemented.

As of 2010, a director of Sustainable Development in the US Department of Agriculture (USDA) has a mission to advance the principles and goals of sustainable development. The USDA encourages management plans that are appropriate to both place and scale, especially across landscapes, supply chains and markets. USDA also supports the principles of 'reduce, reuse, and recycle' in relation to efficient product handling, processing, transportation, packaging, trade, consumption and waste management (USDA, 2007). However, there is no co-ordinated or explicit campaign to encourage local governments to undertake a comprehensive, multi-issue planning and implementation process analogous to LA21 work seen in other countries.

Nevertheless, several local governments have undertaken processes somewhat similar to LA21, invoking terms like 'sustainability' and taking divergent pathways towards sustainable development. In addition, local government has become engaged in LA21 type processes through indirect routes, whereby planning and day-to-day operations have

---

Box 8.9

The green economy, Portland, Maine, USA

Portland provides a good example of policies designed to support the green economy, especially during times of economic downturn.

In the face of the recent financial crisis (2007–2009+), the city formulated an Economic Development Strategy (2009–14). This focuses on green jobs, clean technology clusters and sustainable urban planning and management. The Strategy rests on three pillars:

1 Sustainable job growth (economic sustainability);
2 Sustainable innovation with Eco-Districts (environmental sustainability);
3 Inclusive prosperity (social sustainability).

This builds upon Portland's long history of promoting innovative approaches to urban sustainability, beginning with the 'Urban Growth Boundary' in 1979 that promoted land use density necessary for a functioning public transport system. This contrasts sharply with the urban sprawl characteristic of other US cities. It was also the first US city to adopt a carbon emissions reduction plan. More recently, it has pioneered 'Eco Districts'. A Bureau of Planning and Sustainability was established in 2009 to provide institutional support.

Between 1990 and 2011 Portland achieved GHG emissions reductions of 3 per cent, compared with an increase in the USA of 7.3 per cent roughly during the same period. It has improved low carbon transport options, as well as air and water quality and waste management. Portland's quality of living has attracted a highly educated and creative workforce.

Some of the city's successful policies, for example in relation to mass transit systems, bicycle infrastructure, tax credits for alternative energy use, green building codes, and land use ordinances, have been up-scaled to the state and federal levels.

(*Sources*: ICLEI, 2012)

---

incrementally evolved to include all the attributes of sustainable development comprised in LA21. Although they have not adopted a plan or policy statement that is labelled 'sustainable development', their operations have been modified over several years to the point where they integrate sustainable development objectives (ICLEI, 2012b).

However, many of the projects initiated at the local level in the USA deal only with a single issue, such as economic regeneration, transportation, energy efficiency, a specific environmental concern or the redevelopment of a particular site or district. These programmes do not include integral, long-term approaches to environmental protection and economic and social development. In addition, many do not involve local governmental agencies, nor do they include significant community and stakeholder involvement. Such policies can also be driven by traditional economic development rationale, based on a belief that a more liveable community attracts more businesses and that local clean energy and energy efficiency projects also create local jobs (ICLEI USA, 2014). Such beliefs are reflected, for example, in the Portland case above. Given this range of limitations, it is difficult to consider that these initiatives, by themselves, constitute an LA21 process, especially when judged by the Lafferty and Eckerberg criteria. Sustainable development is clearly here more akin to ecological modernisation. In addition, for a small but vocal minority of Americans, terms like 'sustainable development' are synonyms for central restrictions on natural resource use and limits on the freedom of individual actions, and thus often viewed with hostility and fear (ICLEI, 2012b: 22).

Latin America and the Caribbean

The bulk of LA21 initiatives in Latin America are in Brazil, Chile, Ecuador and Peru. Most have occurred without campaigns at the national level, although the Caribbean example of Cuba provides an exception. The most common priority is community development, and links to the Millennium Development Goals are common. However, while Latin America has a very high rate of stakeholder involvement, ethnic minorities are under-represented (ICLEI, 2002a; ICLEI, 2012b). In addition, many countries lack the capacity to support LA21 activities. It is for this reason that the UNDP Capacity 21, mentioned above, targeted fifteen countries in Latin America in a programme that lasted from 1993 to 2002. The idea was to scale-up successful local experiences into national level policy.

Following the WSSD Plan of Implementation, a Capacity 2015 Latin America project was established to build on best practices and lessons learned from the earlier phase. Focus in particular was placed on developing capacity to implement the MDGs at the local level. This, in turn, is seen to require more integration between the local- and national level policies. Capacity 2015 currently works with eight countries in the region, where knowledge exchange is facilitated through an Information Learning Network.

Box 8.10

Capacity 2015 in Latin America

Capacity 2015 concentrates on four core areas of intervention.

1 **Sustainable local economies**: this seed-funds locally formulated investment and employment strategies.
2 **Leadership for local development**: this helps develop individual capacities for local leadership.
3 **Platforms for local governance – building institutional capacities**: this assists municipalities to create and/or improve platforms where local stakeholders can plan for development.
4 **National policy for local development – societal capacity development**: this works with key policy- and decision makers to strengthen policy support for local development.

(*Source*: UNDP, 2004)

Case study: Brazil

In 2000, Brazil published its national *Agenda 21 Brasileira: Bases para Discussao* (Brazilian Agenda 21: Bases for Discussion). In July 2003, a revised Developing Local Agenda 21 initiative was launched, highlighting the importance of expanding activities. Since then, LA21 initiatives have been established throughout Brazil, with regional networks exchanging experiences. Many initiatives follow the guidelines set out by the Ministry for the Environment in its *Passo a Passo da Agenda 21 Local* (Local Agenda 21: A Step-by-Step Guide). More recently, however, LA21 initiatives have begun to falter, with the period 2002–2009 seeing a reduction of 37.7 per cent in the number of LA21 initiatives. This reflects experience elsewhere (Garcez de Oliveira Padilha and de Souza Verschoore, 2013).

Much of the LA21 activity reviewed to date in this chapter has been based on regional and local, even bottom-up initiatives. In contrast, Cuba provides an example of strong, top-down engagement in support of local sustainable development.

Case study: Cuba

Following its revolution in 1959, Cuba was officially constituted as a socialist state. Immediately after the Earth Summit of 1992, Cuba

Box 8.11

Curitiba, Brazil

Curitiba is recognised internationally as a 'sustainable city'. The key to its success lies in its integrated approach to urban and transport planning and city management, adopted as early as the 1960s. Success is also due to the leadership of Jaime Lerner, an architect, urban planner and a three-time mayor of Curitiba. He was heavily involved in developing Curitiba's Master Plan, which promoted the well known 'Bus Rapid Transit' system. The existence of an independent Institute of Urban Planning and Research ensured continuity in implementing the Plan following the departure of Lerner to become governor of the State of Parana.

Curitiba has the highest rate of public transport use in Brazil and enjoys the lowest rates of urban air pollution. It has also invested in large parks as ecological assets for flood prevention and recreation, retaining green spaces in the city despite substantial population growth. There are also community housing and small business assistance programmes and an innovative waste collection and recycling programme through which the poor can exchange collected waste for transport coupons and food.

Curitiba's ideas have been replicated by many South American cities, including Bogotá, Quito, Guatemala City and Mexico City. It provides good examples of how to link environmental, social and economic objectives.

(*Sources*: ICLEI, 2012b; World Bank, 2009; Gustafsson and Kelly, 2012)

modified Article 27 of its Constitution to widen its understanding of the environment and its relationship to sustainable economic and social development. This was followed in 1997 by the National Environmental Strategy, which established the socialist principles that drive environmental policy, identified the main environmental problems of the country and proposed solutions aimed at improving the protection of the environment through the rational use of natural resources. Cuban environmental policy is based on the principles of sustainable social and economic development, whose primary objective is to generate healthy food supplies for the population while protecting the environment. In 1997 the National Assembly also approved Law 81 on the Environment that reaffirmed national ownership of the country's natural resources. Law 81 also defines sustainable development as a process of equitable and sustained rise in the quality of people's life, providing economic growth and social improvement, in harmony with the protection of the environment. Other important aspects of Cuba's sustainable development policies are poverty eradication, health, culture, education, sports, and equal rights for women (IAEA, 2008). In 1999, the World Wildlife Fund called Cuba the only country in the world to have achieved sustainable development. The WWF used both the UN Human Development Index and ecological footprint analysis, discussed in Chapters 1 and 11, to show energy and resources consumed in each country. Only Cuba passed in both areas, a success credited to the high level of literacy, long life expectancy and low consumption of energy (WWF, 1999).

Box 8.12

LA21 Bayamo, Cuba

LA21 first began in the city of Bayamo in 2001. Bayamo has 142,000 inhabitants, with an economy based on food processing, construction and the provision of social and administrative services.

Consultation with stakeholder groups led to the identification of four priorities: (1) the collection and recycling of solid waste; (2) degradation of the local Bayamo River that limits its uses; (3) an insufficient urban transport service; and (4) insufficient public spaces in some neighbourhoods.

In 2002, the Bayamo Urban Environmental Profile ('Diagnostico Urbano Ambiental de Bayamo') was produced and institutional support provided. This led to: clean-up and water quality monitoring of the Bayamo River; reorganisation of solid waste collection and selective collection for recycling to support urban agriculture; creation of new routes for horse carts and cycle paths and reorganisation of points of traffic conflict; and rehabilitation of public spaces.

Based on the experience in Bayamo, in 2004 the government decided to initiate a National Replication Programme in the three other cities. This programme is jointly supported by UN-Habitat and UNEP.

(*Source*: UN-Habitat, 2008)

The need for gender representation and supporting the role of women in achieving sustainable development has been addressed in many LA21 processes, discussed in Chapters 6 and 11, where matters in relation to women and sustainable development are detailed. For instance, the Korean LA21 Network established a nationwide Gender Network in 2008, dedicated to building gender-sensitive local sustainable development. Similarly, in 2011 the African Local Elected Women's Network was established in Tangier, Morocco with the mandate to strengthen the role of women in local development (ICLEI, 2012b).

Box 8.13

Women's Action Agenda 21 in Brazil

In Brazil, Women's Action Agenda 21 has helped identify local priorities and actions. These often involve initiatives to reduce poverty by tackling environmental problems. In some cities, these experiences have been integrated into municipal environmental policy. Porto Alegre, a city of three million situated in the south of the country, provides a good example. Church groups helped start a co-operative for those working as garbage scavengers. The recycling of paper, cans and glass became a major source of income for women and their children. More than 300 families live on income generated by recycling garbage. The project has improved the self-esteem of these women, including through public recognition of the contribution their work makes to the sustainability of the city. The women are frequently invited to speak in schools and to help organise similar processes in other cities in Brazil.

Similarly, the project 'Mãos Mineiras' was initiated in the early 1990s to promote income generation for women through handicrafts made from garbage. The project is now being replicated in several other places in Brazil.

(*Source*: Corral, 2009)

LA21 in the Asia-Pacific region

One of the noticeable features of LA21 in this region is the strong emphasis on environmental protection. Within that, the most common priority is natural resource management, followed by air quality, water resources and energy management. The presence of a national campaign proved particularly important in stimulating activity in the more developed countries in this region, such as Australia, Japan and South Korea. In developing countries, stakeholder groups share in LA21 decision making to a much greater degree, and the priorities they set also differ.

Case study: Viet Nam

By 2010, political and economic reforms launched in 1986 had transformed Viet Nam from one of the poorest countries in the world to a lower middle income country. The ratio of population in poverty has fallen from 58 per cent in 1993 to 14.5 per cent in 2008, and most indicators of welfare have improved. Viet Nam has already attained five of

---

Box 8.14

LA21 in Vinh, Viet Nam

The first LA21 initiative began in 1998 in the city of Vinh. The city (population 230,000) was entirely destroyed during the Viet Nam War (1959–1975). Its economy is based on light industries, commerce, education and administrative services. The city suffers from severe yearly flooding. A harsh climate, frequently affected by the dry and hot 'Laos' wind and life-threatening storms from the South China Sea, coupled with poor quality soil for agriculture, marks Vinh as one of the poorest cities in the nation.

The Vinh LA21 initiative identified four priorities: (1) revitalisation of the Quang Trung public housing scheme; (2) environmental upgrading of the Vinh market; (3) development of the Lam River front; and (4) waste management.

Consultation on public housing took place in 1997, followed by a consultation on urban strategies. In 1998, a solid waste action plan was prepared and a demonstration project started. In 1999 a consultation was organised on 'Imagining Vinh's Future Identity', leading to the first 'Urban Pact' in 1999, with Urban Pact 2 started in 2001. In 2000, rehabilitation of the Quang Trung estate began. In 2002, an integrated solid waste management project was launched. The same year, community development activities started in the Quang Trung estate and an Urban Pact 3 was adopted, institutionalising the LA21 projects within local government. The People's Committee of Vinh City co-ordinates activities.

(*Source*: UN-Habitat, 2003a)

---

its ten MDGs and is expected to attain two more by 2015 (World Bank, 2014). The Socio-Economic Development Strategy (2011–2020) confirms the country's commitment to sustainable development. Despite this, however, economic development and recent industrialisation have brought problems of pollution, including degradation in air and water quality. Natural resources, especially water and biodiversity resources, including forest resources, have seriously degenerated. Civil society and participatory processes are weak (Socialist Republic of Vietnam, 2012).

India

India faces sustainable development challenges similar to other emerging economies such as China, discussed in Chapter 12. Rapid urbanisation has outpaced the ability of most Urban Local Bodies (ULBs) to provide even the most basic services, such as water and waste

---

Box 8.15

Jawaharlal Nehru National Urban Renewal Mission

Most of the reforms under JNNURM incorporate the principles of sustainable development and of LA21. Reforms include upgrading social and economic infrastructure in cities and providing basic services to the urban poor. These are accompanied by devolution of responsibility to local authorities and reform of planning legislation.

Public participation has also improved following the enactment of the Community Participation Law and the Public Disclosure Law. Both laws have boosted transparency and accountability in the governance of Indian cities.

Since its launch, 530 projects have been approved, including for water supply, solid waste management and public transport. The programme takes a combined top-down and bottom-up approach, with guidelines, reforms and funds from the top and ULB implementation at the local level.

The Peer Learning and Knowledge Sharing Network enables cities to share experiences.

(*Source*: ICLEI, 2012c)

---

management. In response to these challenges, in 2005 the government began an ambitious urban reform programme: the Jawaharlal Nehru National Urban Renewal Mission (JNNURM) to last until 2014.

While the combination of top-down steering and bottom-up engagement has proved beneficial in the Indian case, experience in Korea shows how shifting national priorities are impacting at the local level, resulting in tensions between state steering and local ownership of sustainable development processes.

Korea: from LA21 to Green Growth

The LA21 movement started in 1995 in the city of Pusan and was given renewed impetus with the establishment of the Presidential Commission on Sustainable Development in 2000 and the enactment of a Framework Act on Sustainable Development in 2007. This Act requires cities and provinces to establish a Local Sustainable Development Council and develop a related Strategy. The newly created Councils are closely linked to ongoing LA21 processes, steered by LA21 councils and supported financially by the local governments. As of December 2010, almost 90 per cent of Korean local governments had an active LA21 process.

In 2008, President Lee Myung-bak proclaimed 'Low Carbon, Green Growth' as Korea's new national vision, with three major objectives: promoting eco-friendly growth; enhancing quality of life; and contributing to international efforts to combat climate change. Local Green Growth Committees were mandated, alongside the appointment of Chief Green Officers in cities and provinces, as well as within ministries and governmental agencies. In 2010, Korea passed the internationally praised Framework Act on Low Carbon Green Growth, translating the vision into binding national regulations. Since then, seven Korean provinces and nine cities have developed Five-Year Implementation Plans for Local Green Growth. However, the new vision pays less attention to stakeholder participation and existing LA21 priorities. In addition, newly established Local Green Growth Committees do not have strong relations with existing LA21 secretariats, and citizens are seen more as the target of sustainable lifestyle campaigns than partners in policy discussions (ICLEI, 2012b).

Australia

In Australia, sustainable development policies tend to focus on the ecological dimension, reflecting the importance given to protecting the country's unique ecosystem. However, efforts to conserve biodiversity are delivered through other programmes with objectives such as land and fresh water health, vegetation conservation, or marine protection. Land clearance for urban development and agriculture continues to deplete native species and ecosystems, while roads dissect the remaining wildlife corridors and animal road kill rates are high. Salinity, caused by land clearing and irrigation, and other pressures on water quality threaten not only plants, animals and ecosystems but also the future of agricultural and other industries in many parts of Australia, and even urban water supplies in some cities. The growth in private car use has led to congestion in metropolitan areas and reduced urban air quality, and the country has a poor record in addressing climate change through reductions in GHG emissions. In addition, prioritisation of traditional economic growth, the constraints of a federal system of government and concerns that the promotion of sustainable development is financially burdensome have combined to limit responses (Australian Government, Department of the Environment, 2002).

Box 8.16

Sustainable planning in Melbourne

Melbourne is pursuing sustainable urban development in the context of an increasing population, a high standard of living, and economic growth. The effects of climate change on natural resources and ecosystems and the need for adaptation strategies are keenly felt.

Following a stakeholder and public consultation process, in 2007 the City produced the Future Melbourne Plan. The Plan seeks to make Melbourne one of the world's most sustainable cities by 2020.

The Eco-City component of the Plan aims to achieve zero net GHG emissions by 2020, adapt to climate change impacts, promote resource efficiency and reduced energy consumption, and increase urban density, while protecting water resources. To achieve this, legislative and statutory changes have been implemented, and new financing mechanisms introduced, including the Sustainable Melbourne Fund. The Fund invests in local sustainability projects in water, energy and waste, or in business ventures and new technologies that deliver environmental and economic benefits.

A Buildings Program launched in 2010 has retrofitted commercial buildings to reduce GHG emissions. In 2008 the sustainable water management strategy was updated to include targets for 2020.

(*Source*: ICLEI, 2012c)

Promoting LA21 in Least Developed Countries: LA21 in Africa

Local governments in Africa face challenges that are often well beyond their capacities, including conflict between ethnic groups that often results in violent civil unrest. Demographic pressures, the AIDS crisis, inadequate infrastructure and very limited resources for service delivery and planning add to these difficulties. In many countries, urban environmental management has been added to a long list of municipal responsibilities. The rapid growth of many cities is increasing the human risks associated with poor housing conditions, uncollected solid waste, and over-consumption of fresh water, untreated wastewater and urban air pollution.

Despite the difficult conditions under which local authorities work, LA21 processes are widespread. Economic development remains an overriding concern, with a specific focus on poverty alleviation. LA21 is characterised by strong stakeholder involvement, although the participation of women and ethnic minorities is limited.

Only South Africa has established a national campaign to facilitate local LA21 activities. In addition to the lack of financial and political support, there is also a lack of expertise (ICLEI, 2012b).

Governments across the Maghreb region of North Africa are struggling to address a wide range of socio-economic and political grievances that sparked popular uprisings throughout the Arab Spring of 2011. The problems of the region are rooted in political systems that have been marred by corruption and the marginalisation of large swathes of the population, including young people (Center for Strategic and International Studies, 2012). Despite significant change since 2011, the region is still struggling to introduce political reforms aimed at promoting the rule of law, curbing corruption and overhauling the judiciary. Engagement in LA21 is difficult in the context of such instability. Nevertheless, the Arab countries of the region have had some success with LA21, including Morocco, a country that has managed the economic fallout of regional upheaval relatively well.

Box 8.17

Louga LA21 project, Senegal

Louga is situated in a semi-arid region and vulnerable to drought. The economy is based on trade, agriculture and cattle rearing. Rural influxes caused by deteriorating climatic conditions has seen the settlement grow rapidly and the population now stands at around 125,000.

Louga has several environmental problems, including:

1. Sanitation, with limited solid waste management, liquid waste management or storm drainage;
2. Desert encroachment, resulting in a deterioration of local livelihoods;
3. Mushrooming spontaneous dwellings, resulting in poor provision of services like health and education.

The Louga LA21 project was initiated in 2001. Following consultation, three priority issues were identified: (1) community-based solid waste management; (2) improvement of urban mobility; and (3) HIV/AIDS prevention.

In 2002, working groups were created to prepare action plans. These were discussed in community forums and in 2003 priority projects were presented to a donor's roundtable. In 2004, the Senegal National LA21 Programme began supporting the replication of the Louga project in four other cities.

(*Source*: UN-Habitat, 2003b)

Case study: Morocco

In the last decade, the Moroccan state has launched a series of highly publicised investment projects in infrastructure and transport, which provide some justification for the country's image as the most reforming in the Maghreb (Colombo, 2011). Nevertheless, Morocco faces many sustainable development challenges, including in relation to water management and the growth of its cities, where many people who move to the cities in search of a better life end up living in informal settlements without access to basic services, such as clean water and sanitation, education and health care.

In 2010, during the celebration of the fortieth anniversary of Earth Day, Morocco announced a National Charter for Environment and Sustainable Development. This is the first environmental charter of its kind in Africa as well as in the Arab and Muslim world. The Charter forms the framework for national environmental laws and policy (Kingdom of Morocco, Ministry of Energy, Mines, Water and the Environment, 2011).

Box 8.18

Essaouira, Morocco

Essaouira is a coastal town with 70,000 inhabitants. Its economy relies mainly on tourism and handicraft production. The previous economic base of fisheries is in crisis. Urban growth poses serious threats to the local environment, especially coastal dunes, which provide flood defence and other benefits. On the one hand, the geographical and ecological setting of the city induces severe development constraints and calls for limiting city growth. On the other hand, economic stagnation and population increase calls for enhancing urban development, which can further jeopardise the ecosystem. The lack of local capacity to respond to these conflicting demands prompted the initiation of LA21 in 1996.

Essaouira's LA21 projects address four priority issues: (a) cultural heritage management; (b) creating a green belt and an urban park; (c) strengthening urban planning; and (d) rehabilitation and revitalisation of the Mellah neighbourhood. The development of an urban park was designed to limit further housing development into the dunes, and plans for protecting the wetlands of Ksob river mouth were also proposed.

LA21 projects contributed to the 2001 inscription of Essaouira on the UNESCO list of World Cultural Heritage sites. However, progress has been hampered by limited local resources, although it is supported by the regional LA21 Project.

(*Source*: UN-Habitat, 2003c)

In July 2011, a newly-adopted Constitution established the 'right to sustainable development'. Despite this commitment, however, the country has struggled to integrate sustainable development in other policies, as evidenced by the negative ecological impacts of its economic development strategies (UNEP, 2012).

Limitations of LA21

At one level, LA21 can be seen as concerned with procedural issues, in particular, the integration of sustainable development into local authority planning processes. However, local authorities do not act in isolation and their capacity to shape policy and its outcomes is influenced by a number of factors. To begin with, account has to be taken of the various levels of governance (local, national and international) through which economic, social and political processes interact to shape prospects at the local level. National policy frameworks, for example, can have a major influence on the local planning process, as do statutory obligations, statutory municipal development plans and national budget priorities. National taxation policy can also be highly influential, especially given subsidies and other tax incentives that encourage unsustainable practices, for example in energy and resource use. In many countries, central government maintains control of local budgets, which makes it difficult to co-ordinate national investment plans and local LA21 priorities. The ability of local government to generate revenues is also regulated and restricted by national policies. These factors all combine to shape the capacity of local authorities to successfully implement LA21 initiatives (Gilbert et al., 1996).

While effective LA21 initiatives require local authorities to gain control over their development, globalisation is accelerating investment- and growth-orientated activities by external actors, such as transnational corporations. Often, these have only minimal incentive to be accountable and committed to local development strategies (Gilbert et al., 1996). Thus, the ability of local authorities to structure their development in a sustainable way is limited by the interplay between entanglement of local governance structures in the national political and economic system, and their engagement with systems operating at the international level. Local authorities in industrialised countries, for example, struggle to find ways to deal with waste generated by consumer products and packaging. While this accounts for a large portion of the local solid waste stream, local governments have few direct controls over the products that are sold and used in their jurisdictions and over how they are packaged. In contrast, in the Third World, basic life needs, for health, sanitation and water, structured the priorities of LA21 activities, yet the very international conditions that make these a problem are also the barriers to their effective resolution. The global distribution of political and economic power often blocks the construction of local development models based around LA21 principles. As such, adapting economic, ecological and social interests to the concerns of sustainable development is only possible in a limited fashion at a local level.

The focus on the urban dimension of sustainable development provides a particularly good example of the dangers of divorcing the urban from the other levels through which environmental governance is mediated. The rise of 'new localism' in particular risks the danger that the local can become a 'black box' disconnected from the global, international and national contexts within which localities are framed (Bulkeley and Betstill, 2005; Marvin and Guy, 1997).

Participation

Agenda 21 and LA21 both stress the importance of engaging society and of ensuring the involvement of 'major groups' in national and inter-national efforts to promote sustainable development. LA21 is typically promoted through governance processes that share responsibility across local authorities, economic actors and civil society. Ever since the publi-cation of the Brundtland Report (1987), the participation of both eco-nomic and social actors has come to be seen as a necessary quality of sustainable development governance. The enhanced culture of public participation is often seen as one of the most remarkable achievements of local sustainability processes worldwide. It is also seen as in keeping with the principles of good governance, discussed in Chapter 3.

Participation occurs through institutional settings that bring together various actors at some stage in the policy making process. Practice can vary considerably from country to country and indeed even from one policy issue to another. Countries are best seen as located along a continuum, ranging at their extremes from allowing only a minor, consultative role for non-state actors to more deliberative processes in which actors have a major say in shaping policy goals through dialogue and social learning (van Zeijl-Rozedma et al., n.d.). Thus, LA21 can exhibit a variety of different practices, ranging from very strong steering, particularly from local authorities and business elites, to more open and flexible forms of participatory practice (Lafferty and Eckerberg, 1998).

There is a range of well developed arguments in support of the claim that the promotion of sustainable development is best undertaken through the enhancement of practices of participatory democracy (Dryzek, 2005; Eckersley, 2004). These include instrumental arguments that participation improves the knowledge base for decisions, leading to better policy choices, and enhances implementation prospects (lending so-called 'output legitimacy' to policy), and more ideologically based claims that they help to improve democratic practice (lending 'input' legitimacy). Policy legitimacy is seen to reduce the risk of conflict over policy priorities, decisions and outcomes. Participation, it is argued, is also necessary because promoting sustainable development raises issues of an essentially 'subjective and value-laden' character (Paehlke, 1996). Agreement on the objectives of policy is more likely when these objectives are reached through participatory practices. This reiterates the point that promoting sustainable development is best seen not as a technical task but as a political project. Finally, not only is participation seen as a means of building understanding and commitment for collective policy making, but can also be seen as an end in itself, that is, as an aspect of the necessary and richer alternatives to lives centred on material consumption (Kemp et al., 2005: 16; Micheletti and Stolle, 2012).

This brief encapsulation shows that there is a range of complex and well developed arguments in support of the claim that the promotion of sustainable development is best undertaken through governance arrangements that involve public participation and in particular the enhancement of practices of participatory democracy. However, participatory practices are often weak in terms of traditional political accountability and representation (Hallstrom, 2004). They can also become skewed by the unequal distribution of power, including financial resources, between the participants, for example, between corporate interests and environmental activists. Furthermore, local interests can capture participatory processes, intent on the spatial and temporal displacement of environmental problems to other regions or to future generations, the so-called NIMBY ('not in my back yard') phenomenon. They can thus become an arena for the expression of narrow self-interest. This can weaken their ability to reflect wider, collective interests that, in turn, can undermine their credibility as forums for the governance of sustainable development. In addition, there are also no *a priori* reasons to believe that participation leads to decisions that drive a long-term, strategic approach to promoting sustainable development (Baker and Eckerberg, 2008).

The capacity for participatory processes to provide a meaningful forum for deliberations is heavily dependent upon the type of formal access to policy making that is given; the stage in the policy making process in which participation is allowed; the 'opportunity structures' that exist within the policy process to influence policy making; and the institutional constraints that are placed both on them and that are present more generally (Hallstrom, 2004). The government, for example, can limit the role of 'outsider' stakeholders in policy making. In studies of transition societies in Eastern Europe, participatory processes often became a means of disseminating EU decisions, garnering public support for the EU and increasing public knowledge about the EU, while doing little to enhance societal engagement with decision making, particularly policy formulation (Hallstrom, 2004; Baker, 2008).

Conclusion

Promoting sustainable development has become part of everyday activity for thousands of local governments worldwide. Local initiatives take many forms, reflecting the values and priorities of local communities. Many of these have moved beyond the original LA21 agenda, to inspire a range of complex engagements.

LA21 is both a procedural quest (in relation to planning) and a highly political process, being as much about democratic reform as about planning procedures. LA21 initiatives are designed to facilitate participation from society actors. Extensive public involvement is an integral part of the LA21 process, justified by both instrumental arguments and claims that the legitimacy of policy is better guaranteed with expanded participation. However, LA21 raises the thorny issues of the ability of normal citizens to take effective part in decision making, the capacity of planning process to function effectively under conditions of participation, and the legitimacy of expanded stakeholder democracy. There is also the fact that LA21 activity is limited by the wider economic, political and institutional processes operating at the national and international levels. These constrain the ability of the local level to promote sustainable development in isolation from higher level engagements.

Summary points

- LA21 can be distinguished from general environmental policy because LA21 plans are built upon a commitment to sustainable development.
- LA21 gives local authorities a pivotal role in promoting sustainable development.
- LA21 draws attention to the local scale and the need for actions that are responsive to the specific needs of a local place and its community.
- LA21 links planning, including in relation to land use and transport, and the promotion of sustainable development, particularly in urban settings.
- LA21 helps to focus on the level of social organisation where the consequences of environmental degradation are most keenly felt and where successful intervention can make an immediate difference to the quality of people's lives.
- Case studies show that LA21 activities have been growing throughout the world, but most noticeably in Europe. Priorities differ, with African countries more concerned about poverty alleviation, while industrialised countries are more preoccupied with dealing with energy and waste.
- LA21 processes often lack effective support from national authorities, encounter low political priority and face institutional capacity constraints at the local level.
- There are both instrumental and normative arguments for the participatory practices that form the core component of LA21 practices.
- There are limits to the role that local authorities can play in promoting sustainable development in isolation from actions across all levels of governance.

Further reading

Coenen, F. (2009), 'Local Agenda 21: "Meaningful and Effective" Participation?' in Frans H. J. M. Coenen (ed.), *Public Participation and Better Environmental Decisions: The Promise and Limits of Participatory Processes for the Quality of Environmentally Related Decision-Making* (New York: Springer), 165–182.

For information on local actions, regional engagement and the work of ICLEI, see: www.iclei.org/

References

Australian Government, Department of the Environment (2002), *WSSD: Australian National Assessment Report* (Canberra: Department of the Environment), available online at www.environment.gov.au/node/13078

Baker, S. and Eckerberg, K. (2008), 'In Pursuit of Sustainable Development at the Sub-National Level: The New Governance Agenda', in S. Baker and K. Eckerberg (eds), *In Pursuit of Sustainable Development: New Governance Practices at the Sub-National Level in European States* (London: Routledge, 2008), 1–26.

Baltic Local Agenda 21 Forum (2009), *Baltic Local Agenda 21 Forum BLA21F*, available online at www.bla21f.net/site/index.php?id=15

Bell, D. R. (2004), 'Sustainability through Democratization? The Aarhus Convention and the Future of Environmental Decision Making in Europe', in J. Barry, B. Baxter and R. Dunphy (eds), *Europe, Globalization and Sustainable Development* (London: Routledge), 94–112.

Bryner, G. C. (2000), 'The United States: Sorry – Not Our Problem', in W. M. Lafferty and J. Meadowcroft (eds), *Implementing Sustainable Development: Strategies and Initiatives in High Consumption Societies* (Oxford: Oxford University Press), 273–302.

Bulkeley, H. and Betsill, M. M. (2005), 'Rethinking Sustainable Cities: Multilevel Governance and the 'Urban' Politics of Climate Change', *Environmental Politics*, 14. (1): 42–63.

Center for Strategic and International Studies (2012), *Building Stability through Economic Growth in the Maghreb*, available online at http://csis.org/files/publication/120905_Malka_MaghrebConferenceReport.pdf

Colombo, S. (2011), *Morocco at the Crossroads: Seizing the Window of Opportunity for Sustainable Development* (Brussels: Centre for European Policy Studies, MEDPRO Technical report no. 2/April), available online at www.iai.it/pdf/mediterraneo/medpro/medpro-technical-paper_02.pdf

Corral, T. (2009), 'Regional Studied Development Review: Brazil – Women's Perspectives on Sustainable Development in Brazil', in *UNESCO and Encyclopaedia of Life Support Area Studies – Regional Sustainable Development: Brazil* (Paris: Eolss Publishers), vol. 1, 147–171, available online at www.eolss.net/ebooks/Sample%20Chapters/C16/E1-58-22.pdf

Dorset Country Council (2013), *Dorset Agenda 21 (DA21)* (Dorchester: Dorset County Council), available online at https://www.dorsetforyou.com/387628

Dresner, S. (2002), *The Principles of Sustainability* (London: Earthscan).

Dryzek, J. S. (2005), *The Politics of the Earth: Environmental Discourses* (Oxford: Oxford University Press, 2nd edn).

Eckerberg, K. (1999), 'Sweden: Combining Municipal and National Efforts for Quick Progress', in W. Lafferty (ed.), *Implementing LA21 in Europe: New Initiative for Sustainable Communities* (Oslo: Prosus), 13–26.

Eckerberg, K. and Forsberg, B. (1998), 'Implementing Agenda 21 in Local Government; The Swedish Experience', *Local Environment*, 3 (3): 333–347.

Eckersley, R. (2004), *The Green State: Rethinking Democracy and Sovereignty* (Cambridge, MA: MIT Press).

Evans, B., Joas, M., Sundback, S. and Theobald, K. (2008), 'Institutional and Social Capacity Enhancement for Local Sustainable Development: Lessons from European Urban Settings', in S. Baker and K. Eckerberg (eds), *In Pursuit of Sustainable Development: New Governance Practices at the Sub-national Level in Europe* (Abingdon/New York: Routledge/ECPR Studies in European Political Science), 74–95.

Franquesa, T. (2012), *Barcelona's Agenda 21: 10 Years of Citizen Commitment to Sustainability*, available online at www.sostenibilitatbcn.cat/attachments/article/413/TFranquesa_A21L_Barcelona%20LA21Experience_Educating%20Cities.pdf

Garcez de Oliveira Padilha, L. and de Souza Verschoore, J. R. (2013), 'Green Governance: A Proposal for Collective Governance Constructs Towards Local Sustainable Development', *Ambiente y Sociedade*, 16 (2), http://dx.doi.org/10.1590/S1414-753X2013000200009

Gilbert, R., Stevenson, D., Giradet, H. and Stern, R. (1996), *Making Cities Work: The Role of Local Authorities in the Urban Environment* (London: Earthscan).

Gustafsson, H-R. and Kelly, E. (2012), *Urban Innovations in Curitiba: A Case Study* (New Haven, CT: Yale Law School), available online at www.law.yale.edu/documents/pdf/News_&_Events/LudwigGustafssonKellyCuritibaReport.pdf

Hallstrom, L. K. (2004), 'Eurocratising Enlargement? EU Elites and NGO Participation in European Environmental Policy', *Environmental Politics*, 13 (1): 175–193.

ICLEI (2002b), *The Local Government Declaration to the World Summit on Sustainable Development*, available online at www.dlist.org/sites/default/files/doclib/Local_Government_Declaration_to_the_WSSD.pdf

ICLEI (2003–2011), *The European Sustainable Cities and Towns Campaign*, available online at www.sustainablecities.eu/cities/european-sustainable-cities-and-towns-campaign/

ICLEI (2011), *Charter* (Bonn: ICLEI World Secretariat), available online at www.iclei.org/fileadmin/user_upload/ICLEI_WS/Documents/Governance/Charter_approved_FINALforCOUNCIL20110912.pdf

ICLEI (2012a), *ICLEI at Rio+20* (Freiberg: ICLEI Sectretariat), available online at http://local2012.iclei.org/fileadmin/files/ICLEI_at_Rio_20.pdf

ICLEI (2012b), *Local Sustainability 2012: Taking Stock and Moving Forward. Global Review* (Bonn: ICLEI World Secretariat), available online at http://local2012.iclei.org/fileadmin/files/LS2012_GLOBAL_REVIEW_www.pdf

ICLEI (2012c), *Local Sustainability 2012: Showcasing Progress. Case Studies* (Bonn: ICLEI World Secretariat), available online at http://local2012.iclei. org/fileadmin/files/LS2012_CASE_STUDIES_www.pdf

ICLEI USA (2014), *Sustainability and Local Governments* (Oakland, CA: ICLEI USA), available online at www.icleiusa.org/

International Atomic Energy Authority (IAEA) (2008), *Cuba: A Country Profile on Sustainable Energy* (Vienna: IAEA), available online at http://www-pub. iaea.org/mtcd/publications/pdf/pub1328_web.pdf

International Council for Local Environmental Initiatives – Local Governments for Sustainability (ICLEI) (2002a), *Local Governments' Response to Agenda 21: Summary Report of Local Agenda 21 Survey with Regional Focus* (Canada: ICLEI).

Kemp, R., Parto, S. and Gibson, R. B. (2005), 'Governance for Sustainable Development: Moving from Theory to Practice', *International Journal of Sustainable Development*, 8 (1/2): 12–30.

Kingdom of Morocco, Ministry of Energy, Mines, Water and the Environment (2011), *The National Charter for the Environment and Sustainable Development* (Rabat: Ministry of Energy, Mines, Water and the Environment), available online at www.chartenvironnement.ma/

Lafferty, W. and Eckerberg, K. (eds) (1998), *From the Earth Summit to Local Agenda 21: Working towards Sustainable Development* (London: Earthscan).

Lucas, K., Ross, A. and Fuller, S. (2001), 'Working Paper 1: Literature Review, Centre for Sustainable Development' (London: University of Westminster, Research project funded by the Joseph Rowntree Foundation), available online at http://wwwedit.wmin.ac.uk/cfsd/reports/JRF_LA21_Literature_review.pdf

Lucas, K., Ross, A. and Fuller, S. (2003), *What's in a Name? Local Agenda 21, Community Planning and Neighbourhood Renewal* (London: Joseph Rowntree Foundation), available online at www.jrf.org.uk/sites/files/jrf/185935081x.pdf

Marvin, S. and Guy, S. (1997) 'Infrastructure Provision, Development Processes and the Co-Production of Environmental Value', *Urban Studies*, 34 (12): 2023–2036.

Micheletti, M. and Stolle, D. (2012), 'Sustainable Citizenship and the New Politics of Consumption', *Annals of the American Academy of Political and Social Science*, 644: 88–120.

Ministère de l'Écologie, du Développement Durable et de l'Énergie (20014a), *Agenda 21: French National Framework* (Paris: Ministère de l'Écologie, du Développement Durable et de l'Énergie), available online at www.agenda-21france.org/agenda-21-de-territoire/french-national-framework.html

Ministère de l'Écologie, du Développement Durable et de l'Énergie (20014b), Comité 21, *History and Process of French local Agenda 21*, available online at www.agenda21france.org/agenda-21-de-territoire/history-of-local-agenda-21.html

Paehlke, R. C. (1996), 'Environmental Challenges to Democratic Practice', in W. M. Lafferty and J. Meadowcroft (eds), *Democracy and the Environment: Problems and Prospects* (Cheltenham: Edward Elgar), 18–38.

PCSD (The President's Council on Sustainable Development) (1996), *Sustainable America: A New Consensus for Prosperity, Opportunity, and a Healthy Environment for the Future* (Washington, DC: US Government Printing Office).

Putnam, R. (2001), *Bowling Alone: The Collapse and Revival of American Community* (New York: Simon & Schuster).

Rees, W. E. (1999), 'The Built Environment and the Ecosphere: A Global Perspective', *Building Research and Information*, 27 (4–5): 206–220.

Rowe, J. and Fudge, C. 2003, 'Linking National Sustainable Development Strategy and Local Implementation: A Case Study of Sweden', *Local Environment*, 8 (2): 125–140.

Socialist Republic of Vietnam (2012), *Implementation of Sustainable Development in Vietnam: National Report at the United Nations Conference on Sustainable Development (Rio+20)* (Hanoi), available online at http://sustainabledevelopment.un.org/content/documents/995vietnam.pdf

Tuts, R. (2002), *Urban Poverty Reduction through Good Urban Governance: How Can Local Agenda 21 Initiatives Contribute? Lessons Learned from UN-Habitat's experience* (Nairobi, Kenya: UN-Habitat), available online at http://ww2.unhabitat.org/programmes/agenda21/documents/LA21&governance.pdf

UN (2002a), Johannesburg Declaration on Sustainable Development (New York: UN, A/CONF.199/20), available online at www.un-documents.net/jburgdec.htm

UN (2002b), *World Summit on Sustainable Development: Johannesburg, 2002* (New York: UN Department of Economic and Social Affairs, Division for Sustainable Development).

UNDP (2004), *Capacity 2015: A Continual Process – Capacity 2015 in Latin America Regional Strategy: Implementing the MDGs at a local level* (New York: UNDP), available online at www.latinamerica.undp.org/content/dam/rblac/docs/Research%20and%20Publications/Poverty%20Reduction/Capacity2015-.doc

UNEP (2012), *Morocco* (New York: UNEP Advisory Services), available online at www.unep.org/greeneconomy/Portals/88/documents/advisory_services/countries/Morocco%20final.pdf

UNESCO (2014), *Education for Sustainable Development* (Paris: UNESCO), available online at www.unesco.org/new/en/education/themes/leading-the-international-agenda/education-for-sustainable-development/

UN-Habitat (2003a), *Localising Agenda 21 Projects: Vinh*, available online at http://ww2.unhabitat.org/programmes/agenda21/Vinh.asp

UN-Habitat (2003b), *Localising Agenda 21 Project: Senegal*, available online at http://ww2.unhabitat.org/programmes/agenda21/projects.asp

UN-Habitat (2003c), *Localising Agenda 21 Projects: Morocco*, available online at http://ww2.http://ww2.unhabitat.org/programmes/agenda21/Essaouira.asp

UN-Habitat (2008), *Improving Urban Planning through Localizing Agenda 21: Results Achieved in Bayamo, Cuba* (Nairobi, Kenya UN-Habitat, SCP

Documentation Series, vol. 6), available online at http://cn.unhabitat.org/downloads/docs/2905_73810_Bayamo.pdf

US Department of Agriculture (USDA) (2007), *Sustainable Development* (Washington, DC: Office of the Chief Economist), available online at www.usda.gov/oce/sustainable/index.htm

van Zeijl-Rozedma, A., Cörvers, R., Kemp, R. and Martens, P. (n.d.), 'Governance for Sustainable Development: A Framework', *Sustainable Development*, 16 (6): 410–421.

World Bank (2009), *Curitiba: Planning for Sustainability – An Approach All Cities Can Afford*, available online at http://web.worldbank.org/wbsite/external/topics/exturbandevelopment/extuwm/0,,contentMDK:22446108~pagePK:210058~piPK:210062~theSitePK:341511,00.html

World Bank (2014), 'Vietnam Overview', available online at www.worldbank.org/en/country/vietnam/overview

WWF (1999), *The Other Cuban Revolution*, available online at http://wwf.panda.org/?1944/The-other-Cuban-revolution

Part III
Sustainable development in different contexts

High consumption societies
The promotion of sustainable development in the European Union

Key issues

- Legal and declaratory commitment
- Environmental Action Programmes
- Sustainable Development Strategy
- Environmental policy integration
- Tension between growth strategies and sustainable development
- Distinctive EU understanding of sustainable development.

Introduction

This chapter explores the promotion of sustainable development in the European Union (EU).[1] The EU is an economic and political union founded amidst the ruins of post-war Europe. Through a series of enlargements, it has grown to encompass twenty-eight member states that share a single market and common policies on a variety of issues, ranging from agriculture, climate change, employment, environment, foreign and security affairs, regional and industrial development and transport, and where the majority of members also share a common monetary union and a common currency, the euro.

Since the end of the Second World War, Western Europe has enjoyed unprecedented levels of stability and prosperity. However, this has come at the expense of a healthy environment. Continued economic growth, fuelling high consumption lifestyles, has resulted in ever greater demand for products and services. The social consequences of over-consumption are evident. Approximately 60 per cent of adults and over 20 per cent of children are overweight or obese and coronary heart diseases, often associated with fatty foods and smoking, are the most common death cause in the EU (CSCP, 2012). When judged from a planetary perspective, modern European lifestyles are also unsustainable. The EU's ecological footprint has been increasing almost constantly since 1961 and its ecological deficit has continued to grow (EEA, 2010a). As a result, levels of biological resource use and waste emissions are well above Europe's biological capacity, and it cannot provide the resource inputs needed to supply its consumption demands from within its own borders, as discussed in Chapter 2.

> Biological capacity is the capacity of ecosystems to produce useful biological materials and to absorb waste materials generated by humans, using current management schemes and extraction technologies.

This means that the EU has to draw down the resources of other regions of the globe in order to meet its own demands. If the current patterns of high consumption and production continue, it is estimated that global resource use by the EU will quadruple within twenty years (CEC, 2009a). As the European Environment Agency (EEA), an agency of the EU that provides information to support policy, has warned:

> In a world that is already in ecological overshoot, Europe's ecological deficit contributes to the diminishing amount of renewable natural resources available in the future, adds to overall waste accumulation and puts regional and global ecosystems at greater risk of degradation.
>
> (EEA, 2010a)

Efforts made by the EU to promote sustainable development and how successful these prove to be are thus important, not only from the point of view of the health and well-being of Europe's people and of its ecosystems, but also for dealing with the principles of inter- and intra-generational equity, and ensuring planetary health now and into the future.

This chapter begins by looking at the EU's strong declaratory and legal commitment to the promotion of sustainable development. It then examines how this is put into practice through a series of environment action programmes. Attention is also paid to the policy and strategy documents that frame EU efforts. Examining the problems encountered in integrating sustainable development into other Community policies, the chapter explores the influence of economic growth policies. The concluding section brings together the chapter's findings on the promotion of sustainable development in the EU. Readers can also look at Chapter 6, which discusses the EU's climate change policy, and Chapter 7 that deals with EU biodiversity policy.

Treaty mandate

EU initiatives in the field of environmental protection began after 1972, a time when member states were influenced by growing domestic and international concerns about the environment. The EU declared that economic expansion is not an end in itself, but should result not merely in improvements in the standard of living but also in the quality of life (Baker, 2000). However, it was not until the Single European Act of 1986 that the EU's role in environmental protection was formally recognised. By this time the Commission of the European Communities (the Commission), the main body that initiates new policy proposals in the EU, feared that any strengthening of environmental legislation by member states would act as a barrier to European free trade. As a result, the Commission, in particular the administrative division responsible for the environment, the Directorate-General for the Environment (known as DG Environment), was keen to ensure that environmental policy was more fully Europeanised.

Gradually, there was a shift of policy focus from general environmental protection measures to the promotion of sustainable development. This shift was reflected in modifications to the founding Treaty of Rome (1957). The Treaty of European Union (Maastricht Treaty) (1992) did not fully reflect sustainable development principles, but rather set the objective of promoting sustainable growth while respecting the environment. The Treaty of Amsterdam (1997) went further, calling for 'balanced and sustainable development of economic activities', and adopted environmental policy integration as a key means of achieving this. More importantly, the Treaty of Amsterdam made sustainable development one of the *objectives* of the Community, along with economic and social progress. This makes it applicable to the general activities of the EU, not just its activities in the sphere of the environment. As a result, sustainable development has the status of a guiding principle of the European integration process. The Treaty of Nice (2000) confirmed this. The Lisbon Treaty (2009) reinforced the Union's pledge to pursue this objective both within and beyond its borders. The Treaty of Lisbon states that one of the Union's objectives is to work for the sustainable development of Europe based, in particular, on a high level of protection of, and improvement in, the quality of the environment. Sustainable development is also affirmed as one of the fundamental objectives of the Union in its relations with the wider world, requiring it to contribute to the 'sustainable development of the Earth'. To this end, the Union is expected to work towards the adoption of international measures to preserve and improve the quality of the environment and the sustainable management of global natural resources. The Treaty also makes combating climate change at an international level a specific objective of EU environmental policy. This recognises that the EU has a leading role to play on the world stage in addressing climate change.

Because of these Treaty modifications, there is probably no single government or other association of states with as strong a 'constitutional' commitment to sustainable development as the EU. Sustainable development is now a *norm* of EU politics, both domestically and internationally (Baker and McCormick, 2004).

Environment action programmes

Environment Action Programmes (EAPs) have guided the development of EU environmental policy since the early 1970s. These are drawn up by DG Environment, providing the strategic focus needed to turn declaratory and legal commitments into policy actions. The Fifth EAP (1993–2000) explicitly engaged with the task of promoting sustainable development, although it was strongly influenced by the previous four EAPs.

The First EAP (1973–1976) acknowledged that economic growth was not an end in itself (CEC, 1973), while the Second EAP (1977–1981) referred to the limits to growth stemming from natural resource availability (CEC, 1977). The Second EAP also stressed that neither economic development nor the 'balanced' expansion of the Community could be achieved without environmental protection, and affirmed that 'economic growth should not be viewed solely in its quantitative aspects' (CEC 1977: 8). The Third EAP (1982–1986) forged links between environmental policy and the Community's industrial strategy, arguing that environmental protection measures could stimulate technological innovation (CEC 1983). This argument was to prove decisive, and since the Third EAP environmental protection has continued to be seen as contributing to the competitiveness of the EU's economy. This view displaced the earlier 'limits to growth' argument in favour of a belief in continued economic growth based on efficiency in resource use. The Fourth EAP (1987–1992) further developed this idea, drawing upon, while also helping to promote, the principle of ecological modernisation, in particular the twin focus on efficiency and technological innovation, as discussed in Chapter 2 (CEC, 1987).

Promoting sustainable development

The Fifth EAP, *Towards Sustainability* (1993–2000) made the first explicit policy pledge to the promotion of sustainable development in the EU (CEC, 1992). This was the result of two immediate pressures. The first was recognition of the increased burdens that European integration, particularly the completion of the Single European Market (SEM), posed for natural resources, the environment and the quality of life within the Union. The SEM increases trade between member states, putting more pressure on the environment, for example, from the transport sector. This heightened tensions between deepening economic integration and the EU's ever growing environmental agenda.

The second pressure stemmed from international engagement, particularly the Community's participation in UN-led policy developments. The Fifth EAP was drawn up in parallel with preparations for the Rio Earth Summit and, in the period up to 2001, actions taken under the Fifth EAP came to represent the EU's main response to the obligations it incurred at Rio. When viewed from this international perspective, it is worth noting that the EU was also driven by a sense of moral obligation, in particular acceptance of the principle of common but differentiated responsibilities. As the Commission was subsequently to argue:

> As Europeans and as part of some of the wealthiest societies in the
> world, we are very conscious of our role and responsibilities ...
> Along with other developed countries, we are major contributors to
> global environmental problems such as greenhouse gas emissions
> and we consume a major, and some would argue an unfair, share of
> the planet's renewable and non-renewable resources.
>
> (CEC, 2001a: 11)

Acknowledging the influence of Brundtland, the Fifth EAP defines sustainable development as continued economic and social development without detriment to the environment and natural resources, on the quality of which continued human activity and further development depend (CEC, 1992). The Programme prioritised seven areas for action, including climate change, biodiversity and the urban environment. It also targeted five key sectors for environmental policy integration, namely industry, energy, transport, agriculture and tourism (CEC, 1992). It also called for the use of a wide range of policy instruments, including fiscal and voluntary measures, to support implementation.

The Fifth EAP has been subject to extensive reviews, including by DG Environment and the European Environment Agency (EEA). However, evaluation did not prove an easy task. First, the lack of clarity in the term 'sustainable development' made it difficult to measure progress. This required the adoption and refinement of sustainable development indicators by the EEA so as to help measure progress. In addition, the Fifth EAP did not always set clear implementation targets; for example, it set no quantitative targets for the manufacturing sector. To add to these difficulties, the EAPs are non-binding policy documents and, as such, member states have no legal obligation to report on their implementation.

Nevertheless, there are strong arguments to support the claim that, while there was some decrease in the negative pressures exerted on the environment, progress under the Fifth EAP was very limited (EEA, 1995). There was no reversal in economic and social trends harmful to the environment, particularly in relation to the transport, energy and tourism sectors. Environmental policy integration remained weak (EEA, 1995). In global terms, the record showed similar, depressing trends. The EU is responsible for between 15 and 20 per cent of the world's consumption of resources and this figure remained unchanged during the period. Faced with these problems, the official review of the Fifth EAP, the *Global Assessment*, concluded that 'unless more fundamental changes are made, the prospects of promoting sustainable development remain poor' (CEC, 1999).

The Sixth EAP, *Our Future, Our Choice* (2001–2010), attempted to address some of the previous shortcomings by developing a more strategic and targeted approach (CEC, 2001a).

Detailed measures were set out in seven Thematic Strategies covering soil protection, the marine environment, pesticides, air pollution, the urban environment, natural resources, and waste. The Strategies were

Table 9.1 Priorities of the Sixth EAP

Climate change
Nature and biodiversity
Environment and health
Resource and waste management

used to identify proposals for further legislation. When the Commission was developing the Strategies it consulted with a broad range of stakeholders, including the European Environmental Bureau (EEB). The EEB is Europe's largest federation of environmental organisations, with over 140 member organisations. This participatory process went some way to addressing one of the shortcomings of the Fifth EAP, namely the lack of a sense of 'ownership' of the Programme by those outside DG Environment. It also helped increase the legitimacy of the Strategies among those involved in their implementation (Withana et al., 2010). The Sixth EAP also included a number of explicit international objectives relating to the EU's role in global environmental governance regimes and the integration of environmental considerations in external policies. The Programme also made connection between health and the environment and called for the full application of the precautionary principle, especially when it comes to the impact of poor environmental quality on the health of vulnerable groups, such as children and the elderly.

The Precautionary Principle states that where there are threats of serious or irreversible damage, lack of full scientific certainty shall not be used as a reason for postponing cost-effective measures to prevent environmental degradation.

(UNEP, 1972)

Principle 15 of the Rio Declaration on the Environment and Development commits signature states to the application of the Precautionary Principle. The Principle is widely used in environmental policy, and forms a key component of EU climate change policy, as discussed in Chapters 5 and 6.

The Sixth EAP helped steer environmental policy for over ten years, during which time a number of new environmental policies were adopted, such as in relation to the marine environment, soil, urban development and resource use, and new environmental legislation was introduced. However, there were several weaknesses in the Sixth EAP. There is, for example, a lack of policy coherence between external and internal policies and between different external EU policies. For example, we see a lack of coherence between the EU's internal policies on the promotion and use of biofuels and its external environmental objectives; and between the biofuel policy and the EU's external policies on aid, trade and development co-operation. EU biofuel policy has been subject to ongoing controversy as it has been shown to contribute to a rise in global hunger, deforestation and to land grab (Oxfam, 2013; United Nations, 2013), as discussed in Chapter 11. The Commission later admitted that there are problems of policy coherence between the different strands of EU policy (CEC, 2011a).

The failure to set concrete targets compromised the Sixth EAP, as did limited monitoring and reporting mechanisms. Thus, while the Sixth EAP was important in that it kept environmental issues on the EU policy agenda, implementation was weak. The Natural Resources Thematic Strategy, for example, did not put forward any concrete legal targets or timetables to decouple economic growth from resource use. Furthermore, it focused on 'achieving more sustainable use of natural resources' rather than suggesting a shift to sustainable consumption patterns (Withana et al., 2010). Similarly, the Forest Action does not contain any qualitative or quantitative targets with respect to forests, nor outline any concrete legislative measures to be taken.

Further problems with implementation also arose because reliance had to be placed on member states to voluntarily implement proposed measures. They were also left with the task of determining all the accompanying measures necessary to implement the Programme. As the Environment Commissioner at the time, Janez Potočnik, stated:

> We have obtained good results – but not always as good as we hoped for. Better implementation of EU rules by Member States is needed to close the gap between the Sixth EAP's legislative ambitions and its end-results.
>
> (CEC, 2011b)

This reveals the interdependencies that exist within the EU's system of multi-level governance, a system that involves input from both EU actors and those operating at the member state level (Baker, 2010). Here framework policies and objectives are set at the EU level, while roles and responsibilities for implementation are distributed through member states and down to sub-national, regional and local authorities. Often member states, in particular their sub-national levels of government, have limited capacity to act in pursuit of sustainable development.

Throughout the ten-year life of the Sixth EAP, pressures on the natural environment continued to grow. Transport's environmental burden remained unremitting and environmental pressures from unsustainable consumption and production grew (CEC, 2011a). Over-exploitation of the marine environment continued. In 2010 the EEA published *The European Environment: State and Outlook 2010* (SOER) (EEA, 2010b), a comprehensive analysis of the state and trends of the European environment. It showed that more needs to be done to implement environmental policy and to tackle today's more complex, interlinked systemic challenges in a global context. Furthermore, the SOER 2010 also noted that EU is not on track for addressing pressure on ecosystems (EEA, 2010b).

The SOER 2010 revealed several challenges confronting the EU in its commitment to the promotion of sustainable development. At one level, these can be seen as implementation challenges, that is, the need to confront the reluctance of member states to actually implement the legislation that has been drawn up at the EU level. In addition, efforts to implement the Sixth EAP often had to confront opposition from those that sought to promote economic growth, leading to delays in the adoption of specific implementation measures and in some cases resulted in a lowering of their level of ambition (Withana et al., 2010).

Prioritising growth

In 2013, a Seventh EAP, 'Living Well, within the Limits of Our Planet', was launched to guide environment policy up to 2020 (CEC, 2012a). Consideration had been given to not having a successor to the Sixth EAP, but that most likely would have been perceived as a negative signal, indicating a downward shift in the status of environmental policy, relative to its previous status and in relation to other EU policies. In addition, the EU has set out a set of key objectives and targets in environmental policy and legislation for the period 2010–2050 in support of a green economy. A Seventh EAP can lend coherence to these objectives and targets (EEA, 2013).

The Seventh EAP has to operate under circumstances that in many respects are very different to those functional under the Sixth EAP. The EU is now operating in a very different political and legal framework following its enlargement to twenty-eight members and the entry into force of the Lisbon Treaty, discussed below. Many new member states prioritise economic growth and they have limited track records in dealing with environmental issues (Baker, 2012). With enlargement, the danger is that new members will exert a downward pressure on EU sustainable development policy. This could make the implementation deficit that has long plagued the EU's commitment even more marked and more challenging to confront. Furthermore, across Europe, political priorities have also shifted for the most part towards the prioritisation of economic and social issues, rather than environmental concerns, a trend further exacerbated by the post-2008 financial and economic crises.

One of the striking features of the new Seventh EAP is the extent to which it is expected to support other EU policy priorities, particularly those relating to economic growth and energy efficiency. In particular, the Commission wants the Seventh EAP to contribute to the *Europe 2020 Strategy* of obtaining 'smart, sustainable and inclusive growth' for 2020 and beyond (CEC 2010). Here we see how the EU's commitment to sustainable development has mingled with other influences, such as the state of the European economy and underlying political currents in the member states. Europe 2020 builds on a prior and strategically very important strategy, the Lisbon Strategy. The Lisbon Strategy was introduced in 2000 to boost economic growth and employment through a set of structural reforms. It sought to make Europe 'the most competitive and dynamic knowledge-based economy in the world' (CEC, 2005a).

The Lisbon Strategy focused on growth and employment, implying that this would in turn bring about environmental and social progress (CEC, 2005a). It incorporated a limited number of so-called 'win–win' environmental issues, such as energy efficiency, renewables, and the development of environmental technologies, given their potential to increase competitiveness and create employment. When relaunched in March 2005, the Strategy narrowed its priorities to promoting growth and jobs and to increasing EU global competitiveness (CEC, 2005b). The Europe 2020 Strategy builds on these priorities.

Over this period, the EU has increasingly committed itself to developing a green economy, seen as an economic model that increases prosperity by using resources efficiently, while maintaining the resilience of the natural systems that sustain societies (EEA, 2013). Thus, for example, as part of the *Europe 2020 Strategy*, the EU has a flagship initiative, *A Resource Efficient Europe* (CEC, 2011c) that promotes sustainable growth through developing a resource efficient, low-carbon

---

Box 9.1

Europe 2020

Aims

Europe 2020 is the EU's growth strategy aimed at making the EU a smart, sustainable and inclusive economy.

Priorities

Europe 2020 is built on three priorities:

1 Smart growth: developing an economy based on knowledge and innovation;
2 Sustainable growth: prompting a more resource efficient, greener and more competitive economy;
3 Inclusive growth: fostering a high-employment economy delivering social and territorial cohesion.

These three mutually reinforcing priorities are to help the EU and the member states deliver high levels of employment, productivity and social cohesion.

(*Source*: CEC, 2010)

---

economy. An efficiency roadmap defines medium- and long-term objectives and means for achieving these (CEC, 2011d). However, as discussed in Chapter 2, the focus on efficiency is a feature of an ecological modernisation approach to environmental protection. As a tool for economic growth, its underlying rationale is at odds with the basic principles of sustainable development, particularly in its failure to tackle planetary limits (Baker, 2007). In the EU case, efficiency in resource use is seen to bring economic opportunities, improve productivity and drive down costs. It is also seen as a route to boost industrial competitiveness (CEC, 2011d). However, while the roadmap is designed to reduce environmental impacts, it does not cover all environmental issues, nor does it constitute an overarching environment policy strategy.

The Seventh EAP has been used by the Commission as a means to address the implications of the *Europe 2020 Strategy* for environmental policy, including through the development of legislation. To give work under the Seventh EAP a clear direction, the Commission sets out its vision of where it wants the Union to be by 2050:

> In 2050, we live well, within the planet's ecological limits. Our prosperity and healthy environment stem from an innovative, circular economy where nothing is wasted and where natural resources are managed in ways that enhance our society's resilience. Our low carbon growth has long been decoupled from resource use, setting the pace for a global sustainable economy.
>
> (Potočnik, 2012)

The Seventh EAP differs from previous action programmes in that it is designed to address the increasingly interlinked nature of environmental, economic and social challenges. It aims to enhance Europe's ecological resilience and transform the EU into an inclusive and sustainable green economy. It also highlights the need to take a leadership role in international environmental governance, in particular with respect to climate change. This reflects the fact that the Treaty of Lisbon added the support of international action for fighting climate change to the list of objectives defining EU environmental policy. This requires, in turn, that the increased growth in the demand for natural resources and the impact this has for the environment is addressed both internally within the EU and externally in the Union's international engagements.

The Seventh EAP identifies nine priority objectives up to 2020, including: protecting nature and strengthening ecological resilience; boosting sustainable, resource-efficient, low-carbon growth; and addressing environment-related threats to health. The Programme sets out a framework to support the achievement of these objectives through, *inter alia*, better implementation of EU environment law; enhancement in scientific knowledge; securing the necessary investments in support of environment and climate change policy; and improving the way in which environmental concerns are reflected in other policies. The programme also aims to boost efforts to help EU cities become more sustainable (CEC, 2012a). Currently, nearly 75 per cent of Europeans live in cities and urban areas, and by 2020 this is expected to rise to 80 per cent. Cities face several environmental challenges, including ensuring security of food supply while at the same time reducing their environmental impact; and trying to balance the need for green spaces for both healthy living and the maintenance of biodiversity amidst the ever growing demand for land required by city expansion.

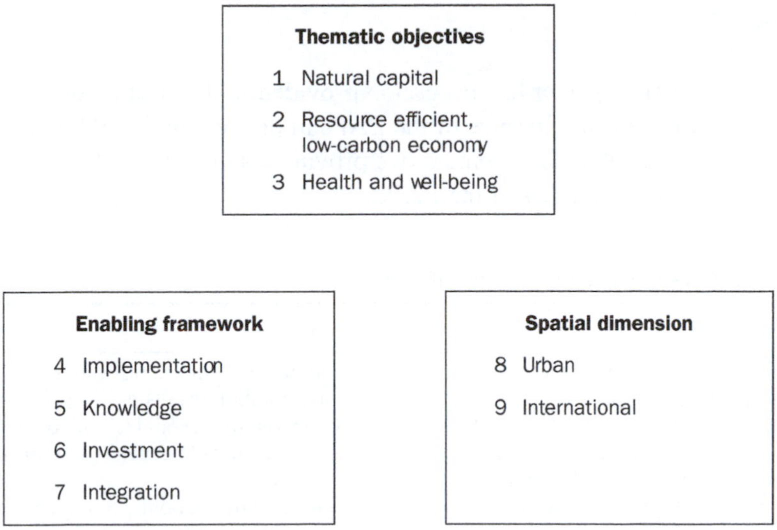

```
┌─────────────────────────────────┐
│      Thematic objectives        │
│                                 │
│   1  Natural capital            │
│                                 │
│   2  Resource efficient,        │
│      low-carbon economy         │
│                                 │
│   3  Health and well-being      │
└─────────────────────────────────┘
```

```
┌────────────────────────────┐   ┌────────────────────────────┐
│     Enabling framework     │   │      Spatial dimension     │
│                            │   │                            │
│   4  Implementation        │   │   8  Urban                 │
│                            │   │                            │
│   5  Knowledge             │   │   9  International         │
│                            │   │                            │
│   6  Investment            │   │                            │
│                            │   │                            │
│   7  Integration           │   │                            │
└────────────────────────────┘   └────────────────────────────┘
```

Figure 9.1 *Structure of the Seventh EAP.*

Source: CEC, 2012.

Although it is still early days, concerns have been already expressed by the EEB about weaknesses in the Seventh EAP, including the lack of specific targets and clear commitments to further binding measures. Such criticisms already plagued the Sixth EAP. However, the EEB was encouraged by the fact that the Seventh EAP has begun to target unsustainable consumption and production. As Jeremy Wates, EEB secretary-general, said:

> The costs of maintaining Europe's excessive consumption through our use and abuse of nature's resources have become prohibitively high – both for the economy and for the environment. The Seventh EAP can and must provide a way forward out of the economic crisis which is environmentally sustainable. The Commission's proposal takes us part of the way there but leaves many elements to be decided later.

> (EEB, 2012a)

However, the requirement that the Seventh EAP serve as a tool for the realisation of the objectives of the *Europe 2002 Strategy* of promoting traditional economic growth may severely limit the capacity of the Programme to contribute to the realisation of the EU legal and declaratory commitment to the promotion of sustainable development.

Summary

To tie together the discussion provided in this section on the history, role and significance of the EAPs in promoting the EU's commitment to sustainable development, we provide a summary of the main strengths and weaknesses of the EAPS.

Table 9.2 *Strengths and weaknesses of EAPS*

Strengths	*Weaknesses*
Responds to treaty mandate to promote sustainable development	Inappropriate tool in fast-moving policy area
Represents a formal political commitment	Long duration results in low political priority
Forces debate on priorities of policy	Hampered by overall lack of coherence within EU sustainable development of other policy areas
Keeps environment on EU agenda	
Maintains focus on wider agenda beyond climate change	Restricted by economic (competitiveness, growth) agenda
Helps develop strategic, planned approach to sustainable development	Scope often too broad and objectives too ambitious
Supports adoption of sustainable development measures, despite threat from competitiveness concerns	Lack of concrete policy proposals
	Failure to include measurable, achievable and time-bound targets
Enables DG Environment to prioritise and focus its efforts	Lack of additional financial resources to support implementation
Provides benchmark to evaluate policy and stimulates ongoing improvements	Poor reporting and monitoring
Participatory process helps to broaden the debate and ensure buy-in among different stakeholders	Lack of appropriate legal support, and limited by reliance on voluntary implementation by member states
	Participatory process can lead to high expectations as to what can be delivered

Putting sustainable development into practice: using laws and soft policy tools

As mentioned above, many of the EAPs developed by the Commission have led to a series of related items of legislation. One of the key ways in which this is done is through the use of EU directives.

> EU directives lay down certain end results that must be achieved in every member state. Each directive specifies the date by which the national laws must be adapted. National authorities have to adapt their laws to meet these goals, but are free to decide how to do so.
>
> (*Source*: http://ec.europa.eu/eu_law/directives/directives_en.htm)

The EU has an array of environmental directives dealing with such issues as air quality, environmental liability, emissions trading, biodiversity, public involvement in environmental decision making, production and product standards, waste, and water. Many of these directives are highly complex and technical in nature, such as the Ambient Air Quality Directive (96/62), the Water Framework Directive (2000/60) or the IPPC Directive (1996/61). Directives play an important part in the promotion of sustainable development not least through regulating pollution, safeguarding biodiversity and dealing with the negative externalities arising from consumption and production. However, while regulations are clearly an important policy tool, as discussed in Chapter 3, many directives are not properly understood, badly transposed into national law and are hampered by national administrations ill-equipped to deliver their objectives (EEB, 2005). Some of these problems can be seen in the EU Water Framework Directive, regarded as a key EU environmental directive yet one of the most complex and difficult to implement.

Case study 9.1

EU Water Framework Directive

Background

The EU has a long history of regulating water quality, with the first wave of water legislation in 1975 setting standards for rivers and lakes used for drinking water abstraction. In 1980, new binding quality targets for drinking water were set. Legislation was also introduced on quality objectives for fish waters, shellfish waters, bathing waters and groundwater.

A second phase of water legislation began in the 1990s, with the Urban Waste Water Treatment Directive (1991), the Nitrates Directive (1991) addressing water pollution by nitrates from agriculture, a new Drinking Water Directive (1998) and a Directive for Integrated Pollution and Prevention Control (IPPC) (1996) addressing pollution from large industrial installations.

WFD

Pressure for a more coherent approach to water policy and management led to the 2000 EU Water Framework Directive (WFD). The main aims of the WFD are to:

(i) Expand the scope of water protection to all waters, surface waters and groundwater;

(ii) Achieve 'good status' for all waters by a set deadline;

(iii) Base water management plans on river basins;

(iv) Adopt a 'combined approach', using both emission limit values and quality standards.

RBMP

The WFD introduced a management system based on river basins. This involves using the natural geographical and hydrological unit instead of administrative or political boundaries as a base for policy. For each river basin district, some of which will traverse national frontiers, a River Basin Management Plan (RBMP) is required, to be updated every six years.

The Directive sets out a number of objectives, including protection of the aquatic ecology, of unique and valuable habitats, of drinking water resources, and of bathing water. All these objectives must be integrated for each river basin into the RBMP.

The WFD deals with pollution at source, through the application of existing technology. On the source side, it also sets out a framework for developing further technological controls. This is then combined with the objectives of good quality status for all waters. The RBMP has to provide a detailed account of how the objectives set for the river basin, including ecological status, quantitative status, chemical status and protected area objectives, are to be reached within the timescale required. It also requires that an economic analysis of water use within the river basin be carried out. This is to enable discussion on the cost-effectiveness of the various possible measures.

Significance

One advantage of the WFD is that it has rationalised the Community's water legislation by replacing seven of the early directives and integrating them into a coherent whole.

The WFD was a major landmark which established new requirements for integrated river basin planning in order to achieve ecological objectives. It has helped tackle the negative impacts on aquatic ecosystems from physical changes, in particular dams and weirs.

However, the WFD is plagued by implementation problems and many member states have yet to adopt RBMPs. The RBMP requires substantial policy integration, across a range of sectors, including agriculture. Furthermore, nutrient pollution control has not been fully introduced. In addition, it is also common for Plans to require considerable cross-border co-operation, as many of Europe's major rivers are trans-boundary in nature, such as the Danube, Rhine, Elbe, Tisza, and Meuse.

In response to concerns about weak implementation, the Commission has produced a Water Blueprint, which outlines actions to improve implementation. The Blueprint synthesises policy recommendations and introduces new water-related green infrastructure measures such as reforestation, flood plain restoration, soil management, and sustainable urban drainage systems. The Blueprint also explains how to internalise costs for water use and water pollution, so as to apply economic instruments. The Commission also plans to develop a methodological framework for the application of payments for ecosystem services. In addition, the Blueprint sets out ways to integrate water management issues into the Common Agricultural and Cohesion policies.

(*Sources*: CEC 2012b; 2012c; EEB, 2012b)

The Water Blueprint is closely related to the EU's 2020 Strategy and the 2011 Resource Efficiency Roadmap. It highlights that preserving water is not only about environmental protection, health and well-being, but also about economic growth and prosperity. This includes ensuring that all economic sectors have the water they need for creating growth and jobs.

Environmental policy integration

As can be seen from the discussion to date, one of the key challenges facing the EU is to integrate environmental considerations into its other policies, particularly at the sectoral levels.

> At the simplest level, environmental policy integration means the integration of environmental considerations in the design and implementation of policy.
>
> (Lafferty and Hovden, 2003)

In the 1960s and 1970s, environmental policy was conceived and implemented as a 'standalone' policy area, largely independent of policies in other sectors. While this specialisation helped in the development of more targeted policy responses to particular problems (Kemp et al., 2005: 19), it also led to the neglect of broader considerations. Often solutions to one set of problems displaced the problem either to other sectors, to future generations or to other countries. However, by the 1990s the limitations of specialisation and differentiation were increasingly evident, resulting in new awareness of the need to take environmental considerations into account in a wide range of sectoral policies. Growing commitment to the promotion of sustainable development also required an integrated approach. The term 'environmental policy integration' (EPI) is used in the literature to capture the need to enhance coherence between sectoral, economic and environmental policies and between them and sustainable development policies. Lafferty and Hovden (2003) provide a comprehensive understanding of environmental policy integration:

> the incorporation of environmental objectives into all stages of policy-making in non-environmental policy sectors, with a specific recognition of this goal as a guiding principles for the planning and execution of policy; – accompanied by an attempt to aggregate presumed environmental consequences into an overall evaluation of policy, and a commitment to minimise contradictions between environmental and

sectoral policies by giving principled priority to the former over the latter.

Lafferty and Hovden argue that *principled prioritisation* of environmental policy objectives forms the core of EPI. To argue that sustainable development merely implies that the essential needs of the world's poor and of future generations (the two key target groups of Brundtland) should merely be 'balanced' with a myriad of other societal goals misses, in their view, the fundamental premise of the Brundtland Report and its follow up within the UN. As understood by Lafferty and Hovden, because it requires a principled and consequential integration of environmental considerations into all sectoral activity, EPI involves a significant break with the traditional, capitalist, economic development model. At its most radical, EPI involves prioritising, over and above sectoral interests, the sustainable development implications of specific policies, programmes and activities (Lafferty and Hovden, 2003).

EU approach to EPI: the Sustainable Development Strategy

A key means used by the EU to promote EPI has been the adoption of *A Sustainable Europe for a Better World: A Strategy for Sustainable Development* in 2001 (CEC, 2001b). EPI has a central role in this Strategy, which states that 'sustainable development should become the central objective of all sectors and policies' (CEC, 2001b: 6). The Sustainable Development Strategy was introduced in preparation for the WSSD, Johannesburg, 2002. The Strategy was renewed in June 2006 following enlargement of the EU (CEC, 2005a). Because it was adopted in Gothenburg, it is sometimes called the Gothenburg Strategy.

The Strategy aims to dovetail policies for sustainable development over the long term. Specific quantitative targets and measures were also laid down, for example with respect to GHG emissions and energy efficiency. These targets and objectives are, to a large extent, based on EU and member state commitments at the time, for example under the Kyoto Protocol, the Johannesburg Summit and the MDGs; or reflect policy developments at the time, such as on biofuels (Withana et al., 2010). The external dimension is also addressed in the renewed Strategy. The revised Strategy seeks to strengthen the fight against global poverty, improve monitoring of global sustainable development and ensure greater compliance with international commitments. To achieve this, the EU has to improve the coherence of its own development aid policies.

Box 9.2

Objectives of EU Sustainable Development Strategy

Aim	Objective	Indicative actions
Limit **climate change**	Meet commitments under Kyoto Protocol and European Strategy on Climate Change.	Enhance energy efficiency and make more use of renewable energy.
Reduce adverse effects of **transport**	Break links between economic growth and transport growth; more environmentally friendly transport, that is also conducive to health.	Develop alternatives to road transport and vehicles that produce less pollution and use less energy.
Promote more sustainable modes of **production and consumption**	Break links between economic growth and environmental degradation; pay attention to how much ecosystems can tolerate.	Use green public procurement; set environmental and social performance targets for products; expand distribution of environmental innovations and technologies; provide information about and appropriate labelling of products and services.
Ensure sustainable management of **natural resources**	Avoid overexploitation; improve efficiency of resource use; recognise value of ecosystem services; halt loss of biodiversity.	Complete *Natura 2000* network; define and implement actions to protect biodiversity; support recycling and reuse.
Limit major threats to **public health**	Ensure food safety and quality; remove food threats to health and environment posed by chemicals by 2020; research links between health and pollutants.	Tackle issues relating to epidemics, resistance to antibiotics, and lifestyle; prepare for possible pandemic and combat HIV/AIDS.
Combat **social exclusion and poverty**	Ensure active ageing and viable pension and social protection systems; bring better situation for families, especially children; achieve equality between men and women.	Integrate legal migrants and develop a community immigration policy.
Strengthen fight against **global poverty**	Better monitoring of global sustainable development and compliance with international commitments.	Increase amount of aid to less favoured countries; improve international governance; improve development aid policies.
Make more use of **financial and economic instruments**	Build a market that offers less polluting products and services; change consumer behaviour.	Make prices reflect actual environmental and social costs; apply fiscal measures to energy and resource consumption and/or pollution; co-ordinate financial support from European funds to optimise efficiency.
Improve **communication**	Enhance dialogue between businesses and consumers.	Different parties to establish partnerships among themselves.

We can see from this that the Sustainable Development Strategy is both ambitious in its objectives and also wide in its scope. However, it lacks clear guidelines as to how to integrate environmental concerns in sector policies (Withana et al., 2010). In addition, some of the integration objectives include vaguely formulated ambitions, for example to decouple economic growth from environmental degradation (CEC, 2007; 2009b). Decoupling remains one of the most important yet challenging issues for the Union. To address decoupling, revisions need to be made to the traditional hierarchy of policy objectives, which tends to give precedence to economic growth over environmental considerations. Critics have also suggested that there is a tendency to reduce sustainable development to its economic and environmental dimensions, while disregarding social aspects (Withana et al., 2010).

In addition, attention needs to be paid to implementation (Persson, 2004). To date, the EU has had very limited success with EPI. EU transport policy in Eastern Europe provides a good example, which has prioritised road-building (CEE Bankwatch and FoEE, 2007). The EU-funded Via Baltica transport corridor, for example, has generated considerable controversy, and came before the European Court of Justice because the Polish section ran through the Rospuda valley, a protected EU *Natura 2000* site. Similarly, the destruction of the Drava wetlands, due to motorway construction along the Budapest (Hungary)–Sarajevo (Bosnia and Herzegovina)–Ploče (Croatia) corridor, provides another example of sectoral conflict (Baker, 2012).

Furthermore, as climate change and energy issues move to centre stage on the EU's policy agenda, another concern arises. At first sight, it may seem paradoxical to suggest that actions to address climate change have shifted the EU away from the promotion of sustainable development. It could be argued that climate change is a key result of unsustainable models of development and that addressing climate change lies at the heart of promoting more sustainable futures. However, discussions on growth and competiveness now routinely include considerations of how to move to a low-carbon economy and create green jobs, emphasising the 'win–win' potential of addressing climate change. As such, addressing climate change is explicitly linked with the achievement of traditional, developmentalist objectives, of growth, economic competitiveness and employment. If sustainable development was truly regarded as the overarching objective of the EU, this would imply a principled and consequential application of this objective, across all EU policies. From such a perspective, it would be more logical to clarify the overarching economic, social and environmental objectives of the EU's sustainable development agenda; and only then set short-term operational objectives, such as in relation to energy efficiency, competitiveness and employment. However, exactly the opposite has happened (Withana et al., 2010), and the commitment to sustainable development is framed within the need for traditional economic goals of promoting economic growth and European competiveness.

Distinctive EU understanding of sustainable development

Taken together, the seven EAP and the key strategy documents represent the policy framework of the EU, within which it puts its legal obligations (declaratory intent) to promote sustainable development into practice. They frame the context within which actions, secondary legislation, specific programmes and funding are structured. Their analysis can reveal a great deal about the EU's understanding of sustainable development. At a general level, it reveals that throughout the First EAP, the imperative of economic growth continued to take precedence over environmental protection. By the Fifth and Sixth EAPs, the Union had evolved an environmental policy that intertwined the twin imperatives of economic development and environmental protection in a new way: both came to be seen as compatible, mutually reinforcing aims of policy. However, by the time the Sixth EAP was completed, the pursuit of traditional economic growth objectives, as evidenced by both the Lisbon and Europe 2020 Strategies, had come to dominate. The result is that sustainable development is now seen to be dependent upon the prior achievement of economic growth, rather than acting as a fundamental driver of policy choices. Through this historical progression, a distinctive EU understanding of sustainable development has emerged. These are summarised below.

The characteristics of the EU's understanding of sustainable development bear several resemblances to the understanding formulated by Brundtland. These include a concern for the global dimension, acknowledgement of the differentiated responsibilities of the North and the recognition of the social dimensions of sustainable development. However, the European approach prioritises European concerns, neglecting to embed these in the wider global context. First, the traditional emphasis of economic growth is maintained. Less attention is given to issues of development, in particular what the Brundtland call for necessary social

---

Box 9.3

EU understanding of sustainable development

1 Promoting sustainable development is strongly linked with the stimulation of economic growth.
2 Sustainable development within Europe arises from the achievement of efficiencies within the 'green economy'.
3 Sectoral policy integration is a key component of sustainable development.
4 Sustainable development has social dimensions and is linked to issues of health and food safety.
5 The promotion of sustainable development is a shared responsibility, including for consumers.
6 Promoting sustainable development is linked with the resolution of global environmental problems.
7 At a global level, promoting sustainable development is a moral obligation for Europeans.

---

and economic improvement in the Third World might mean for European growth aspirations. Second, discussion on the social aspects of sustainable development increasingly stresses issues of concern to Western consumers, such as food safety and health. Absent from this discussion are issues of inter-and intra-generational equity, particularly in relation to the need to modify the high consumption and resource use patterns of Western consumers. The idea that the promotion of sustainable development can rely upon partnership is premised upon the existence of a developed civil society and of responsible entrepreneurship, or at least an economic sector that has come to accept the benefits of ecological modernisation. This idea has not taken hold at the global level and, as discussed in Chapter 4, may not be capable of having a global remit.

Conclusion

The EU has made strong legal and declaratory commitments to sustainable development, and it now serves as an objective of the European integration process. In addition, a comprehensive range of legislation, strategy documents and action programmes frames that commitment. New issues continue to arise on the EU's policy agenda and over time these have stretched concerns beyond the EU border, particularly in the face of global environmental change, especially climate change. Recent years have seen the EU play an ever greater role in ensuring more effective international environmental governance in the face of global environmental change.

However, when analysis moves from exploration of the Union's constitutional and declaratory commitments to its implementation efforts, especially at the sectoral level, a different and altogether more pessimistic picture emerges. This is all the more so given the current prioritisation of economic growth over and above considerations of sustainable development. The focus on efficiency and competitiveness, including in the energy sector, can only serve to heighten this negative profile.

These negative reviews, however, do not mean that the EU's commitment to the promotion of sustainable development has been a policy failure. Rather, its commitment is important for several reasons. First, the EU has set out an ambitious vision for sustainable development for Europe. Its acceptance by the EU is of deep, symbolic importance. In addition, this vision is backed by legal, treaty-level obligations. Second, the term sustainable development now permeates the official discourse of the Union. In its official programmes, legal commitments and public discourse, the EU has moved from an earlier phase characterised by disregard for the environmental consequences of policies, to a new realisation that the environment matters. Third, medium-term policy frameworks and strategies have been put in place. These show clear evidence of policy learning, as seen by the willingness of DG Environment in particular to expose its policies to ongoing evaluation and to search for new ways to improve practice and outcomes. Fourth, even if policies fall far short of promoting sustainable development (and they most certainly will), they provide important environmental and development criteria against which the EU integration process can be appraised. A failure to realise the visions embedded in declaratory intent can be expected, not least because the promotion of sustainable development is a long-term process of social, cultural, political and economic change. What is important is that the EU has launched Europe on a *path* towards sustainable development, even if it is faltering along the way. Finally, the EU's commitment is important for shaping the negotiating position and behaviour of the EU at the international level. Beyond the borders of the EU, the commitment shapes the EU's identity by marking it out as *different* from other actors.

However, in order to nurture what is positive about this difference, the focus on achieving growth through efficiencies needs to be critically reviewed. The promoting of sustainable development on a planet characterised by finite resources and by planetary limits has to accept limits to growth, over and above any advantages that may accrue from efficiency gains.

Summary points

- Declaratory commitment to the promotion of sustainable development is high in the EU. This is important, as we should never underestimate the power of ideas in politics.
- For the EU, the promotion of sustainable development involves decoupling economic growth from environmental destruction. It is primarily a technical, managerial task of decoupling through eco-efficiency. This understanding does not challenge the Western economic development model either by limiting growth or changing existing patterns of high consumption.
- International developments within the UN strongly influenced the EU's commitment to sustainable development. However, the EU's understanding of sustainable development differs in fundamental respects from that presented by the Brundtland Report.
- Efforts made to date by the EU are not sufficient to promote sustainable development.

Note

1 For the sake of simplicity, the term EU is used throughout this chapter, although the Union has undergone several name changes from its original inception in the 1950s as the European Coal and Steel Community and then the European Economic Community.

Further reading

Critical commentary on EU sustainable development policy

Baker, S. (2007), 'Sustainable Development as Symbolic Commitment: Declaratory Politics and the Seductive Appeal of Ecological Modernisation in the European Union', *Environmental Politics*, 16 (2): 297–317.
Jordan, A. and Adelle, C. (eds) (2013), *Environmental Policy in the EU: Actors, Institutions and Processes* (London: Earthscan, 3rd edn).

Environmental policy integration

Jordan, A. J. and Lenschow, A. (2010), 'Environmental Policy Integration: A State of the Art Review', *Environmental Policy and Governance*, 20 (3): 147–158.

Lafferty, W. M. and Hovden, E., 2003, 'Environmental Policy Integration: Towards and Analytical Framework', *Environmental Politics*, Vol. 12, no. 3, 1-22.

Web resources

Official EU Web: http://europa.eu

References

Baker, S. (2000), 'The European Union: Integration, Competition, Growth – and Sustainability', in W. M. Lafferty and J. Meadowcroft (eds), *Implementing Sustainable Development: Strategies and Initiatives in High Consumption Societies* (Oxford: Oxford University Press), 303–336.

Baker, S. (2007), 'Sustainable Development as Symbolic Commitment: Declaratory Politics and the Seductive Appeal of Ecological Modernisation in the European Union', *Environmental Politics*, 16 (2): 297–317.

Baker, S. (2010), 'The Governance Dimensions of Sustainable Development', in P. Glasbergen and I. Niestroy (eds), *Sustainable Development and Environmental Policy in the European Union: A Governance Perspective* (The Hague: Open University Press), 21–59.

Baker, S. (2012), 'Environmental Governance: EU Influence beyond Its Borders', in I. Gladman (ed.), *Central and South-Eastern Europe 2013* (London: Europa Publications, 13th edn), www.routledge.com/books/details/9781857436501/

Baker, S. and McCormick, J. (2004), 'Sustainable Development: Comparative Understandings and Responses', in N. J. Vig and M. G. Faure (eds), *Green Giants: Environmental Policies of the United States and the European Union* (Cambridge, MA: MIT Press), 277–302.

CEE Bankwatch and FoEE (2007), *EU Cash in Climate Clash: Comparative Analysis of the 2007–2013 Structural Funding Allocations for Energy and Transport in the New Member States* (Brussels: CEE Bankwatch and FoEE), available online at www.bankwatch.org/documents/EU_cash_climate_clash_bw.pdf

CEC (Commission of the European Communities) (1973), 'First Programme of Action on the Environment', *Official Journal*, C112, 20 December.

CEC (1977), 'Second Environmental Action Programme, 1977–1981', *Official Journal of the European Communities*, C139, 13 June.

CEC (1983), 'Third Environmental Action Programme', *Official Journal of the European Communities*, C46, 17 February.

CEC (1987), 'Fourth Environment Action Programme', *Official Journal of the European Communities* C328, 7 December.

CEC (1992), *Towards Sustainability: A European Community Programme of Policy and Action in Relation to the Environment (1992–2000)* (Brussels: Commission of the European Communities, COM(92) 23 final).

CEC (1999), *Global Assessment – Europe's Environment: What Directions for the Future?* (Brussels: Commission of the European Communities, COM(1999) 543).

CEC (2001a), *Environment 2010: Our Future, Our Choice – the Sixth Environment Action Programme 2001–2010* (Brussels: Commission of the European Communities, COM(2001) 31 final).

CEC (2001b), *A Sustainable Europe for a Better World: A European Union Strategy for Sustainable Development* (Brussels: Commission of the European Communities, COM(2001) 264 final).

CEC (2005), *Review of the Sustainable Development Strategy – A Platform for Action* (Brussels: Commission of the European Communities, COM(2005) 658 final).

CEC (2005a), *Common Actions for Growth and Employment: The Community Lisbon Programme* (Brussels: Commission of the European Communities, COM(2005) 330 final).

CEC (2005b), *Working Together for Growth and Jobs. A New Start for the Lisbon Strategy* (Brussels: Commission of the European Communities, COM(2005) 24 final).

CEC (2007), *Progress Report on the Sustainable Development Strategy 2007* (Brussels: Commission of the European Communities, COM(2007) 642 final).

CEC (2009a), 'Sustainable Consumption and Production: A Challenge for Us All', Fact sheet (Brussels: Commission of the European Communities Environment), available online at http://ec.europa.eu/environment/eussd/pdf/brochure.pdf

CEC (2009b), *Mainstreaming Sustainable Development into EU Policies: 2009 Review of the European Union Strategy for Sustainable Development* (Brussels: Commission of the European Communities, COM(2009) 400 final).

CEC (2010), 'EUROPE 2020: A Strategy for Smart, Sustainable and Inclusive Growth' (Brussels: Commission of the European Communities, COM(2010) 2020 final).

CEC (2011a), 'The Sixth Community Environment Action Programme, Final Assessment' (Brussels: Commission of the European Communities, COM(2011) 0531 final).

CEC (2011b), 'Final Assessment of the Sixth Environment Action Programme Shows Progress in Environment Policy – But with Shortfalls in Implementation',

Press release (Brussels: Commission of the European Communities, available online at http://europa.eu/rapid/press-release_IP-11-996_en.htm

CEC (2011c), 'A Resource Efficient Europe' (Brussels: Commission of the European Communities, COM(2011)).

CEC (2011d), *Final Roadmap to a Resource Efficient Europe* (Brussels: Commission of the European Communities, COM(2011) 571), available online at http://ec.europa.eu/resource-efficient-europe/

CEC (2012a), 'Proposal for a Decision of the European Parliament and of the Council on a General Union Environment Action Programme to 2020, "Living Well, within the Limits of Our Planet"' (Brussels: Commission of the European Communities, COM(2012) 710 final).

CEC (2012b), 'A Blueprint to Safeguard Europe's Water Resources' (Brussels: Commission of the European Communities, COM(2012) 0673 final).

CEC (2012c), *A Commission Report to the European Parliament and the Council on the Implementation of the Water Framework Directive: River Basin Management Plans* (Brussels: Commission of the European Communities, COM(2012) 670).

CSCP (Centre on Sustainable Consumption and Production) (2012), Emerging Visions for Future Sustainable Lifestyles 2050', European Policy brief (Wuppertal, Germany: CSCP), available online at ec.europa.eu/research/.../policy-briefs-spread-november-2012_en.pdf

EEA (European Environment Agency) (1995), *Environment in the European Union – 1995 – Report for the Review of the Fifth Environmental Action Programme* (Copenhagen: EEA), available online at www.eea.europa.eu/publications/92-827-5263-1

EEA (2010a), *Ecological Footprint of European Countries (SEBI 023) – Assessment* (Copenhagen: EEA, May), available online at www.eea.europa.eu/data-and-maps/indicators/ecological-footprint-of-european-countries/ecological-footprint-of-european-countries

EEA (2010b), *The European Environment: State and Outlook 2010* (Copenhagen: EEA), available online at www.eea.europa.eu/soer

EEA (2013), *Towards a Green Economy in Europe: EU Environmental Policy Targets and Objectives 2010–2050* (Copenhagen: EEA Report no. 8/2013), available online at www.eea.europa.eu/publications/towards-a-green-economy-in-europe

EEB (European Environmental Bureau) (2005), *EU Environmental Policy Handbook: A Critical Analysis of EU Environmental Legislation* (Brussels: EEB).

EEB (2012a), 'EEB Welcomes Respect for Planetary Limits in Seventh EAP Proposal But Misses Concrete Targets', 29 November, available online at www.eeb.org/index.cfm/news-events/news/eeb-welcomes-respect-for-planetary-limits-in-7th-eap-proposal-but-misses-concrete-targets/

EEB (2012b), 'Ten Years of the Water Framework Directive: A Toothless Tiger? A Snapshot Assessment of the EU Environmental Ambitions'

(Brussels: EEB), available online at www.eeb.org/index.cfm/activities/biodi-versity-nature/water/water-what-is-the-eeb-doing/

Kemp, R., Parto, S. and Gibson, R. B. (2005), 'Governance for Sustainable Development: Moving from Theory to Practice', *International Journal of Sustainable Development*, 8: 12–30.

Lafferty, W. M. and Hovden, E. (2003), 'Environmental Policy Integration: Towards an Analytical Framework', *Environmental Politics*, 12 (3): 1–22.

Oxfam (2013), 'What Are Biofuels? What's the Problem with Them?' available online at www.oxfam.org/en/grow/campaigns/what-are-biofuels

Persson, A. (2004), 'Environmental Policy Integration: An Introduction', Background paper, PINTS – Policy Integration for Sustainability (Stockholm: Stockholm Environment Institute), available online at www.sei-international.org/mediamanager/.../Policy.../pints_intro.pdf

Potočnik, J. (2012), 'Living Well, Within the Limits of the Planet', presented at High Level IEEP [Institute for European Environment Policy] Conference on the Future of EU Environmental Policy, 4 December (Brussels: Commission of the European Communities, European Commissioner for Environment, speech, 12/900), available online at europa.eu/rapid/press-release_speech-12-900_en.doc

UNEP (United Nations Environment Programme) (1972), 'Rio Declaration on the Environment and Development', in *Report of the United Nations Conference on the Human Environment, Stockholm, 5–16 June 1972*, available online at www.unep.org/documents

United Nations (2013), 'Note on the Impact of the EU Biofuel Policy: On the Right to Food' (New York: United Nations Human Rights, Office of the High Commissioner, 23 April), available online at www.srfood.org/fr/special-rapporteur-urges-phase-out-of-eu-biofuel-incentives

Withana, S., Farmer, A., Pallemaerts, M., Hjerp, P., Watkins, E., Armstrong, J., Medarova-Bergstrom, K. and Gantioler, K. (2010), *Strategic Orientations of EU Environmental Policy under the Sixth Environment Action Programme and Implications for the Future*, Final report (London: Institute for European Environmental Policy), available online at www.ieep.eu/assets/556/Strategic_Orientations_of_6EAP_-_Revised_report_-_May_2010.pdf

Changing times in Eastern Europe

Key issues

- Transition
- Environmental legacies of communist rule
- New environmental pressures
- EU, conditionality and Eastern enlargement
- Environmental justice
- Soft security.

Introduction

This chapter explores the prospects for, and the barriers to, the promotion of sustainable development in the countries in transition in Central and Eastern Europe (CEE), including the Baltic States and the Balkans. It focuses on the tensions between the development demands of transition and the promotion of sustainable development. It also looks at the EU enlargement process, which created an ideal opportunity for the EU to influence how transition states manage their environment, putting its own commitment to sustainable development into practice. The Eastern enlargement of the EU has created a large geopolitical block, a powerful trading body and the world's largest internal consumer market. Therefore, what happens to the pursuit of sustainable development as this region transitions from communist rule to new forms of economic and political development is of global significance.

The chapter begins by exploring the environmental legacies of communist rule, and then examines the influence of marketisation and democratisation. This is followed by an exploration of sustainable development through the lens of EU enlargement. Throughout the chapter, case studies are given and country profiles presented.

Understanding transition

The communist regimes in Eastern Europe collapsed in 1989 and since then countries in the regions have undergone a complex process of transition. This involves political democratisation and the introduction of market economies. Transition is far from being a simple process (Smith and Pickles, 1998: 15–16). While countries adopt new institutions of governance, they *adapt* them to suit their particular country context, as seen for example in the variation in the types of environmental protection agencies established across the region. Second, transition is not taking place in a vacuum, but rather involves complex reworking of old social and economic relations, as countries construct new forms of capitalism on, and with, the ruins of their old communist system. Privatisation policy provides a good example, where countries sold off their state-owned enterprises, often resulting in complex forms of mixed public–private ownership unique to the region (Stark, 1997).

> Transition in Eastern Europe involves democratisation, marketisation and the adoption of new institutions of governance and their adaptation to suit the particular country in which they are adopted. Transition takes place in the context of a complex reworking of old social and economic relations.

Transition can be expected to be varied, given the cultural, religious and ethnic diversity across countries in the region. They also had different experiences under communism. For example, levels of industrialisation and the degree of economic centralisation varied, with Hungary and Bulgaria providing contrasting examples. The use of the single term 'Eastern Europe' during the period of communist rule masked this political and economic diversity. Furthermore, countries experienced different political reorganisations after 1989, especially with respect to the extent of reforms, degree of public support, and the stability of subsequent regimes. The post-communist situation in the Balkans is less stable than in the *Visegrad* countries, that is, Poland, Hungary, the Czech Republic and Slovakia. The Balkans region has experienced high turnover rates of governments, low public acceptance of change and poor commitment to reform, as reflected in the continuous strength of the communist successor parties. The wars that followed the break-up of the former Yugoslavia also contributed to instability in the Balkans, as did political unrest further afield, especially in Ukraine. Since 1989, these factors have contributed to the emergence of a highly differentiated transition process between the countries of the Balkans region and those elsewhere in Eastern Europe (Baker, 2012). The legacies of the old regime shaped the capacity of the countries to respond to the challenges of the post-1989 period and are thus key factors influencing the promotion of sustainable development in the region.

Promoting sustainable development in transition societies

Since 1989, various efforts have been made to address the environmental legacies of communist rule, while countries have also put in place new political, economic and administrative structures as part of their transition to democratic, market economies.

Addressing environmental pollution

While retaining large tracts of unspoilt land, often possessing rich biodiversity, most countries have inherited a heavy legacy of pollution from the communist era (EEA, 1999). The contrast between heavily polluted environmental 'hot spots' and ecologically rich areas has been attributed to the communist system of central planning. This concentrated industrial production, particularly heavy industry, in a small number of regions, close to cheap, and often dirty, sources of energy. Much of the area between these agglomerations remained relatively untouched (von Homeyer, 2001).

One of the hallmarks of the old Soviet period was its much publicised environmental mistakes and disasters. Environmental problems include poor air and water quality, inadequate treatment and disposal of industrial waste (including hazardous and nuclear waste), soil deterioration and contamination of land (including agricultural land and land used for Soviet military installations). Most countries also lacked a comprehensive waste management strategy and effective legislation. The most notorious area for air pollution was the so-called 'black triangle', covering the Czech and Slovak republics, Poland and the former German Democratic Republic (East Germany). Other areas of especially high concentrations of pollution include the Black Sea and the Danube river basin (Carter and Turnock, 2002; EEA, 1999).

Other regions of the world, including Western Europe, also experienced similar pollution problems. However, from the 1970s onwards, Western countries began to adopt pollution amelioration policies and many firms underwent a process of ecological modernisation, as discussed in Chapter 2. In contrast, while environmental legislation was strengthened during the 1970s and 1980s in Eastern Europe, in some cases setting standards above those in the West, limited attention was paid to implementation. The capacity of Eastern countries to deal with their pollution was limited, not least because their environmental infrastructure suffered chronic neglect. Furthermore, under the old regime firms had to give priority to reaching production quotas as opposed to meeting environmental standards. Factory managers were often closely associated with the local political elites and, as the major employer in a town or region, they commanded considerable power at the local level. Firms, particularly large firms, were often put in charge of their own environmental monitoring. Environmental fines were set at very low levels, encouraging them to pay fines rather than install costly pollution prevention measures. In other words, under the old system there was close relationship between the economic and political elites and the system of public administration, resulting in weak responses to the growing environmental problems that industrialisation was causing (Baker, 2012; Pickvance, 2004).

The environmental situation, however, has not remained static since 1989. First, governments have undertaken environmental clean-up, particularly in heavily polluted industrial zones. The collapse of production in many of the large state-owned industrial enterprises also led to improvement in ambient quality, such as air quality in the Czech Republic (Carter and Turnock, 2002); similarly, industrial restructuring has discouraged the high levels of resource use, especially energy inputs, typical of the old system.

However, there is no simple correlation between the end of communist rule and improvements in environmental quality. The period since 1989 has witnessed a reduction in some forms of environmental pressure, only to find the emergence of new environmental problems. Examples include growing consumer waste, as well as problems associated with the rise of road transport and the use of private cars. Furthermore, environmentally insensitive land restitution policies have opened up hitherto protected state forests to commercial logging interests.

Improving environmental management

The legal, administrative and institutional capacity of states to address the environmental consequences of economic activity has also

Country profile 10.1

Bulgaria

Profile

Democratic consolidation is weak in Bulgaria and there are ongoing problems with the rule of law, political corruption and organised crime.

Bulgaria has significant problems with food safety and public health, both air and water quality is poor, and the country has some heavily polluted 'hot spots'. Between 1992 and 2002 Bulgaria lost 15 per cent of its forests because of changes in forestry policy and incorrectly conducted land restitution, although 34 per cent of the territory is now under *Natura 2000* legislation and the Ministry of Environment and Water has established a national biodiversity monitoring system.

Improved governance

In 2005, several new appointments were made at the Ministry of the Environment and Water and other environmental bodies. Furthermore, several new environmental bills and legislative amendments have been introduced, including in relation to water quality, GMOs, noise, waste management and renewable energy, and a biodiversity act was introduced in 2012.

There has also been a plethora of action plans, including on the environment (2006), aspects of waste management (2011), climate change up to 2020 (2012), and energy efficiency (2011); and a resource efficiency roadmap was introduced in 2012. A National Environmental Strategy and Action Plan (2006–2015) and a Sectoral Operational Programme (2007–2013) have also been formed. These developments have taken place alongside plans to provide tax relief and rebates for recycling initiatives under the National Reform Programme (2011–2015) and a tax on plastic bags came into operation in 2011. The Better Regulation Programme (2010–2013) hope to facilitate better public participation in planning decisions.

Regional co-operation has also improved, as seen in the West Stara Planina project with Serbia and the development with Romania of a joint system for the management of air and water quality. Under the Carpathian Convention (2006), Bulgaria is also involved in the protection and sustainable development of the Carpathian region.

Ongoing problems

Recent concerns about bogus projects and corrupt networks diverting Community money have led the EU to freeze funding to Bulgaria, including for its environmental programme. This has resulted in the suspension of many environmental projects. Administration capacity to implement programmes and plans and to adhere to legislation remains weak, as does political will.

Bulgaria has received several warnings about its failure to comply with nature conservation legislation, including in the protected areas of the Rila and Pirin Mountains, and in relation to the destruction of important habitats because of road works undertaken along the Vaya River in the protected area of Emine-Irakli. In 2011 the European Commission carried out a site visit to the Pirin and Rila Mountains, following numerous petitions. Furthermore, since Pirin National Park is a World Heritage Site, UNESCO and IUCN have urged the Bulgarian government to reconsider its development plans for this region, as they could mean adding Pirin to the 'World Heritage in Danger' list.

(*Sources*: Baker, 2012; CEC, 2012; 2014b; Vesela Valkova and Andonova-Rangelova, 2012)

improved. This is driven by the desire of countries to become members of the EU, discussed below. These changes are occurring alongside new strategies for regional co-operation, especially between neighbouring states that share a common ecological feature, such as a river. Examples include co-operation among countries in the Danube river basin and those bordering the Black and Caspian seas, co-operation that is financially aided by the EU (Baker, 2012).

Promoting ecological modernisation

In addition to these direct, state-level responses, there has been a parallel process of ecological modernisation of the economy. Much of this is taking place within industrial firms, particularly their production processes. The privatisation of state-owned companies has provided an important conduit for the transfer of ecological modernisation practices. In order to privatise, many countries have had to clarify their laws and practices on environmental liability. Furthermore, privatisation, especially when involving foreign ownership, has been followed by plant modernisation, which has helped improve air quality, reduce the energy intensity of production, and advance waste management and recovery practices. The Belgian company Union Minière, for example, which bought a controlling stake in the Bulgarian Pirdop metallurgical plant, installed new dust abatement technology and modernised waste storage and management at the plant (US Geological Survey, 2000).

Privatisation has also resulted in new partnership arrangements between the public and the private sectors. The establishment of public–private partnership schemes has proved particularly popular with respect to waste management and landfill (Slovakia), water and sewage schemes

(Czech Republic) or municipal water (Sofia, Bulgaria) (Baker, 2011). This ensures that the private sector plays a role in the management of the environment, which, as we have seen in Chapter 3, brings both advantages and disadvantages for the pursuit of sustainable development.

Ecological modernisation is especially important, as under the old regime the energy intensity of the economies of Eastern European states was more than three times the EU average (EEA, 2003). There is strong synergy between the reduction in energy intensity, the achievement of greater energy efficiency and an enlarged Europe meeting its Kyoto Protocol targets designed to address climate change. However, the benefits of improvements may be outweighed by other developments, such as the steady rise in transport emissions, a rebound effect discussed in Chapter 2.

The energy production sector also provides an example of ecological modernisation, including through the modernisation of power plants, transmissions and grid systems and the application of new coal technology. Old and inefficient coal fired plants contribute almost half of all CO_2 emissions from the region. Here there is real potential for reducing GHG emissions and also improving environment quality at a local level. It is also important for the switch from reliance on unsafe nuclear technology.

Box 10.1

Rebound effects in transport

The transport sector is the fastest growing consumer of energy and producer of GHGs in the EU. More than two thirds of transport-related GHG emissions are from road transport. Transport is also responsible for other negative environmental externalities such as air pollution, particulate matter and noise. Transport infrastructure also has a negative impact on landscapes and biodiversity.

State investment strategies in the new EU member states prioritise road transport infrastructure. This has fuelled higher transport demand, which has also fed into increased ownership of private cars, and urban sprawl.

There has also been a rapid increase in passenger aviation, in part due to the growth of low-cost airlines in the region.

While the EU has played a major role in forcing improvements in fuel efficiency, and in setting CO_2 emission targets for cars, these gains have been offset by increases in passenger and freight transport demand across the region.

(*Sources*: EEA, 2013; CEC, 2014a; Eurostat, 2013)

Case study 10.1

Nuclear power and nuclear safety

High risk facilities

Following the 1986 Chernobyl reactor accident in Ukraine, the safety of Soviet-designed reactors became a major international concern.

A 1998 EU-sponsored study of nuclear safety in CEC found that six reactors were operating at high levels of risk. These include the Kozloduy reactor in Bulgaria, the Ignalina reactor in Lithuania and the Jaslovské-Bohunice reactor in Slovakia. There was deep concern about the Ignalina reactor because it is of the same design and uses the same technology as in Chernobyl. A leak in April 2003 from Hungary's Paks nuclear power plant has added this facility to the list of high risk sites. Neighbouring Austria remains deeply concerned about the safety of the nuclear power plants at Jaslovské Bohunice and Mochovce.

EU conditionality

As part of their accession treaties to the EU, Lithuania, Slovakia and Bulgaria agreed to close several nuclear reactors. Kozloduy was to close units 1 to 4, Ignalina units 1 and 2 and Bohunice V1 units 1 and 2. Bulgaria agreed to close the Kozloduy units in exchange for a substantial loan from the European Commission to modernise the remainder of the nuclear plant. Slovakia had nuclear reactors at Jaslovské Bohunice and Mochovce decommissioned between 1984 and 1999.

Substantial EU assistance was given to fund decommissioning and waste management, to mitigate the closure consequences by, for example, supporting replacement capacity for electricity production and energy efficiency measures, and to mitigate social consequences. Despite this investment, the two nuclear reactors at Ignalina remain unsafe and the European Parliament has levelled criticism at inefficient procedures during the decommissioning processes at all three power plants.

Significance

Nuclear energy provides an important example of the complex changes occurring in the region. The closure of the nuclear plants was accompanied by the construction of new plants and the expansion of existing units, in some cases with Russian or French assistance.

Poland is developing its nuclear power capacity, possibly through a co-operative project with Lithuania, Estonia and Latvia. Russia is co-operating with Finland to supply a new nuclear station. However, plans to build a nuclear power plant in the Shkoder border region of Albania have been postponed, due to concerns following the Fukushima disaster in Japan as well as territorial disputes with Montenegro. Similarly, in Poland, while the energy utility company PGE (Polska Grupa Energetyczna) withdrew from the Baltic nuclear power plant at Visaginas, the parliament passed legislation in 2011 to build two nuclear power plants by 2020 and 2030 respectively.

The development of nuclear power in the region could be seen as an effort, at least in part, to link nuclear energy output with a regional electricity market. This is supported not only by the countries themselves, but also by private companies seeking to export energy supplies.

(*Sources*: Baker, 2012; CEC, 1998; 2011a; 2011b; New Europe Online, 2012; World Nuclear Association, 2012; WNN, 2014)

Ensuring nuclear safety is an essential feature of the pursuit of sustainable futures.

Ecological modernisation is a key element in the EU's approach to the industrial sector of the new member states. It is seen as necessary for industrial competitiveness, especially for fledgling export-orientated firms (CEC, 2001; 2002). In order to be competitive in world markets, industries in the transition states have to reduce the cost of production, especially through reduction in energy and raw materials input and introduction of resource recovery methods. However, while ecological modernisation can make an important contribution, the promotion of sustainable development involves a broader range and deeper set of social, economic and political changes. In addition, there is no guarantee that marketisation will result in ecological improvements. In some countries, particularly in Southeastern Europe, including in Bulgaria, state assets have often been sold off into the hands of the old *nomenklatura*, which has a history of eschewing environmental regulations. Here privatisation has enabled the old political elites to gain new, economic, power. It has also allowed the political 'embeddedness' of institutions of governance, industrial production and environmental regulation to endure in the post-communist period (Baker, 2012).

Enhancing democratic participation

The transition process has also brought profound political and social changes. The formation of political parties, holding of regular, free elections and the establishment of several new state-level environmental bodies and institutions all allow environmental concerns to be routed through the democratic system (Tickle and Welsh, 1998).

Democratisation has also helped to introduce greater transparency into the sustainable development policy making process. Decentralisation of public administration has given local authorities a new voice in sustainable development policies, while at the same time exposing their poor resources and weak administrative capacity. The democratisation of political life has also opened up new arenas for participatory forms of public engagement. This has contributed to the enhancement of the roles of environmental NGOs, and thus to the strengthening of civil society

Box 10.2

Environmental NGOs

Background

Under the communist regimes, the strength of independent opposition varied considerably between countries. Only in Bulgaria did the environmental movement prove capable of challenging the communist monopoly of power.

Early engagement

The role played by the environmental movement in the collapse of the communist regimes proved important in shaping their strength after 1989. As a result, the ability of the movement to participate in the political developments of the post-communist period varied widely. Existing opportunities also differed, particularly given differences in the extent of democratic reforms.

Evolution of groups

Mass mobilisation around environmental issues declined throughout the region after 1989. There was a marked shift from protest to more moderate forms, such as lobbying, negotiation and discussion. Environmental concern was increasingly expressed through participation in professional, goal-orientated environmental NGOs.

There is considerable debate about the benefits of these changes and the significance of the strong influence exercised by foreign groups and funding agencies. Criticism of the increasing dependence of the region's environmental NGOs on foreign aid and support has pointed to their shift in priorities to those topics most likely to receive foreign support. Funding has nonetheless helped to enhance the capacity of environmental NGOs to engage.

Impact of democratisation

The process of democratisation had given new opportunities for environmentalists to engage in more institutional forms of political activity. Examples include the involvement of Bulgarian NGOs in the European Environment Agency; the institutionalisation of consultation between the environment ministry and environmental NGOs in Poland; and NGOs' involvement in the creation of a National Development Plan in Hungary and their representation on the Hungarian National Environmental Council. The Romanian 2001 Law on NGOs is considered one of the most liberal of such laws in the region, allowing NGOs considerable latitude with regard to their activities, rights and responsibilities. In contrast, however, Estonia provides an example of a country where the role of NGOs remains weak.

Influence of EU

The EU exercises strong influence over NGOs and sets a significant proportion of their agendas and shapes their activism. The CEE Bankwatch Network, for example, an international NGO based in Prague, Czech Republic, spends much of its time monitoring the impacts of EU funding.

The limited role played by NGOs reflects the EU style of interest mediation, which favours input from technical and scientific advisers. The fact that many CEE countries themselves favour this style tends to further limit participation from civil society groups. In this context, there is a real danger that the gap between EU environmental policy and the concerns of citizens will widen further.

(*Source*: Baker, 2012)

(Tickle and Welsh, 1998). This is taking place alongside their increased involvement with international organisations, including the UN.

Despite the fact that transition has brought democratisation of political life and some ecological modernisation of industry, the social conditions necessary for the promotion of sustainable development have, in some respects, deteriorated. There is a growing gap between those sections of the community benefiting from the opportunities of transition and those that are marginalised by it (Baker and Welsh, 2000). In addition, the emergence of a class of nouveau riche and of organised crime has the potential to threaten the gains won by transition, since these constitute the least public-spirited segments of society. Such groups demonstrate a preference for private gain over the public good and their continued strength retards the formation of strong sustainable development norms in the region. Furthermore, civil society remains weak and centralised administrative structures endure, alongside closed and highly politicised bureaucratic cultures (Smith and Pickles, 1998). These characteristics are not in keeping with the type of structures, institutions and processes needed to promote sustainable development. At the same time, marketisation and the growth of consumerism has given rise to a new wave of environmental problems associated with consumer waste and the growth of private car ownership.

In sum, following the collapse of the communist regime, the policy making process has become more open to environmental interests, despite marked differences in practices among countries. This has provided new opportunities for environmental groups and strengthened the pressure on governments to integrate environmental considerations into policy making. When coupled with increases in administrative capacity, efforts to address environmental risks posed by pollution and unsafe nuclear plants, and the trend towards ecological modernisation, these developments have proved significant. Collectively they help integrate environmental considerations into the normal business of government and lead to the acceptance of environmental protection as a legitimate task for those charged with the governance of transition. However, while these developments are important for the pursuit of sustainable development, we recall from discussions in Chapter 2 that sustainable development requires engagement with a wider set of policy imperatives, particularly those that relate to matters of equity. The next section deals with this issue, looking in particular at the issue of environmental justice.

Environmental justice

The term 'environmental justice' was coined to represent the observation that members of ethnic minorities, communities of lower socio-economic status, the least educated and otherwise marginalised segments of society disproportionately suffer from environmental harm, as discussed in Chapter 6.

There are several pressing issues of environmental injustice in Eastern Europe. Refugees, for example, including those displaced by the wars that followed the break-up of the former Yugoslavia and the ongoing ethnic tensions in the Balkans, tend to suffer disproportionately from poor environmental conditions. One of the most explicit manifestations of environmental injustice relates to the treatment of the Roma, a minority group within Europe that frequently experiences discrimination and marginalisation.

Case study 10.2

Environmental injustice among the Roma

Background

The Roma people constitute the largest ethnic minority in Europe. The EU is home to approximately 10–12 million Roma, with the largest concentrations in Romania, Bulgaria, Hungary, Slovakia, the Czech Republic and Spain. The majority of Roma suffer from social exclusion, discrimination, segregation and deep poverty.

Situation in Romania

The Roma population of Romania experiences prejudice and racism. They are poor, often geographically segregated and have limited life chances. Restricted access to housing is compounded by eviction from, and demolition of, housing they occupy. Issues of environmental injustice are starkly revealed when housing policy is examined.

Evicted Roma are often rehoused near waste disposal sites and various industrial areas on the periphery of cities. Public facilities, transportation and employment opportunities are minimal at best in these areas. For example, in Dorohoi-Centrul Vechi the local authorities demolished a housing unit occupied by fourteen to fifteen Roma families. The families were subsequently moved to social housing located directly in an industrial zone. In Constanta, twenty new houses were built for Roma in an area solely occupied by factories. Transportation is minimal and certain basic community facilities, such as schools, are absent. In another case in Bucharest, thirty-five Roma families were moved to outer circles of the city where there were no basic shelter provisions and no public utilities, including drinking water, sewerage or electricity.

In 2004, in the municipal region of Harghita-Miercurea Ciuc, about forty to fifty Roma were evicted from the centre of town by the local authorities to a site on the other side of a fence surrounding a water filtering station that emits toxic gases. Warnings regarding the toxic emissions from the filtering station are posted on the fence surrounding the station. Residents have asserted that two infants have died because of exposure to the toxic air generated by the plant, but medical experts have not confirmed this.

In Piatra Neamt, again Roma were transferred to the periphery of the city into a precarious situation near a garbage dump and shooting grounds. In Salaj Silmeul Silvaniei, the Roma live in environmentally risky conditions where there is no access to water, heat, or sewerage. In Episcopia Bihor, a Roma settlement was built on the municipal rubbish dump of Oradea. Most of the Roma in the settlement live off the recycled garbage. They are exposed to toxic materials and smog and air pollutants from incinerated waste.

Wider problems

Access to quality and affordable housing has proved more difficult in the post-communist period in Romania. Those in the lower socio-economic stratum of Romanian society have few opportunities to avoid environmentally dangerous living conditions. Roma face similar problems of environmental injustice in other transition and EU member states.

(*Source*: Steger, 2007)

The situation of the Roma demonstrates how public policy contributes to the unequal distribution of environmental benefits and harm. From this case we can see that Roma disproportionately suffer two main forms of environmental injustice. First, they suffer from exposure to environmental hazards due to their proximity to hazardous waste sites, factories, abandoned industrial facilities and other sources of pollution; second, they are denied environmental benefits such as potable water, sewage treatment facilities, sanitation, and access to natural resources (Harper et al., 2009). The status of Roma minority communities and the poor quality of the environmental conditions under which they reside and work is an indicator of how unsuccessful European society is at recognising and integrating marginalised people. Displacing environmental harm to other groups in society breaches one of the fundamental principles of sustainable development, that of intra-generational equity. It is to this issue of sustainable development that we now turn.

Sustainable development and EU membership

The EU's engagement in Eastern Europe can be seen as an attempt to control the negative consequences of the political and economic upheavals that followed the collapse of the old Soviet system. European support aims at the promotion of democracy and marketisation and the application of principles of good governance. Over time, these concerns have been consolidated as 'soft' security issues, which include justice and home affairs, the environment, and energy. The collapse of the old regimes shifted focus to the EU's perceived vulnerability in relation to soft security issues, including environmental matters, nuclear risk, transboundary pollution and, more recently, climate change.

The term 'soft security' was originally used to distinguish military issues from other security issues. Subsequent widening of the notion of security has added environmental, water and energy concerns.

Membership aspirations

Eight Eastern European countries joined the EU in May 2004 (the Czech Republic, Hungary, Poland, Slovakia, Slovenia and the three Baltic states of Estonia, Latvia and Lithuania). In 2007, membership was extended to Bulgaria and Romania and to Croatia in 2013. Increasingly, the EU's influence has extended into Southeastern Europe, to the western Balkans (Albania, Bosnia and Herzegovina, Kosovo, the Former Yugoslav Republic of Macedonia (FYRM), Montenegro and Serbia). Turkey, Serbia, the FYRM and Montenegro are EU candidate countries and the EU has a potential candidate country list that includes Albania, Bosnia and Herzegovina, and Kosovo. Membership is not currently seen as a necessary outcome of the current expansion of EU interest to the south, or indeed, further east among the former Soviet states, including Ukraine.

EU policy towards the former communist bloc can be described as a policy of 'conditionality' – a bargaining strategy involving reinforcement by reward. Over time, conditionality moved from a narrow preoccupation with the legal transposition of EU rules and regulations to broader concerns with implementation. As a result, attention has increasingly focused on reforming institutions and administrative practices, upgrading monitoring capabilities, and ensuring that the formerly highly centralised states start to devolve implementation tasks to subnational, regional and local levels of government. The rewards given by the EU for compliance with conditionality clauses vary, ranging from financial and technical assistance through to the establishment of institutional ties. Institutional ties can themselves range from trade and co-operation agreements (Association Agreements) to full membership of the EU (Baker, 2012).

EU conditionality refers to a strategy by which the Union sets rules as conditions that non-member states have to fulfil in order to get EU rewards, such as assistance and institutional ties.

(*Source*: Schimmelfennig and Sedelmeier, 2004: 663)

Establishing closer ties with the EU became the determining factor shaping environmental policy in the post-1989 period in transition states. As a result, the prospects for the promotion of sustainable development became closely linked with these ties. However, in exploring the links we need to be aware that there are other external and internal influences in operation. Both the UNDP and the Organisation for Security and Co-operation in Europe (OSCE), for example, are involved in the Environment and Security Initiative in Southeastern Europe, an initiative that was relaunched in May 2011 as the Environment and Security Network. This network aims to encourage co-operation, especially on implementation, to promote networking, raise awareness and share information on environmental and security issues (UNEP, 2006; 2013). Furthermore, while the EU has played a considerable role in shaping transition, this process of change is also 'path dependent', involving a complex reworking of social, political and economic relations inherited from the period of communist rule.

Nevertheless, because the EU is exercising a predominant influence in the region, attention is turned to exploring the promotion of sustainable development in transition states within the context of their relationship with the EU. The question we ask is: Has the EU had a positive or negative influence on the prospects for sustainable development in the region?

Setting policy within a sustainable development framework

To begin with, the EU requires all new or potential member states to make explicit commitment to the promotion of sustainable development, which is enshrined in EU treaties and forms part of the environmental *acquis*, as discussed in Chapter 9. This has led states to pass sustainable development acts, such as the 1995 Estonian Act on Sustainable Development. The EU also requires that the principle of sustainable development underpin all public policy. New member states are also expected to establish a national council for sustainable development, alongside formulating national sustainable development strategies (NSDS), aimed at putting their legal commitment into practice. Hungary, for example, adopted a NSDS in 2007 and subsequently established its Council in 2008. Similarly, Romania launched its NSDS in 2008 and Croatia in 2009, while 2010 saw Latvia adopting its latest NSDS, providing long-term strategic planning to 2030 and laying out plans for the integration of sustainable development considerations into sectoral policies.

Strengthening environmental legislation

As part of their preparations for membership, countries have to adopt the *acquis communautaire* of the EU, that is, the entire body of EU

Case study 10.3

NSDS Romania

Background

Romania launched an NSDS in 1999 and in 2008 produced the revised *Strategy 2013–2020–2030*. The production of a revised strategy was an obligation arising from Romania's EU membership. The UNDP supported Romania in devising the Strategy.

Aims

The NSDS plans to move Romania toward a new model of development, geared to the continued improvement of the quality of life and human relationships in harmony with the natural environment.

Growth orientated

The Strategy is based on the belief that sustainable development can be achieved through attaining high rates of economic growth. Growth will result in a significant reduction in the social and economic disparities between Romania and the rest of the EU. The Strategy focuses on how natural capital can be used as a foundation for social development, economic growth and dealing with the impacts of climate change.

Targets

The strategy has set specific targets for the short (2013), medium (2020) and long term (2030), all involving moving Romania closer to the average performance of the EU member states, using sustainable development indicators.

Critique

The Strategy is strong in terms of its focus on the social pillars of sustainable development. However, its chief weakness lies with its over-emphasis on economic growth as the tool for achieving social sustainability. The complex issues of how to address potential trade-offs between economic growth and the ecological and social pillars of sustainable development are not addressed. Implementation capacity is also weak and the country lacks both administrative and financial resources to put the Strategy into action. Political will is also lacking.

(*Sources*: Romanian Government, 2008; European Sustainable Development Network, 2013; Kohlmann et al., 2009)

legislation, treaties and case law. Countries have to negotiate with the EU to agree dates for transposition of legislation and timetables for implementation, although typically the EU has to concede transitional periods in several areas, including in relation to 'investment intensive' environmental directives. These include directives relating to Integrated Pollution Prevention and Control, the control of emissions from large combustion plants and the treatment of urban wastewater, because these require substantial investment in training and often also in new, physical infrastructure.

The *acquis* contains an impressive body of legal acts covering an array of environmental matters, as discussed in Chapter 9. Making the adoption of the environmental *acquis* a condition of membership gives the EU an important way in which it can exercise its influence over environmental governance in the region. This is particularly the case given the differences between the EU environmental *acquis* and the environmental legislation inherited from the communist era. The *acquis* covers a wider range of environmental matters, such as in relation to recycling and waste management. Furthermore, it deals not just with environmental processes, such as civil society participation in planning procedures, but also environmentally related product standards. It uses a wider range of implementation tools, including voluntary agreements, and requires new monitoring and reporting systems and standards. There is also new pressure to achieve more effective policy implementation, often only weakly addressed under the old communist system. All of these improvements have the capacity to support efforts to promote sustainable development.

Despite the benefits, challenges remain and these are especially acute in the Balkan region, where economic decline, combined with social and political divisions and conflict, hamper progress. However, implementation could enable the Eastern European states to avoid the mistakes that the EU member states have made in the past, enabling them to make a smoother transition to a sustainable development path.

We can see from the example of Albania that adoption of the environmental *acquis* through its transposition into national legislation is but one step – legislation has also to be implemented and enforced. This presents major challenges, as environmental legislation is both expensive and technically complex to implement. The new member states have to enhance their system of public administration, including, for

Box 10.3

Potential benefits to sustainable development from implementing the EU environmental *acquis*

- Better public health as exposure to pollution is reduced, resulting in decreases in respiratory diseases and premature deaths;

- Less damage to forests, buildings, fields and fisheries through reduction in acid rain and other forms of pollution, leading to wider economic benefits and reduced costs;

- Lower risk of (irreversible) damage to natural resources, such as groundwater aquifers;

- Better protection of natural ecosystems and (endangered) species;

- Promotion of tourism as a result of a cleaner environment (forests, bathing waters, nature reserves);

- Reduced risk of water-related illnesses as a result of better bathing water and drinking water quality;

- Increased economic efficiency and higher productivity, supporting industrial competitiveness;

- Lower production and maintenance costs to industry through availability of cleaner water, reducing pre-treatment needs;

- Lower consumption of primary material due to more efficient use and higher levels of reuse and recycling;

- Support for employment and benefits for local and regional development;

- Improved awareness of environmental risks and sounder approaches to minimising risks;

- Social benefits through greater learning, awareness, involvement and responsibility with regard to environmental matters.

(*Source*: ECOTEC, 2001)

example, their environmental protection agencies and inspectorates. The lack of administrative and financial resources and technical skills acts as a major barrier to effective implementation of legislation in transition states. In addition, states have to update existing or build new physical installations and infrastructure, such as wastewater treatment plants in order to be able to implement the Urban Wastewater Directive.

Country profile 10.2

Albania

Background

In Albania, democratic consolidation is weak, corruption is widespread and there are problems with human rights and the protection of minorities. In addition to money laundering and drugs trafficking, Albania is a country of origin for human trafficking.

Albania has several major environmental problems, including industrial pollution and poor air quality. It lacks a modern system of waste collection, disposal and recycling, and uncontrolled dumping and burning of waste persists, especially in rural areas. There is no clear strategy for the safe disposal of hazardous waste. Albania does not fully comply with the Euratom Treaty, in particular with regard to environmental monitoring and radiation protection in medical and industrial applications. There are also concerns about illicit trafficking of nuclear and other radioactive materials.

Influence of the EU

Because Albania is keen to join the EU, it has to adopt the environmental *acquis* and develop new strategies to deal with its environmental problems. However, progress has been slow.

A National Waste Strategy was adopted, but there are no procedures in place for the management and control of landfills. Efforts have also been made to address pollution at the Sharra landfill near Tirana, the copper smelter in Rubik, the nitrate fertiliser plant in Fier, the chemical plant in Durrës and the Ballshi oil refinery. Albania has also negotiated a grant from the Netherlands for environmental remediation at the former industrial site of Porto Romano near Durrës, where pollution is directly affecting human health. A project to assess the environmental problems at the Patos-Marinez oilfield began in 2007. Remediation measures are also required at the PVC thermoplastic resin plant in Vlorë. Measures have been implemented to identify and assess other environmental 'hot spots'.

While emissions from vehicles and large industries have been slightly reduced, there is no clear long-term policy on air pollution abatement and emissions reduction. Tirana is now one of the most polluted cities in the world and although a new Tirana air management plan has been prepared with the help of the EU, this has yet to be adopted by the municipality. The European Commission has warned that Albania is facing a humanitarian catastrophe due to growing pollution, which is estimated at ten times above the average tolerance level set by the WHO. Enforcement of existing legislation on vehicle emissions and fuel quality is weak.

The outdated nature of the water supply and sanitation systems is compounded by the poor quality of water. There are only two operational wastewater treatment plants in the country, although four new plants have been completed, but are not yet operational.

While the country's National Environmental Strategy remains in draft form, some improvements in nature protection can be reported. A new law on hunting was adopted in 2009. By 2012, protected areas accounted for about 13.17 per cent of the national territory, although implementation of *Nature 2000* stalls and illegal logging continues. The EU Habitats Directive has been transposed, and a law for implementation of the Convention on International Trade in Endangered Species of Wild Fauna and Flora (CITES) has been adopted. However, training and resources still need to be provided to the staff operating in these areas.

Limited capacity

Environmental protection is hampered by a lack of political will, weak administrative capacity, lack of financial resources, fragmented responsibilities and the weak judicial system. Administrative capacity is particularly weak at the sub-national level, and inter-institutional co-operation and co-ordination is poor. Temporary contracted employees fill at least 20 per cent of posts in the public sector. The National Environmental Guard, for example, is poorly resourced, and fines are pitched at a low level, providing limited incentive for change. Albania does not have a National Climate Change Strategy, and only one staff member covers matters in relation to climate change within the Ministry of Environment.

Furthermore, public participation in the decision making process is restricted. Nonetheless, the 2001 Law on NGOs is considered one of the most liberal of such laws in the region, allowing NGOs considerable latitude with regard to their activities, rights and responsibilities. Environmental education in primary schools and the creation of a network of environmental summer camps should also be noted.

The basic administrative capacity and political will to implement commitment to sustainable development is not present.

(*Sources*: Baker, 2012; CEC, 2011c)

Mixed messages

The EU provided financial assistance in order to help transition states address environmental matters, now provided through the Instrument for Pre-Accession Assistance (IPA). However, EU funding provides a prime source of policy inconsistency. The IPA has been criticised for prioritising large-scale investment projects, even in cases where smaller, localised development would have been more appropriate. This includes financial support for large-scale sewage and wastewater treatment projects, rather than the low technology, low-cost biological treatment that is more suitable for the plethora of small, rural communities in the region (REC, 2003).

The now-suspended Special Accession Programme for Agriculture and Rural Development (SAPARD) funded the extension of the Common Agricultural Policy (CAP) to the east. It has been heavily criticised for promoting intensive agricultural methods and displacing the environmental advantages associated with low-intensity farm management practices. Poland, for example, is now confronting serious environmental threats from agricultural intensification. Farm modernisation has also led to a deepening crisis of land abandonment, with a resultant loss of biodiversity. In Latvia, the past decade has witnessed the disappearance of semi-natural meadows, an experience shared by Sweden following its accession to the EU in 1995.

Whither nature?

Because of fewer development pressures, lower-intensity agriculture and their system of forest protection, transition states host species and habitat types that have nearly vanished from Western Europe, including mammals such as the European bison. The last great wilderness on the European continent, the Carpathian Mountains, stretches across seven countries and is one of the last homes for large European carnivores, including bears, wolves and lynx; the Danube Delta is globally significant as a wetland area; the Vistula River in Poland is one of the few large European rivers with major natural features; and the Baltic coast is one of the most important corridors for migrant birds in Europe (WWF, 2003). This rich natural and cultural heritage is very significant since it forms the potential foundation of development strategies based upon the principles of sustainable development (WWF, 2000).

Furthermore, a relatively high proportion of forest territory has protected area status. The period of communist rule proved particularly good for the maintenance of rich forest landscapes across the region, resulting in high levels of biodiversity. For example, in both Hungary and Slovakia approximately 20 per cent of the forest is within protection areas. In some countries legislation is stronger than that provided by the EU *acquis*. This could have been viewed as a positive asset in negotiation between the EU and the candidate countries. Unfortunately, this has not been the case as the post-1989 period has seen rising vulnerability to non-sustainable forms of commercial forest exploitation. Part of the problem lies in the fact that the EU has currently no adequate policies to promote sustainable forestry practices and it also lacks a clear strategy on how to handle forests within an enlarged EU. When account is also

Box 10.4

Opportunities for sustainable agriculture

Traditional practices

Unlike their Western European counterparts, agricultural practices in the CEE largely worked in co-operation with nature and the landscape. Mixed farming and low intensity agriculture enabled the region to maintain a rich and biologically diverse landscape, which ranges from coastal meadows and wet grasslands in the Baltic region, to the strip-land farming landscape of southern Poland and small-scale livestock rearing in the Carpathian Mountains.

Sustainable development potential

Agricultural and rural areas represent significant contributions in terms of natural capital, cultural heritage and social cohesion, which the new member states bring to an enlarged Europe. Building upon this natural and cultural capital could provide an excellent stepping stone for sustainable development.

The low-intensity agricultural practices of Eastern Europe also offer an opportunity for the transfer of best practice to the West. However, despite several reforms that brought environmental considerations into the CAP, the logic of EU agricultural policy remains production-orientated and a switch towards more ecologically responsible land and landscape management has yet to take place. The drive to agricultural intensification continues to be reflected in current agricultural strategies towards Eastern and Central European (ECE) countries.

Lost opportunity

Transition challenges the EU to reform its agricultural policy and to value nature as a key component of the wealth of rural Europe. This could support Western Europe as it struggles to de-intensify agriculture in the face of the rising financial costs of the CAP, agri-chemical pollution, food safety scares, growing concerns about animal welfare and husbandry practices, and loss of biodiversity. Here, there is synergy between CAP reform and the creating of a new model of rural sustainable development in an enlarged EU.

However, with current policy priorities, the opportunity to draw upon the practices in ECE countries to help construct new models of sustainable agriculture in Europe will be lost.

(*Sources*: FoE and CEC Bankwatch, 2000; Grodzinska-Jurczak and Cent, 2011; WWF, 2002a and b; 2003)

taken of the policies of privatisation and of land restitution, it is clear that the forests of Eastern Europe are increasingly threatened. Lithuania, for example, has just agreed new logging concessions as commercial forestry takes hold in the country.

However, EU policy in relation to biodiversity protection is reasonably well developed. Two key pieces of legislation frame EU nature protection policy: the 1970 Directive on the Conservation of Wild Birds (the Birds Directive) and the 1992 Directive on the Conservation of Natural Habitats and of Wild Fauna and Flora (the Habitats Directive). The Birds Directive is the EU's oldest piece of nature legislation and one of the most important, creating a scheme of protection for wild bird species. The main aim of the Habitats Directive is to promote the maintenance of biodiversity by requiring member states to take measures to maintain or restore natural habitats and wild species. These two directives form the legal base for the *Natura 2000* programme, aimed at the creation of a comprehensive, linked network of protected habitats across the nine bio-geographical regions of the EU.

However, the eastern extension of the *Natura 2000* programme has proved difficult. Historically, nature conservation policy has been plagued by disputes over whether the EU should have competence in this area or whether this is best left to the member states (Baker, 2003). Consequently, the implementation of the *Natura 2000* programme has been subject to considerable delay in existing member states. The eastern expansion of *Natura 2000* has exacerbated this problem, not least because the programme had to be applied to a much larger territory than before.

When identifying sites for inclusion in the *Natura 2000* network there is a common tendency for countries to reselect areas that are already protected. This means that site identification often makes little reference to wider ecological regions or to the preservation of biodiversity in Europe as a whole (WWF, 2002b). Furthermore, there is insufficient independent control over the site selection process. In addition, all countries have acknowledged that they lack the administrative and institutional capacity to implement the *Natura 20002* programme. Another major difficulty is created by the mismatch between the timing of investment funds for traditional development activities, such as road building programmes, and those designed to help nature conservation. As a result, when the nature conservation systems are finally in place, they will only help conserve the biodiversity that has managed to survive large-scale investment

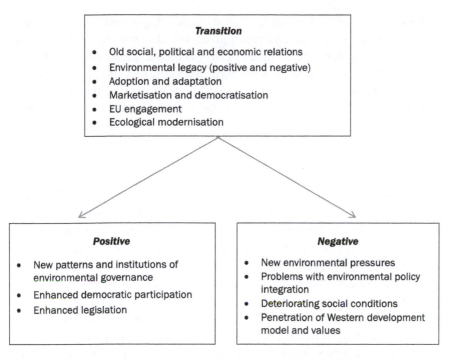

Figure 10.1 *Promoting sustainable development in transition societies.*

programmes aimed at promoting traditional forms of development (WWF, 2000: 5). Older member states made the same mistake in the past, especially those in peripheral regions, such as Ireland, that received large amounts of EU Sructural Funds and are now only able to implement nature conservation strategies to preserve what is left over after modernisation and development has taken its toll. This is despite the realisation that the promotion of sustainable development requires application of the principle of environmental policy integration, that is, the integration of environmental considerations into all stages of the policy process.

Conclusion

This chapter explored the prospects for, and barriers to, the promotion of sustainable development in transition countries in Eastern Europe. The environmental dimensions of transition have to be viewed within the context of the complex interface that is evolving between the social, political, cultural and administrative legacies of the old regimes and the new systems of environmental management being introduced across the region. On the one hand, there is a great deal of continuity with the past, especially the low priority given to environmental protection over economic development. At the same time, marketisation and the growth of consumerism has given rise to a new wave of environmental problems associated with consumer waste and the growth of private car ownership. The role played by environmental NGOs in public policy formation and implementation, although expanded since 1989, remains limited by ongoing weaknesses in civil society and the continuation of centralised administrative structures alongside closed and highly politicised bureaucratic cultures (Smith and Pickles, 1998). These characteristics are not in keeping with the type of structures, institutions and processes needed to promote sustainable development. On the other hand, transition has seen the emergence of new features of environmental governance, in part driven by external influences. Countries in the region have become anchored into the system of international environmental governance. This has helped spread a commitment to the principles of sustainable development, at least at the level of declaratory politics. Capacity enhancement, including decentralisation of public administration, has given local authorities a new voice in environmental management, while at the same time exposing their poor resources and weak administrative capacity.

At times, it is difficult to distinguish the impact that membership of the EU is having on the environment from the more general impacts of transition, which is bringing modernisation, a growth in consumerism, market liberalisation and economic restructuring. However, in relation to transport and agricultural policy, there is a clear link between the policies of the EU and enhanced pressures on the environment, particularly upon natural habitats.

Like their counterparts within the developed world, governments in the transition states give low priority to environmental protection, nature conservation and, more generally, the promotion of sustainable development when faced with the pull of modernisation and growth. Economic growth is already changing consumer habits and generating more waste. Transportation is expected to increase by as much as seven times its current level and most of it will be road traffic, one of the most polluting forms of transport. Public transport networks are declining and in countries which once had good systems of recycling, an explosion of waste is clogging landfills. These trends, combined with expanding tourism, more intensive farming and forestry and the expected increases in energy consumption, have the potential to destroy the invaluable natural environment that exists in the region. The tragedy of this development is that it is funded and aided by the EU, despite the fact that the EU has both a declaratory and legal commitment to the promotion of sustainable development. There is a danger here that transition will vindicate the traditional, capitalist form of development, and that the end of communism will result in the triumph of consumerism.

However, a different way to view the process of enlargement is to see eastward expansion in terms of opportunity. Enlargement can bring enormous additional natural capital and biodiversity for the enlarged European Union to cherish, enjoy and safeguard for future generations. It provides the social and ecological conditions for Europe to embark on a more sustainable path. From a global perspective, an enlarged EU could have a stronger positive influence on efforts to address global environmental problems, such as climate change and biodiversity loss.

> Enlargement offers the European Union an opportunity to put its paper commitments to sustainable development into actual practice.
>
> (WWF, 2003: 10)

However, this chapter has shown that, far from taking advantage of the opportunity that enlargement brings, an enlarged Europe will be an environmentally poorer region.

Summary points

- Since the collapse of the communist regimes in CEE in 1989, countries in the regions have undergone complex processes of transition.
- The legacies of the old regime shape the prospects for, and barriers to, the promotion of sustainable development in CEE. There is much in common with the past: low priority assigned to environmental protection; limited involvement of civil society in public policy; continuation of centralised administrative structures alongside closed and highly politicised bureaucratic cultures.
- Transition is bringing new styles of environmental governance and anchoring the region into the system of international environmental governance.
- The social and ecological conditions for promoting sustainable development have deteriorated, especially with respect to social cohesion and ecological diversity.

- Eastern enlargement has the potential to enrich the EU, ecologically and socially.
- Preparations for enlargement undermined the very significant ecological and social contributions that this region could make to the promotion of a sustainable Europe.

Further reading

Baker, S. (2012), 'Environmental Governance: EU Influence beyond Its Borders', in I. Gladman (ed.), *Central and South-Eastern Europe 2013* (London: Europa Publications, 13th edn), www.routledge.com/books/details/9781857436501/

The European Union Sustainable Development Network provides ongoing information on SD strategies in Europe: www.sd-network.eu/

OECD (2012), 'Green Growth and Environmental Governance in Eastern Europe, Caucasus, and Central Asia', Green Growth paper (Paris: OECD), available online at www.oecd-ilibrary.org/environment/green-growth-and-environmental-governance-in-eastern-europe-caucasus-and-central-asia_5k97gk42q86g-en

Ongoing information on EU policy towards the accession states can be found at http://Europa.eu

The Regional Environmental Center for Central and Eastern Europe has ongoing information on environmental matters from a civil society perspective: www.rec.org/

References

Baker, S. (2011), 'Environmental Governance and EU Enlargement: Developments in New Member States and the Western Balkans', in I. Gladman (ed.), *Central and South-Eastern Europe 2012* (London: Europa Publications, 11th edn), www.routledge.com/books/details/9781857436501/.

Baker, S. (2012), 'Environmental Governance: EU Influence beyond Its Borders', in I. Gladman (ed.), *Central and South-Eastern Europe 2013* (London: Europa Publications, 13th edition), www.routledge.com/books/details/9781857436501/.

Baker, S. (2003), 'The Dynamics of European Union Biodiversity Policy: Interactive Functional and Institutional Logics', *Environmental Politics*, 12 (3): 24-41.

Baker, S. and Welsh, I. (2000), 'Differentiating Western Influence on Transition Societies in Eastern Europe: A Preliminary Exploration', *Journal of European Area Studies*, 8 (1): 79–103.

Carter, F. W. and Turnock, D. (eds) (2002), *Environmental Problems of East Central Europe* (London: Routledge, 2nd edn).

CEC (1998), *Nuclear Safety in Central and Eastern Europe and in the New Independent States* (Brussels: Commission of the European Communities).

CEC (2001), *The Challenges of Environmental Financing in the Accession Countries, Commission* (Brussels: Commission of the European Communities, COM (2001) 304).

CEC (2002), *Industrial Policy in an Enlarged Europe* (Brussels: CEC, COM(2002) 714, final).

CEC (2011a), 'Nuclear Safety: EU Will Give Extra €500 Million for the Decommissioning of Old Soviet Type Nuclear Reactors', Press release (Brussels: CEC, IP/11/1449, 24/11/2011), available online at http://europa.eu/rapid/pressReleasesAction.do?reference=IP/11/1449&format=HTML&aged=0&language=en&guiLanguage=en

CEC (2011b), *On the Use of Financial Resources During 2004–2009 Provided to Lithuania, Slovakia and Bulgaria to Support the Decommissioning of Early Shut-Down Nuclear Power-Plants Under the Acts of Accession* (Brussels: CEC, COM(2011) 432 final), available online at http://ec.europa.eu/energy/nuclear/decommissioning/ndap_en.htm

CEC (2011c), *Albania 2011 Progress Report* (Brussels: CEC, SEC(2010) 1205).

CEC (2012), *Monitoring Member States' Policy Developments on Resource-efficiency/Environment in Europe 2020: Country Profile Bulgaria* (Brussels: CEC), available online at http://ec.europa.eu/environment/ms_policydev.htm.

CEC (2014a), 'Climate Action: Reducing Emissions from Transport', available online at http://ec.europa.eu/clima/policies/transport/index_en.htm

CEC (2014b), *Member States' Policy Developments on Resource Efficiency and Environment: Bulgaria*, available online athttp://ec.europa.eu/environment/ms_policydev.htm

ECOTEC (2001), 'The Benefits of Compliance with the Environmental Acquis for the Candidate Countries', Final report (Birmingham: ECOTEC), available online at http://ec.europa.eu/environment/enlarg/pdf/benefit_long.pdf

EEA (European Environment Agency) (1999), *Environment in the European Union at the Turn of the Century* (Copenhagen: EEA).

EEA (2003), *Europe's Environment: Third Assessment*, State of the Environment Report no. 3/2003 (Copenhagen: EEA), www.eea.europa.eu/publications/environmental_assessment_report_2003_10

EEA (2013), *Final Energy Consumption by Sector* (Copenhagen: EEA, CSI 027/ENER 016), available online at www.eea.europa.eu/data-and-maps/indicators/final-energy-consumption-by-sector-5/assessment

European Sustainable Development Network (2013), *Single Country Profile: Romania*, available online at www.sd-network.eu/?k=country%20profiles&s=single%20country%20profile&country=Romania

Eurostat (2013), 'Transport Energy Consumption and Emissions' (Brussels: CEC, Eurostat, 3 December), available online at http://epp.eurostat.ec.europa.eu/statistics_explained/index.php/Transport_energy_consumption_and_emissions#Further_Eurostat_information

Friends of the Earth Europe and CEE Bankwatch Network (2000), *Billions for Sustainability? The Use of EU Pre-accession Funds and Their Environmental and Social Implications* (Prague and Brussels: CEE

Bankwatch Network and Friends of the Earth Europe, 2000), available online at http://bankwatch.org/documents/BillionsforSustainability2.pdf

Grodzinska-Jurczak, M. and Cent, J. (2011), 'Expansion of Nature Conservation Areas: Problems with Natura 2000 Implementation in Poland?' *Journal of Environmental Management*, 47 (1): 11–27.

Harper, K., Steger, T. and Filcak, R. (2009), 'Environmental Justice and Roma Communities in Central and Eastern Europe', *Selected Publications of EFS Faculty, Students, and Alumni*. Paper 1, available online at http://scholarworks. umass.edu/efsp_pub_articles/1

Kohlmann, R., Preissler, S. and Stengel, M. (2009), 'National Councils for Sustainable Development: An Appropriate Tool of Common Use?' *Economic and Environmental Studies*, 9 (1): 59–73.

New Europe Online (2012) 'Albania Postpones Building of Nuclear Power Plans', available online at www.neurope.eu/article/albania-postpones-building-nuclear-power-plant

Pickvance, C. (2004), *Local Environmental Regulations in Post-Socialism: A Hungarian Case Study* (London: Ashgate).

Regional Environmental Centre (REC) (2003), *Environmental Financing in Central and Eastern Europe 1996–2001* (Budapest: Regional Environmental Centre).

Romanian Government (2008), *National Sustainable Development Strategy of Romania 2013–2020–2030* (Bucharest: Government of Romania, Decision no. 1460, Official Gazette no. 824/8), available online at www.undp.ro/download/files/sndd/sndd_eng_176x235_final.pdf

Schimmelfennig, F. and Sedelmeier, U. (2004), 'Governance by Conditionality: EU Rule Transfer to the Candidate Countries of Central and Eastern Europe', *Journal of European Public Policy*, 11 (4): 669–687.

Smith, A. and Pickles, J. (1998), *Theorizing Transition: The Political Economy of Post-Communist Transformations* (London: Routledge).

Stark, D. (1997), 'Recombinant Property in East European Capitalism', in G. Grabher and D. Stark (eds), *Restructuring Networks in Post-socialism* (Oxford: Oxford University Press), 35–69.

Steger, T. (ed.) (2007), *Making the Case for Environmental Justice in Central and Eastern Europe* (Budapest: CEU Center for Environmental Policy and Law, The Health and Environment Alliance and The Coalition for Environmental Justice), available online at www.cepl.ceu.hu/system/files/ceu_teljes_pdf.pdf

Tickle, A. and Welsh, I. (1998), 'Environmental Politics, Civil Society and Post-Communism', in A. Tickle and I. Welsh (eds), *Environment and Society in Eastern Europe* (Harlow: Longman).

UNEP (2006), *Environment and Security in South Eastern Europe: Improving Regional Cooperation for Risk Management from Pollution Hotspots as Well as the Transboundary Management of Shared Natural Resources*, available one line at nvsec.grid.unep.ch/see/pub/ADA%20ENVSEC%20Mining-Mountains%20in%20SEE.pdf

UNEP (2013), *The First Decade of Partnership for Environment and Security*, Environment and Security Initiative, available online at www.envsec.org/index.php?option=com_content&view=article&id=94&Itemid=176&lang=en

US Geological Survey (2000), *Mineral Yearbook, 2000*, available online at http://minerals.usgs.gov/minerals/pubs/myb.html

Vesela Valkova, V. and Elena Andonova-Rangelova, E. (2012), *Bulgaria: Environment 2013* (Schoenherr Attorneys at Law), available online at www.mondaq.com/x/206920/Waste+Management/Environment+in+22+jurisdictions+worldwide+2013

von Homeyer, I. (2001), 'Enlarging EU Environmental Policy', Environmental Studies Workshop, Robert Schuman Centre for Advanced Studies, European University Institute, Florence, May.

WNN (World Nuclear News) (2014), 'Russia, Finland Agree on More Cooperation', 26 February, available online at www.world-nuclear-news.org/NP-Russia-Finland-agree-on-more-cooperation-2602144.html

World Nuclear Association (2012), *Nuclear Power in Poland*, online at www.world-nuclear.org/info/inf132_poland.html

WWF (2000), 'The "New" European Union', *WWF Agenda for Accession*, December (Brussels: European Policy Office Belgium, World Wide Fund for Nature).

WWF (2002a), *Enlargement and Agriculture: Enriching Europe, Impoverishing our Rural Environment?* WWF European Agriculture and Rural Development team report, October (Brussels: European Policy Office Belgium, World Wide Fund for Nature).

WWF (2002b), 'The "New" European Union', *WWF Agenda for Accession*: an update (Brussels: European Policy Office Belgium, World Wide Fund for Nature).

WWF (2003), *Progress on Preparation for Natura 2000 in Future EU Member States*, available online at assets.panda.org/downloads/n2000progressmailing20030122.pdf

Challenges in the Third World

Key issues

- International development agenda
- Least Developed Countries
- Poverty
- Food security
- Gender
- Trade
- Knowledge
- Finance.

Introduction

A pervasive feature of international meetings and summits taking place under the remit of the UN, and their accompanying declarations and plans, is the attention given to the division between the North and the South, between the rich and the poor, between the high consumption societies of the industrialised world and those in the Third World struggling to sustain basic livelihoods. The Brundtland Report also acknowledged that both the positive and negative consequences of industrialisation and agricultural modernisation are, when viewed from a global perspective, inequitably distributed. Recognition has also grown that the consequences of global environmental change, including climate change and biodiversity loss, are uneven in their impacts.

This chapter looks at the problems facing Third World countries and how these shape their emerging environment and development agendas. Building upon discussions in Chapter 1 on the Western development model, the chapter begins by exploring critiques of development from a Third World perspective. The challenges involved in the promotion of sustainable development in the Third World are then discussed in general terms, before attention is turned to six key themes that are explored in finer detail.

The international development agenda

Many theorists argue that the Western development model sustains inequalities and leads to under-development in the Third World. In the decades since the Second World War many countries have paid a high economic, social, environmental and cultural price for adopting policies aimed at 'catching up' with Western development. Among rural populations in particular, the Western model is seen as having undermined traditional subsistence agriculture and directed resources towards the production of cash crops and away from traditional food crops. While the former command ever-decreasing prices on the world market, the shortage of the latter results in ongoing crises of food insecurity and hunger (Shiva, 2000).

Many critiques of development draw upon neo-colonial perspectives. Imperialism describes 'theories and practices developed by a dominant metropolitan centre to rule distant territories, by force, by political means or by economic, social, and cultural dependence' (Banerjee, 2003: 146). Colonialism, a consequence of imperialism, involves the establishment of settlements on outlying territories. A range of relations, between nation states, international institutions and transnational corporations, structures the process. Colonial rule brought highly destructive patterns of resource extraction and cultural and social disintegration. Using a post-colonial perspective, global institutions, such as the Global Environmental Facility, the World Bank and the World Trade Organisation (WTO) – all of which are discussed below – are seen as important ways to institutionalise the relationships of imperialism and colonialism in the post-colonial period. This allows, for example, post-colonial relations to be played out in trade regimes between the industrialised and Third World. It also allows the development agenda to become part of the way in which the 'South' is constructed as in need of 'development' and 'progress', preferably achieved through adoption of Western marketisation and democratisation models and through the transfer of science and technology. These transfers would allow them to 'catch up' with the developed world.

> Ever since President Harry S. Truman coined the notion of 'under-development' in his inaugural address in January 1949, and promised assistance to the countries of the Southern hemisphere in their efforts to catch up with the North, it has been taken for granted that, first, development could be universalised in space and, second, that it would be durable in time. This belief has proved wrong.
>
> Development has in fact ... deepened the crisis of injustice between North and South, just as it has provoked a manifold crisis of nature which undercuts its prospects for the future. It has revealed itself to be finite in (global) space as well as in time.
>
> (Sachs, 1997: 71)

In addition to the idea that the Third World should 'catch up' with the developed world, more recent efforts at transfer have been driven by the belief that the developing world could 'leapfrog' over certain stages of development. It is hoped that 'environmental leapfrogging' would prevent latecomer countries from going through the same pollution-intensive stages of industrial development as industrialised countries have experienced in the past.

Environmental leapfrogging is the skipping of pollution-intensive stages of development by developing countries.

Environmental leapfrogging has attracted increasing attention in energy and climate debates. Here renewable energy technologies, that were either unavailable or unaffordable in the past, are seen as providing a new opportunity for carbon neutral growth and industrial development in the South. The idea that developing countries might be able to follow more sustainable, low carbon development pathways is particularly attractive in the context of climate change. However, questions arise as to whether developing countries are capable of this uptake and whether or not such pathways are suitable for use in economies characterised by local, small scale producers (Perkins, 2003). Developing countries often lack the capacity to select, absorb and adopt technologies that lie outside the conventional environmentally intense technology paradigm. Research has also found that leapfrogging is dependent upon a country's resources and that it involves much trial-and-error (Sauter and Watson, 2008). Furthermore, there is also growing evidence to suggest that many developing countries are industrialising in the same environmentally destructive manner as the developed world, as seen, for example, in the development pattern of emerging economies, including China, discussed in Chapter 12. Furthermore, as we will see in this chapter, developing countries are also locked into global trade and debt regimes that give them limited room for autonomous manoeuvre.

A post-colonial perspective on development has also been used to criticise the UN's engagement with sustainable development. The discourse on sustainable development is seen to share characteristics of colonising discourse, becoming another example of 'a Western style for dominating, restructuring, and having authority over' the Third World (Said, 1979: 3).

> In the sustainable development discourse, poverty is the agent of environmental destruction, thus legitimating prior notions of growth and development.
>
> (Banerjee, 2003: 159)

Such critics call for a radical reconceptualisation of development itself and its 'meta narratives' of progress. However, adopting a Third World perspective does not necessarily lead to an abandonment of the sustainable development project, especially as formulated by Brundtland. Development remains a prime objective of Brundtland in the face of poverty, disease and hunger. Brundtland argued that such is the overwhelming need of the world's poor that economic development may take precedence over environmental protection considerations. This argument also recognises that economic growth can provide the resources necessary to protect the environment in the poorest countries, not least because the lack of resources causes environmental damage as much as do the pressures of industrialisation. But economic growth in developing countries needs to be accompanied by limits to growth in high consumption societies.

Brundtland recognised that, given the Earth's limited resources, global equity, at current OECD consumption levels, is simply not possible. Thus, the promotion of global sustainable development involves a two-fold task: overcoming the barriers to sustainable development in the developing world and reducing the high consumption levels in the industrialised world. This is what is meant by the claim that sustainable development requires that the principle of equity be inserted into the development paradigm. An equitable development paradigm addresses the inequalities wherein the industrialised countries over-consume while most of the rest of the world consumes at barely subsistence levels. Thus, as discussed throughout this book, while the international,

---

Box 11.1

Inequality of consumption

- Twenty per cent of the world's people consume 86 per cent of its goods while the poorest 20 per cent consume just 1.3 per cent.

- The richest 20 per cent use 58 per cent of all energy and the poorest 20 per cent use less than 4 per cent.

- Twenty per cent of people produce 63 per cent of the world's greenhouse gas emissions while the other 20 per cent produce only 2 per cent.

- Twelve per cent of the world's people use 85 per cent of the world's water.

- The richest 20 per cent consume 84 per cent of all paper and have 87 per cent of all vehicles, while the poorest 20 per cent use less than 1 per cent of each.

(Royal Government of Bhutan, 2012: 53)

---

institutional engagement with this concept can be criticised, at the heart of the Brundtland formulation lies a strong argument for the radical transformation of the structures of political and economic power. This transformation provides the scope for Third World communities to construct varied and relevant development paradigms that reflect their needs, values and aspirations.

The promotion of sustainable development in the Third World has to be undertaken in parallel with actions directed at realising several interrelated development goals. In this context, Third World countries have to overcome multiple, and very pressing, obstacles. Climate change, for example, is expected to deepen the vulnerability of the poor by increasing water stress and reducing food security, as discussed in Chapters 6 and 7. More extreme weather events, coupled with sea level rises, will see the displacement of millions of people and potentially contribute to increased incidence of vector-borne diseases (Pisupati and Warner, 2003: 31). Climate change threatens to undo decades of development and poverty-reduction efforts, undermining opportunities for sustainable development.

This list gives an idea of the daunting nature of the tasks ahead and points to how the promotion of sustainable development is linked, in an integral way, with the resolution of complex social, economic and political problems. While the North also experiences many of these problems, including within the transition states of Eastern Europe, the intensity

Box 11.2

Obstacles to sustainable development in the Third World

- Asymmetrical global power relations;
- Post-colonial legacies;
- Indebtedness;
- Political corruption;
- Armed conflict and civil unrest;
- Racial, ethnic and religious tensions;
- Trafficking in drugs, arms and people;
- Organised crime;
- Limited capacity within the system of public administration;
- Lack of basic education, health provision and infrastructure;
- Growth of megacities and urban slums;
- Population growth;
- Disease, especially AIDS;
- Climate change;
- Loss of biodiversity and ecosystems;
- Limited agricultural productivity;
- Rapid resource depletion;
- Discriminatory attitudes and practices towards women and girls.

with which they are felt marks Third World society. Some of the issues, such as the rise of urban poverty in the megacities of the South, raise new challenges. This urbanisation of poverty is difficult to address, as most research and policy initiatives have focused on rural poverty (Hulme and Turner, 2014: 640). Despite helping to get a sense of the extent of the difficulties, the list reveals nothing about the *causes* of these problems, leaving unresolved the question as to why Third World countries are so socially, politically, economically and ecologically vulnerable. To understand this, we begin by looking at the engagement of Third World countries within the UN.

Engagement within the UN

It is fair to say that countries from the Third World approached the Rio Earth Summit with a mixture of fear and hope (Grubb et al., 1993). The fear was that the people of the South were to be asked to sacrifice their chance to live 'the American dream', while having to endure the environmental and social costs stemming from economic growth in the industrialised world. There was also fear that the North would use its environmental concerns to place new conditions upon overseas development aid (ODA), a 'green conditionality' that might well reflect Northern priorities and be supervised by global financial institutions in which the South has little faith. Yet, there was also hope.

Herein lay the possibility that the environmental concerns of the North may give the South some negotiation room in global politics. If the North wants the South to change their future behaviour and development paths, then they will have to meet Southern demands (Grubb et al., 1993). These demands included debt relief and increased ODA, as well as increased market access and fairer prices for their commodities.

For its part, the North felt that the South had to make some changes. Chief among these was the need to curb population growth and to introduce more stable, transparent and democratic political structures, principles of good governance discussed in Chapter 3. Most Third World countries resist international efforts to discuss population targets and controls and the UN has always found it difficult to find a common ground on which to discuss this topic. It raises sensitive cultural and religious issues and touches upon personal matters of reproduction. Yet, addressing planetary over-population is essential for the promotion of sustainable development, not least because unregulated population growth may push us beyond the carrying capacity of the planet, while simultaneously undermining efforts to raise living standards for the world's poor. However, a distinction needs to be made between population growth and population impact: in the South, population growth is the highest; but the population of the North has the greatest environmental impact. Over-consumption in developed countries contributes more to global, unsustainable patterns of development than population growth in the low-income countries (Goodland and Daly, 1996).

Differentiating the 'Third World'

Nevertheless, while many of the concerns expressed by the Third World in the lead-up and the aftermath to the Rio Earth Summit continue to plague international governance regimes, there have been several important changes since 1992 that call for a more differentiated picture of the 'Third World'. While subsistence livelihoods are a common feature, there are nonetheless significant divisions and differences within the G77 countries, as the group of Third World countries is known. There are differences in levels of economic development, social inequality and life chances, environmental conditions, in the nature of their trade and levels of exports, and in extent of political stability. These differences have been highlighted by the emergence of the BRIC (Brazil, Russia, India and China) countries, an emergence that undermines the notion that Third World countries can be viewed as a unified entity. This is not least because of the economic power BRIC countries are now beginning to possess (Taylor, 2014: 288). The BRICs, for example, already have a greater share of world trade than the USA. We will also see in Chapter 12 how emerging economies are playing a more active role in global governance, and attention in particular will focus on the rising global economic, political and environmental significance of China. While it is uncertain how these emerging economies will interact with established powers, it is nonetheless clear that a shift of influence away from the Western world is occurring. From this perspective, while it was permissible to assert that in the immediate post-war period the G77 possessed common interest, such commonality of purpose has since diminished (Taylor, 2014). In addition, the so-called 'rise of the South' is also to be noted, as Indonesia, Mexico, South Africa, Thailand, Turkey and other developing countries are growing in economic and political importance. The UN's 2013 *Human Development Report* identifies more than forty developing countries that have done better than expected in human development in recent decades, with their progress accelerating markedly over the past ten years (UN, 2013a). In addition, the post-war division into the 'wealthy capitalist world', the communist/socialist world and the large majority 'underdeveloped world' has been transformed by the demise of communism and the transition to marketisation and democratisation in the countries of Eastern Europe, as discussed in Chapter 10. Thus, the current global map displays 'a configuration of nations and communities that fit uncomfortably with inherited categories and divides' (Marshall, 2014: 564). Recognition has also to be given to the fact that there is not 'one set of villains and another of victims' (WCED, 1987: 47). Social inequalities also exist within the Third World, many of which have political and economic elites that enjoy affluent, Western lifestyles. In contrast, the bulk of their populations, especially in rural areas, are struggling to eke out a livelihood. Problems of predatory corruption contribute (Marshall, 2014: 569) in part to these internal inequalities. It is also important not to construct an image of Third World communities as passive victims of wider global processes. The recent growth of resistance movements against globalised agriculture and biotechnology at the local level within many developing states and the spread of anti-globalisation alliances, witnessed in demonstrations in Seattle in 1999 and Genoa in 2001 and that now routinely accompany meetings of both the G8 and G20 countries, provide ample counter-evidence against such a view.

Recognising such differences between countries, many international organisations, including the UN, distinguish least developed countries (LDCs) within the G77. Since 2012, the UN has adopted a broader understanding of LDCs than it had previously, which had confined understanding to matters relating to economic growth. LDCs are now defined as low-income countries suffering from the most severe structural impediments to sustainable development (UN, 2013b). The explicit acknowledgment of the prospects of promoting sustainable development as the criterion of classification reflects the UN's concern to view development across its economic, social and environmental dimensions. The UN sees handicaps to sustainable development in terms of low levels of human resources and high levels of structural, economic vulnerability. In 2014 there were forty-eight LDCs recognised by the UN.

Box 11.3

Criteria used to classify LDCs

- Gross national income per capita.

- Human Assets Index: this provides information on the level of development of human capital. It is a combination of four indicators: two of health and nutrition and two of education.

- Economic Vulnerability Index: this reflects the risk posed to a country's development by exogenous shocks, the impact of which depends on the magnitude of the shocks; and on structural characteristics that determine the extent to which the country would be affected by such shocks, that is, their resilience.

(*Source*: UN, 2013b)

Promoting sustainable development in LDCs is particularly problematic. Poverty is widespread, education and employment opportunities are weak, health care is minimal and life chances are severely restricted, especially for women. Several LDCs will not meet a single Millennium Development Goal (MDG), discussed below. They are among the most vulnerable in the international trading and economic system. Many, for example, are net food importers, which increases their vulnerability to food price increases and drops in export earnings. The food price rises in 2007 and 2008 worsened the food security situation for many countries already experiencing limited coping capacities, as discussed below. They also have the least capacity to respond to global environmental change, including climate change, although these changes will bring significant adverse impacts, as discussed in Chapter 6. Their capacity for environmental management is further weakened as many are caught up in webs of international indebtedness, which often forces countries to deplete their natural resources to service debt repayments. This makes it difficult to manage resources in ways that are relevant to local, long-term needs.

UN engagement with LDCs is structured under Programmes of Action, framed by sustainable development goals. There have been four such Programmes to date. In 2011 the UN undertook an appraisal of its Third Programme of Action (2001–2010), which found that:

> The poorest and most vulnerable countries still have a long way to go to fully achieve economic, social and sustainable development, structural transformation and graduation from least developed country status. The Programme of Action remains unfinished business. There is a need for continued and enhanced support through a new programme of action, as business as usual will not suffice.
>
> (UN, 2011a: 19)

The Fourth UN Conference on Least Developed Countries, which took place in Istanbul in 2011, saw member states agreeing to the *Istanbul Programme of Action for Least Developed Countries for the Decade 2011–2020* (UN, 2011b). However, the UN does not act alone and it is also fair to say that the spread of global social movements, such as the UK Jubilee Debt Campaign and Make Poverty History, alongside the growth of international NGOs such as Oxfam, World Vision and CARE, have transformed international development policy into a complex array of global campaigns, exchanges and actions (Marshall, 2014: 568). The importance of civil society in the promotion of sustainable development is a recurring theme in this book. Its significance is, for example,

reflected in the 1996 decision of the IMF and the World Bank to agree
debt relief through the Highly Indebted Poor Countries Initiative, followed in 2005 by the Multilateral Debt Relief Initiative intended to help
highly indebted, poorer countries achieve the MDGs, discussed below.
Both of these decisions followed intense lobbying by NGOs and other
groups. However, while these initiatives have made some contribution
to reducing the debt burden of the LDCs, often one relief is followed by
another pressure. Concerns are growing, for example, that the rise of
China, with its land grabs, loans and trade deals with African countries,
will cause a new wave of highly indebted poor countries on that
continent.

Understanding the ways in which the myriad international organisations
and financial regimes are involved in shaping the economic, environmental and development trajectories of Third World countries is a complex task. One way to explore these influences is to examine key themes
that are emerging in the sustainable development agenda of the Third
World. Six key themes have been identified as being of particular rele-
vance, and in the following section these themes are used as lenses
through which to explore whether, and in what ways, the global governance regime promoted by the UN is responsive to the needs, values and
aspirations of the Third World.

Sustainable development: themes from the Third World

Theme 1: poverty

The post-war period brought the countries of Asia, Latin America and
Africa into the international system of global macroeconomic management through the work of the Bretton Woods Institutions, that is, the
International Monetary Fund (IMF) and the World Bank. By the 1980s,
the lending programme of the World Bank, designed to enable the Third
World to 'catch up' with development, resulted in a spiral of debt and
poverty when the interest rate on the so-called 'cheap' loans taken out in
the 1970s rose dramatically. When debtor countries were forced to call
in the IMF to avoid complete bankruptcy, the IMF imposed tough
'structural adjustment' conditions on the new round of loans. Structural
adjustment programmes involve the provision of loans conditional on
the adoption of 'structural adjustment' reforms. These reforms are
driven by the neoliberal belief that, in order to achieve long-term or

accelerated economic growth in poorer countries, economies need to be restructured and government interventions reduced. Thus governments are encouraged or forced to reduce their role in the economy by privatising state-owned industries, including the health sector, and opening up their economies to foreign competition (WHO, 2014a). In addition, countries were put under pressure to increase their exports to pay for the debt, a process that encouraged unsustainable exploitation, especially of forest resources (Dresner, 2002; Reed, 1997). These programmes also had gender-specific impacts (Baker, 1994).

The 1990s saw issues of poverty rise on the international agenda, amidst growing evidence about the ways in which the spread of Western models of development deepened poverty. This period saw the World Bank argue that economic development had to be accompanied by social policies. Its 'dollar a day' head count, introduced in 1990, estimated that around 1 billion people in the world lived in extreme poverty (Hulme and Turner, 2014: 634). Although adjusted upwards over time, the 'dollar a day' count has become a commonly used measure for purposes of international comparisons of income poverty. The count, however, saw poverty understood largely in terms of income: to be poor meant that one could not afford the cost of providing a proper diet or home. Similarly, the UN had also measured poverty in terms of the ability to meet a minimum number of calories or to have a minimum level of income to satisfy needs, that is, it was measured in terms of income poverty. A poverty line defined this minimum level and the poor constituted the actual number of people whose incomes or calorie intake was below this line.

However, during the 1990s, the concept of poverty was expanded to take account of non-monetary factors. Inspired by the work of Amartya Sen (Sen, 1999) and the 'capability approach', the United Nations Development Programme (UNDP) introduced measures for progress and for deprivation that focus on poverty from a human development perspective.

Most of the current discussions of poverty, including in international organisations such as the World Bank and the UN agencies, draw upon Sen's approach either explicitly or implicitly. Under the leadership of Kofi Annan in particular, when the UN paid considerable attention to poverty eradication, poverty began to be seen as a denial of choices and opportunities for living a tolerable life (UN, n.d.). Poverty encompasses factors such as isolation, powerlessness, vulnerability and lack of security, such as in relation to land rights, as well as an individual's capacity and capability to experience well-being.

Box 11.4

Capability approach to poverty alleviation

Capability for Sen means the alternative combinations of 'functionings' an individual can achieve. 'Functionings' refer to the various things a person may value doing or being. These can vary from elementary functionings like nourishment and shelter, to complex ones such as self-esteem and community participation.

Underlying the capability approach is a specific conception of what constitutes human well-being. At a very basic level, well-being can be thought of as the quality or the 'well-ness' of a person's being or living, and living itself can be seen as consisting of a set of interrelated functionings. The level of well-being thus depends on the level of those functionings, that is, how well a person can do or be the things he/she has reasons to value. This can include, for example, to what extent a person can be free from hunger or take part in the life of a community, and so on. The concept of 'capability' refers to a person's freedom or opportunities to achieve well-being in this sense.

The capability approach has implications for policy. It requires that policies should not focus on collective outcomes such as the distribution of income, but rather on building individual capabilities, and ensuring that people have the freedom to convert economic wealth into outcomes they desire.

Subsequent work by Ballet and others has extended the notion of capabilities from individuals to also cover societies.

(*Sources*: Lehtonen, 2004; Sen, 1999; UN, 2004)

Sen's capability approach has been important for furthering understanding of the social dimensions of sustainable development. It puts emphasis on improving social conditions from one generation to another, and on the interactions between the three spheres of development, that is, economy, ecology and society. Because it also focuses on meeting basic human needs, such as shelter and nourishment, the

Poverty is the inability to have choices and opportunities, a violation of human dignity. It means lack of basic capacity to participate effectively in society. It means not having enough to feed and clothe a family, not having a school or clinic to go to, not having the land on which to grow one's food or a job to earn one's living, not having access to credit. It means insecurity, powerlessness and exclusion of individuals, households and communities. It means susceptibility to violence, and it often implies living in marginal or fragile environments, without access to clean water or sanitation.

(*Source*: UN, 1998)

capability approach stresses the links between the environmental and the social pillars of sustainable development (Lehtonen, 2004), such as in the links between ecosystem services and food provision.

The capability approach is also supported by the elaboration of the Human Development Index (HDI), a composite measure of health, education and income. The HDI was introduced by the UNDP in its first *Human Development Report* (1990), has since become the most widely accepted and cited measure of its kind, and has been adapted for national use by many countries (UNDP, 1990). It has become an alternative to purely economic assessments of national progress, such as measuring growth in terms of GDP.

Policies to address poverty received a new impetus at the turn of the millennium, an occasion for soul searching and review, and for raising new aspirations for the future, with the 2000 UN Millennium Summit at centre stage (Marshall, 2014). The Summit resulted in the Millennium Declaration and the Millennium Development Goals (MDGs). The Declaration represented a reaffirmation of the UN's vision of a peaceful, prosperous and more just world, while also showing a commitment to new and bolder forms of partnership between different economic and social actors to address the continuing global problem of poverty (Marshall, 2014: 565).

However, to date, many of the MDGs have not been met and the spectre of poverty still haunts humankind. Furthermore, it has been difficult to keep the issue of poverty on the international agenda. This is despite the fact that the economic crisis of 2007 onwards has, according to the FAO, seen an additional 100 million people fall into hunger between 2008 and 2009 (Hulme and Turner, 2014: 635).

The failure to realise the MDG has led critics to claim that: 'for the majority of member states ... the years since 2000 have been "business as usual" at the UN General Assembly – that is, making grand statements about the eradication of poverty without demonstrating commitment through action' (Hulme and Turner, 2004: 635). The year 2015 was set as the date for the achievements of the MDGs, and the UN secretary-general has appointed a System Task Team to make preparations for the post-2015 UN development agenda. This includes strengthening the links between the MDGs and sustainable development after 2015 (UN, 2014). It will be recalled from Chapter 5 that the outcome of the Rio+20 Summit included agreement to establish a set of sustainable development goals. There is broad consensus now within the UN on the

Box 11.5

The UN Millennium Declaration and the MDGs

The Millennium Declaration, agreed at the Millennium Summit in 2000, summarised the agreements and resolutions of the UN world conferences held during the previous ten years. It also led to the MDGs, which serve as benchmarks for measuring actual development. The MDG are a set of development targets that centre on halving poverty and improving the welfare of the world's poorest by 2015.

There are eight MDGs, with the environment as an essential component. The first seven are about poverty reduction and improving health. These goals are directly linked to the promotion of sustainable development. Each contains deadlines and measurable ends, to be achieved by 2015.

Goal 7 is particularly important for the promotion of sustainable development. It has several targets: mainstreaming the environment into policies and programmes (environmental policy integration); reversing the loss of environmental resources and of biodiversity; and improving access to environmental services.

The MDGs aim to halve, by 2015, the proportion of people without sustainable access to safe drinking water. They also seek, by 2020, a significant improvement in the lives of at least 100 million slum dwellers.

Goal 8 stresses that the achievement of these goals requires global partnership for development. The MDGs are reflected in the Johannesburg Plan of Implementation, agreed at the WSSD.

(*Sources*: UN, 2006; UNDP, 2013)

need to have one global development agenda for the post-2015 period, with sustainable development at its centre. In addition, it is expected that more attention will focus on inequalities, especially given the growing gap between the rich and the poor, including in emerging economies.

Although disappointing in terms of actual performance, the MDGs are important for the promotion of sustainable development. The formulation of the MDGs helped to define globally shared objectives for reducing poverty and secure higher levels of human development. The have also helped to differentiate the needs of different countries and regions, recognising, for example, that sub-Saharan Africa experiences the most extreme and multidimensional poverty, while Asia has the highest number of poor people (Hulme and Turner, 2014). They have focused attention on crucial bottlenecks to development, including finance, technical skills and implementation capacity. From a wider perspective, the emphasis on the development of human capabilities, including through education and health, has helped to emphasise quality of life rather than narrow notions of economic prosperity. The MDGs have also sharpened awareness of the impact of environmental degradation and the loss of ecosystem services on the poor. It is now recognised that in all regions, and particularly in sub-Saharan Africa, the condition and management of ecosystem services is a dominant factor influencing prospects for reducing poverty. The degradation of ecosystem services is now understood as a significant barrier to achieving the MDGs, with growing recognition that many of the regions facing the greatest challenges in achieving the MDGs are also those facing significant problems of ecosystem degradation (World Resource Institute, 2005). Since biodiversity impacts on poverty, equality in access to and use of biodiversity resources is fundamental for poverty alleviation and for the achievement of sustainable development (Pisupati and Warner, 2003: 7). This understanding makes strong links to the normative principles of sustainable development, especially those dealing with equity.

> There is nothing about the MDGs as an ideal which is revolutionary, but the combination of their simplicity, moral weight, specificity and coherence as a set of interconnected objectives – and a firm foundation in high-profile, universal assent at international level – gives them a normative power and appeal. ... This is a thin reed, but it is not worthless; it is a feature the MDGs share with the human rights regime: the potential of norm diffusion and embedding to edge the powerful beyond what deliberative self-interest would otherwise entertain.
> (Poku and Whitman, 2011: 189)

Further progress in the understanding of poverty and what constitutes development have also resulted from Rio+20. In the lead-up to the Summit, the International Council for Science (ICSU) published a series of policy briefs that called for a move beyond GDP as a measure of progress, with focus instead placed on inclusive wealth and well-being (ICSU, 2012). Following from the Summit, there are increasing moves to have happiness and well-being concepts included in the potential sustainable development goals mentioned above. As the UN secretary-general, Ban Ki-moon, has stated:

> We need a new economic paradigm that recognizes the parity between the three pillars of sustainable development. Social, economic and environmental well-being are indivisible. Together they define gross global happiness.
> (UN, 2012)

The focus on happiness and well-being as indicators of development is about 'redefining what we mean by progress' (Earth Institute, 2012). The first *World Happiness Report* produced by the Earth Institute showed that, while the least happy are poorer countries, social factors, such as supportive relationships, personal freedoms and the absence of corruption, are more important than income in determining happiness (Earth Institute, 2012). While recognising that the lifestyles of the rich imperil the survival of the poor, the Report also warns of the disorders of development and its related focus on growth:

> Affluence has created its own set of afflictions and addictions. Obesity, adult-onset diabetes, tobacco-related illnesses, eating disorders such as anorexia and bulimia, psychosocial disorders, and addictions to shopping, TV, and gambling, are all examples of disorders of development. So too is the loss of community, the decline of social trust, and the rising anxiety levels associated with the vagaries of the modern globalized economy, including the threats of unemployment or episodes of illness not covered by health insurance. … These phenomena put a clear limit on the extent to which rich countries can become happier through the simple device of economic growth.
>
> (Sachs, 2012: 3–4)

The second *World Happiness Report* (2013) further strengthened the case that well-being should be a critical component of how the world measures its economic and social development (SDSN, 2013). Considerations of human happiness are contributing to a new understanding of what constitutes progress by moving beyond indicators that merely consider the possession of material goods (or the income to buy them). By refocusing attention on quality of life, such an understanding has the potential to contribute to the promotion of stronger forms of sustainable development. The focus on well-being and happiness reveals that the quest for material possessions and the persistent creation of new material 'wants' does not in itself make us better off. This realisation can, in turn, help support value changes, especially in high consumption societies, allowing material goods to be seen as a means to meet human needs, rather than an end in itself.

Theme 2: food security

The issue of food security can help us understand the complexities of sustainable development when viewed through a development lens.

> Food security exists when all people, at all times, have physical, social and economic access to sufficient, safe and nutritious food which meets their dietary needs and food preferences for an active and healthy life.
>
> (FAO, 2003)

Food security is linked to all three pillars of sustainable development. It is linked to the social pillar, especially to matters of health and nutrition. Food security raises issues in relation to whether households get enough food, how it is distributed within the household, and whether that food fulfils the nutritional needs of all its members. It is also linked to the economic pillar of sustainable development, including to issues of trade. International agriculture agreements, for example, are crucial to a country's food security. However, the link is not straightforward. Trade liberalisation, as promoted by the international trade regime of the WTO, might increase the supply of food commodities, reduce prices in food-importing countries and remove some pressure on these countries' natural habitats. At the same time, reductions in trade barriers might also lead to increased production in food-exporting countries, where commercial agriculture could increase vulnerability to deforestation, pests, diseases, and/or natural disasters, and might reduce the availability of ecosystem services (Sachs et al., 2009). Food security is also linked to the environment pillar, including through the close connection that exists between ecosystem services and food supply, as mentioned above. Food security is also critically linked to the maintenance of biodiversity. Biodiversity is essential to food provision, by making wild foodstuffs available and by supporting food production through soil formation and land productivity, pest and disease control in agricultural systems, and pollination. Biodiversity also enhances the adaptation and resistance (resilience) of crops, which helps against future risks associated with climate change. However, poverty alleviation, food security and conservation strategies are not related in a simple, cause-and-effect manner. While synergies do exist between biodiversity and alleviating poverty, oftentimes these prove elusive in practice and difficult trade-offs have to be made, as discussed below (Billé et al., 2012). The links between food security and social, economic and environmental dimensions of sustainable development are thus complex.

Consequently, it comes as no surprise to find that there is a great deal of debate around food security and conflicting arguments abound. These include claims that:

● There is enough food in the world to feed everyone adequately; the problem is distribution.

- Future food needs can – or cannot – be met by current levels of production.
- National food security is paramount – or no longer necessary because of global trade.
- Globalization may – or may not – lead to the persistence of food insecurity and poverty in rural communities.

(WHO, 2014b)

One area that has attracted considerable attention is the link between rising investment in biofuels and food insecurity. This is not least because the rise in the price of basic foodstuffs, in particular during the 2007–2008 food crisis, has been attributed to the conversion of land from agricultural food production to the production of biofuels.

Box 11.6

Biofuels and food security

Background

Biofuels are liquid fuels made from organic matter, typically crops. There are two principal kinds: ethanol, produced from carbohydrates such as sugar cane, sugar beet, corn or wheat; and biodiesel, manufactured from oilseeds, including rapeseed (canola), oil palm, soy and jatropha.

The growth of biofuels is designed to tackle two of the most difficult energy challenges: ensuring security of energy supply and addressing climate change. As a result, many governments have set targets for biofuel production or use, and these are often mandatory. Much of the original attraction of biofuels lay in their perceived GHG neutrality. As crops grow, they fix carbon from the atmosphere. When they are burned (as biofuel), this carbon is simply released back, so that over the life cycle of the fuel, the net impact on atmospheric carbon is neutral.

However, biofuels are no longer seen as GHG neutral, as there are emissions associated with all stages of their lifecycle, particularly if the crops are grown intensively, using nitrogen-based fertilisers and machinery, grown on deforested land, or if the refining process requires large inputs of (fossil) energy.

Food security

The growth of biofuels incentivises the diversion of crops and agricultural land away from food and into fuel production. Biofuels are considered by the UN, World Bank, IMF and OECD as having played a significant role in the food crisis of 2007–8. The FAO has also argued that they have pushed up the price of foodstuffs. Biofuels do not just consume food directly; they compete with it for land, water, and other inputs, pushing up prices. Estimates that the total biofuel consumption is set to increase tenfold between 2004 and 2030 add to these concerns.

Impact on sustainable development

The use of biofuels can deflect attention from the need to reduce consumption in the industrialised world. In the meantime, the social and environmental costs of biofuel policies, in terms of deepening poverty and environmental degradation, are being borne by developing countries. The industry is characterised by high concentration of land and resource use. The growth of biofuels has contributed to deforestation, including for example in Indonesia. It has also been associated with the displacement of rural communities. Secure access to land is also threatened by unregulated biofuel expansion which is pushing vulnerable communities aside and undermining agrarian reform programmes. The industry is also plagued by poor working conditions. In poor countries, biofuels may provide some opportunities for national consumption or for export. However, the opportunity costs associated with diverting feed stocks into biofuel production, rather than selling them on commodities markets, can be significant.

(*Source*: Oxfam International, 2008)

The case of biofuels reveals the complex nature of the trade-offs between developing low carbon energy to address climate change, maintaining biodiversity and ensuring food security.

Theme 3: gender

Prior to the publication of Ester Boserup's seminal book on women and development (Boserup, 1970), it was believed that development processes affected woman and men in the same ways. However, Boserup's research on women in developing countries challenged many of the fundamental assumption of international development policy, including that development would 'trickle down' and that it would result in equitable distribution and equality of benefits. Subsequent research has revealed the extent to which gender inequalities persist. In India, for example, women provide 75 per cent of labour for transplanting and weeding rice, yet fewer than 10 per cent actually own land; during rainfall shortages in India, more girls die than boys, and the nutrition of girls suffers more during periods of food shortage and rising food prices. Similarly, an analysis of credit schemes in five African countries found that women received less than 10 per cent of the amount of credit awarded to male smallholders (IUCN, 2010).

Box 11.7

Women's Environment and Development Organisation

In 1990 the Women's Environment and Development Organisation (WEDO) was established. It was founded specifically to influence the 1992 Earth Summit. WEDO's mandate is to work to ensure that sustainable development policies, plans and practices are gender responsive.

The mission of WEDO is 'to increase the power of women worldwide as policymakers in governance and in policymaking institutions, forum and processes, at all levels, to achieve economic and social justice, a peaceful and healthy planet and human rights for all'.

WEDO acted as a lead voice for women in the run-up to the Rio Earth Summit. Its policy document 'Women's Action Agenda 21' served as a basis for introducing sections on gender equality in Agenda 21 and the Rio Declaration.

WEDO has since participated in all the UNCED summits, as well as the major international UN conferences on development, including at Beijing, Istanbul and Cairo, and the UN Commission on the Status of Women in New York in 2014. The related 'Women's Action Agenda for a Healthy and Peaceful Planet 2015' served as a basis for women's lobbying during the WSSD events. They subsequently formed part of the Women's Major Group for the Rio+20 process.

WEDO has helped to establish the Women's Caucus, which acts as an advocacy group, advancing women's perspectives at the UN and other inter-governmental forums.

(*Sources*: WEDO, 2004–2008; 2014)

The connections between gender and the environment now form a common theme within Third World sustainable development discourse (Buckingham-Hatfield, 2000). That the Brundtland conceptualisation of sustainable development and the global efforts of the UN environment and development programmes now recognise a gender dimension is also due to the activities of women's groups. In particular, it owes much to efforts mounted during the UN International Decade for Women (1976–1985) and the work of the UN International Research and Training Institute for the Advancement of Women. The participation of women's networks in the 1992 Rio Earth Summit and the Women's Environment and Development Organisation (WEDO) policy document 'Women's Action Agenda 21' and the *Planeta Femea* event held at the Global Forum at Rio all helped to infuse a gender perspective into the output of the Rio Summit.

In 2013, the IUCN established the 'Environment and Gender Index' to provide quantitative data on governments' performance in translating the gender and environment mandates in the three Rio conventions into national policy and planning (IUCN, n.d.).

As a gender dimension was added to the study of the development process, a new theoretical approach was needed (Momsen, 2004). This approach has become known as the Women, Environment and Development (WED) perspective. The WED discourse highlights the links between the social and economic position of women in the Third World and environmental degradation. The position of women makes them more vulnerable to the negative effects of environmental degradation than their male counterparts. They are more marginalised, usually work harder, especially if engaged in agricultural labour, have less adequate diets and are often denied a voice in the political, economic and social spheres. It makes an explicit link between gender inequality and the impacts of environmental degradation, drawing attention to how accelerated environmental degradation is making women's daily search for firewood, fodder and water more difficult. The controversial link between population size and environmental degradation also comes to the fore. Thus, for example, the work of CBD now acknowledges that just as the impact of biodiversity loss is disproportionately felt by poorer communities, there are also disparities along gender lines. Biodiversity loss affects access to education and gender equality by increasing the time spent by women and children in performing certain tasks, such as collecting fuel, food and water. Thus, the conservation of biodiversity has to build upon understanding and exposing gender-differentiated biodiversity practices, gendered knowledge acquisition and usage (UNEP, n.d.a).

Box 11.8

Gender-specific impacts of environmental degradation

- Women make up the majority of the world's poor. Poverty is caused by, but also contributes to, environmentally damaging patterns of natural resource use.

- Given their economic and social roles, environmental degradation and biodiversity loss have a disproportionate impact upon the daily lives of women.

- Given their vulnerable economic and social positions, the negative aspects of globalisation (growing inequalities, inequitable distribution of wealth and resources) have a disproportionate gender impact.

- Environmental security and health issues have a gender dimension: given women's reproductive and social roles, environmental hazards have different effects on women and men.

Box 11.9

Women and the promotion of sustainable development

- Because of their domestic, agricultural and cultural roles, women are key agents in the promotion of sustainable patterns of natural resource management.

- Women are holders of knowledge about their local environment (indigenous ecology). This stems from their role in the provision of food and traditional medicine. They are therefore the key to developing appropriate biodiversity preservation strategies.

- Promoting sustainable livelihoods at the community level can be accelerated by giving women the right to inherit land and to have access to resources and credit.

- There is a strong connection between promoting human rights, especially for women, and promoting sustainable development, which is based upon principles of equity and partnership.

- Promoting democratic environmental governance has a gender dimension: participation based on the principle of gender equality of access is more legitimate, democratic and effective.

However, the early years of the WED debate presented women as passive victims of the environmental degradation stemming from global processes. As the discourse shifted from discussion of the environment to that of sustainable development, a new focus emerged. This emphasised women's positive role as efficient environmental resource managers within the development process.

Several of the arguments put forward in support of a gender perspective on sustainable development, especially those that stress the role of women in local promotion efforts, are built upon a claim that women had 'privileged knowledge and experience of working closely with the environment' (Braidotti et al., 1994: 2). However, some have gone further than this, stressing that women have a special relationship with nature, a claim known as ecofeminism.

Ecofeminism has been very important in providing a critique of traditional development models (Baker, 2004). It has helped in particular to further our understanding of multiple, intersecting inequalities and their impact on efforts to promote sustainable development (Cudworth, 2013). The ecofeminist approach has also been used to critique the notion of ecological footprint, discussed in Chapter 2. Ariel Salleh, for example, has argued that the methods used to calculate footprints ignore

Box 11.10

Ecofeminism

Ecofeminism is both an analysis of society–nature relationships as well as a prescription for how these relationships can be transformed.

Ecofeminism draws upon the feminist theory of patriarchy and combines this with insights gained from environmental and peace activism. A central argument is that a common dualistic belief system, rooted in the principle of domination and subjugation, underlies modern, negative attitudes towards both women and nature. To counteract this dominant ideology, ecofeminism aims to reconstruct a new understanding of the place of human beings within the natural world. In particular, it aims to situate women, nature, and sometimes men, in a more balanced and equitable relationship with each other.

Ecofeminism, as political activism, arose from what had hitherto been two different social moments, the environmental movement and the women's movement. From within the latter, it has inherited much from the women's peace and spirituality movements. It seeks to counteract the myriad ways in which the degradation of the natural environment impinges on the daily lives of women, especially in the Third World.

A convention has grown up in the literature that divides ecofeminism into two broad groupings: 'cultural ecofeminism' and 'socialist ecofeminism'.

Cultural ecofeminism

Cultural ecofeminism draws heavily on the tradition of radical feminism. Radical feminist analysis has located women's oppression with men, particularly with male sexuality, regarded as the site of male power. Central to radical feminism is the belief that there exists an innate female nature, which differs from the gendered self of the male. Cultural ecofeminism has extended this central tenet to argue that women, by virtue of their bio-logical capacity, have a closer relationship with the natural world than do men. This is significant for social movement activity: it has allowed women to keep sight, throughout history, of the mutually interdependent relationship that exists between humanity and the natural world. Thus, women are in a unique and advantageous position to engage politi-cally, culturally and socially on behalf, and in defence, of nature. This 'standpoint theory', as it is known, is based on the belief that only those that are oppressed can understand and counteract the relationship of oppression.

The claim that women have an innate nature or essentialist characteristics is called *essen-tialism*. This return to essentialism is the source of considerable disquiet, one could almost say alarm, within the broader contemporary feminist movement. Many contemporary feminists strongly distrust the ecofeminist reconnection of femaleness with the sphere of reproduction and with nature. Such connections are seen as dangerously conservative, representing the antithesis of the aims of the feminist movement.

Acceptance or rejection of essentialism provides a key way of distinguishing the two main tenets of ecofeminism.

Socialist ecofeminism

Crucial to the socialist ecofeminist position is the claim that the exploitation of nature relates to exploitation in society. The reason why women have different experiences of nature than men is not due to their 'essential nature' but, rather, to the fact that we live in a gendered society. Women do not have a natural affinity with nature, but rather the link between women and the environment is socially and culturally constructed.

(*Sources*: Baker, 2004; Buckingham-Hatfield, 2000)

the gendered divide between paid and unpaid work, as well as transport and energy use (Salleh, 2009). This way of measuring environmental impact panders to practices that render women invisible in policy. Research has also been extended to explore the gendered impacts of climate change (Spitzner, 2009; UN Women Watch, 2009).

However, some feminists have severely criticised the argument that women should provide the moral and the practical efforts necessary to reverse environmental deterioration. While promoting sustainable development can be seen as progressive, 'cleaning up' after men is not and conforms to existing stereotypes about women and their role in society. Women's involvement in the provision of Primary Environmental Care, for example, promoted by some Third World agencies, is seen as adding to women's daily burden. While participatory and community based, it nonetheless equates 'community' work with voluntary, unpaid work by women. This form of green participation is not seen as ultimately liberating for women (Agarwal, 1997). Despite these critiques, however, there are numerous examples of activities that have helped to promote gender-sensitive sustainable development trajectories. Some include capacity enhancement activities, while others involve more direct forms of engagement. These latter include campaigns to protect traditional ways of life, reverse ecological damage and undertake ecological restoration projects.

India's Chipko, or tree hugging movement, formed in the 1970s, is among the most famous of the direct action campaigns. The Chipko movement helped develop a 'feminist forest paradigm' that has been influential across the Himalayas and beyond (Shiva, 1989). Similarly, the crisis in women's access to wood and water in Kenya motivated

Case study 11.1

Capacity building for women leaders in sustainable development

In Brazil, the Network for Human Development has tested a capacity building model geared to increase women's effective participation in Local Governance for Sustainable Development.

Capacity building involves a training course, with thirty to forty participants, selected to ensure balance among private and public sector and civil society actors. The training course develops skills on policy, planning and resource allocation; negotiation; computer proficiency; and the use of data resources. Training concludes with the elaboration of a common agenda for allocating responsibilities and identifying needed resources.

(*Source*: WEDO, 2001)

Wangari Maathai to launch the Green Belt Movement (Maathai, 2003). Wangari Maathai was subsequently to receive the Nobel Peace Prize for her contribution to sustainable development, democracy and peace. The multiple roles of ecological restoration are examined in various chapters in this book.

Similarly, the Environmental Movement of Nicaragua explicitly deals with women's issues, in particular the high levels of exposure of women agricultural workers to pesticides.

Theme 4: trade

Another theme relates to the links between trade, especially liberalisation policies or free trade, and sustainable development. The conflicts between free trade and environmental protection have come into ever-sharper focus as the pressures of globalisation, and the related increase in international trade, continue.

Established in 1994, the World Trade Organisation (WTO) is now the major international organisation that upholds trade liberalisation. The WTO is also responsible for enforcing international free trade law. Its mission is to promote:

> The optimal use of the world's resources in accordance with the objective of sustainable development ... Contributing to these

Box 11.11

The Green Belt Movement

Set up in 1977, the Green Belt Movement is an environmental organisation that empowers communities, particularly women, to conserve the environment and improve livelihoods. It was established in response to the needs of rural Kenyan women who reported that their streams were drying up, their food supply was less secure, and they had to walk further and further to get firewood for fuel and fencing.

The Movement has three main areas of activity: Community Empowerment and Education; Tree Planting; and Advocacy. It promotes ecological restoration, encouraging women to work together to grow seedlings and plant trees to bind the soil, store rainwater, and provide food and firewood. Women receive a small monetary token for their work. The goal is to create climate resilient communities through restoration and protection of forest watersheds, and the creation of sustainable livelihoods for communities in Kenya and across Africa. Ecological restoration activities, especially tree planting, help to maintain biodiversity and promote food security.

At the international level, the Movement advocates for environmental policy that ensures the protection of natural forests and community rights, especially communities living close to and in forest ecosystems in sub-Saharan Africa and the Congo Basin Rainforest Ecosystem.

(The Green Belt Movement, 2014)

objectives by entering into reciprocal and mutually advantageous arrangements directed to the substantial reduction of tariff and other barriers to trade and to the elimination of discriminatory treatment in international trade relations.

(Preamble to the WTO Agreement, 1994)

The free trade system upheld by the WTO is an important lynchpin in the global economic system, and its trade regimes play an important role in supporting globalisation (Hoekman, 2014). While it declares commitment to sustainable development, it upholds traditional economic values, particularly the view that free trade can be a route to modernity, especially for the Third World (Sampson, 2005). Free trade, it is argued, encourages a country to specialise according to its comparative advantage and, by exposing countries to competition, also forces more efficient resource use. In addition, an environmentally regulated free trade system can also apply sanctions against countries with low environmental standards, thus promoting higher environmental quality at a global level.

Box 11.12

Women, environment and health: pesticides

UNEP estimates accidental poisoning from exposure to pesticides causes 20,000 deaths and 1 million illnesses worldwide every year. To understand the gender implications it is helpful to consider:

● The differential use of pesticide between men and women during agricultural production;

● The unique health effects on women;

● The extent of information about pesticides available to each gender.

Women farmers and workers are frequently directly exposed to dangerous pesticides. Impacts on women's reproductive health include greater incidences of miscarriage, still-births and increased incidence of birth defects. There are also potential carcinogenic effects. One extreme example is the insecticide dichlorodiphenyltrichloroethane, commonly known as DDT, once widely used for controlling insect pests on agricultural crops. DDT is highly persistent in the natural environment and accumulates through the food chain. It increases the risk of breast cancer, and an infant feeding on breast milk can receive up to twelve times the acceptable limit of DDT. DDT is now illegal in many countries but it is used in many Third World countries, as it is cheaper than less persistent alternatives.

Chapter 14 of Agenda 21 recommends increased awareness of sustainable agriculture methods in women's groups. This includes reducing the use of agricultural chemicals and also making wider use of traditional practices of pest control. Examples of gender-sensitive activity in this area include the Environment and Development Action Network in Africa, and Mama-86 in Ukraine.

(*Sources*: MAMA-86; WEDO, 2004–2008)

Despite these claims, however, many argue that trade regulations restrict the ability of states, particularly those in the Third World, to promote sustainable development. It is not surprising, therefore, to learn that the WTO has received considerable attention from environmental activists, especially from anti-globalisation protesters. Ministerial meetings of the WTO in Seattle in 1999 and in Cancún in 2003 were met with large protests, followed in 2011 by the emergence of the Occupy Wall Street movement. High on the agenda of these protests are concerns about the way in which the international financial and trade systems create inequality.

Critics point to the lack of transparency in the way in which the WTO conducts its business. Trade experts dominate meetings, which remain closed to environmental and civil society organisations, behaviour not in keeping with the principles of good governance. However, the direct relationship between environmental protection measures and free trade policies is also a source of additional conflict. To deal with the conflict, the WTO has established a Committee on Trade and the Environment.

One of the issues that the Committee addresses is the use of trade restriction measures in multilateral environmental agreements (MEAs). Among the 250 MEAs currently in existence, the WTO estimates that twenty of these include environmental trade measures (WTO, 2014). The 1973 Convention on the International Trade of Engendered Species of Wild Flora and Fauna (CITES), the 1987 Montreal Protocol on Substances that Deplete the Ozone Layer and the 1989 Basil Convention on the Transboundary Movement of Hazardous Wastes and their Disposal are all good examples of MEAs that restrict trade.

A second area of concern is the trade disputes that arise when environmentally based distinctions are made between what would otherwise be considered 'like' goods: such as distinguishing 'dolphin safe tuna', fur from animals not caught in leg-hold traps and beef produced without artificial hormones. The WTO abhors such distinctions, seeing them as a form of protectionism. In contrast, these distinctions are seen by many as the backbone of environmental policy (Von Moltke, 1997). Thus, the WTO ruling that the US could not ban shrimp caught with nets that trap turtles became notorious among environmental activists (Dresner, 2002). Similarly, there are disagreements over attempts by governments to restrict the importation of goods that do not conform to certain environmental standards. The WTO sees this as a restriction on free trade, while national actors can see such measures as a way in which they can uphold their environmental standards. This strengthens arguments that trade liberalisation promotes a race to the bottom, as countries come under competitive pressure to lower their environmental standards.

Another area of disquiet is the impact that WTO rulings and agreements have on national environmental protection measures. Under extreme pressure to open up their markets, many developing countries find that free trade agreements have adverse effects on their agriculture, forests and fisheries. Significant and irreversible impacts on the forest sector, for example, have been found in 'biodiversity hot spot' countries, such as Brazil, Indonesia, countries in the Congo basin, and Papua New Guinea, from trade liberalisation (FoE, 2008).

Faced with such criticisms, there have been increased calls for the development of a trade regime that enhances, not destroys, the prospects for sustainable development. This resulted in the WTO meeting in Doha, Qatar in 2001 agreeing the Doha Declaration. A fundamental objective of the Declaration is to improve the trading prospects of developing countries and help the integration of LDCs into the WTO multilateral system. Another aim is to deal with Trade-Related Aspects of Intellectual Property Rights (TRIPS). However, the Doha round of negotiations on free trade has broken down, especially over the 'special and differential treatment' of developing countries, a key normative principle of sustainable development.

Some have argued that the breakdown of the Doha round is due, at least in part, to the growing importance of emerging economies, making it difficult to reach agreement along traditional lines and using established coalitions (Hoekman, 2014: 558). The Declaration, and its related work programme, can also be criticised for its limited understanding of the difficulties faced by the Third World in the global economy. Such difficulties are primarily seen as technical in nature, such as the lack of technical and administrative expertise, and are seen to limit their capacity to derive benefits from international trade. The solution is to support an international programme of enhanced transfer of best practice, technology and financial resources. This approach ignores the possibility that these transfers may increase the structures of dependency that tie the Third World into an inequitable relationship with the industrialised world. In addition, the underlying premise of the Declaration is that trade and the promotion of sustainable development can be mutually supportive. However, green theorists and supporters of strong sustainable development criticise this assumption. They argue that free trade, the development of global markets and the stimulation of economic growth are not consistent with the sustainable development principles of localism and equity (Pepper, 1996). In addition, free trade is seen as having resulted in a devastating pattern of environmental asset stripping, to the benefit of the industrialised world and to the detriment of the developing world.

> The use of 'free trade' to cement global economic integration removes the principle of sustainability from local communities across the globe. It does so by stimulating the trade in natural resources, and their products, without strengthening local communities or encouraging responsible environmental management. Resources are depleted to provide foreign exchange, and sustainable livelihoods are eroded.
>
> (Redclift, 1997: 395)

Theme 5: knowledge

Box 11.13

Chapter 35 of Agenda 21

- Enhance scientific understanding;
- Improve long-term scientific assessments;
- Strengthen scientific capacities in all countries;
- Ensure that the sciences are responsive to emerging needs.

(*Source*: UN, 1992)

Chapter 35 of Agenda 21 addressed the relationship between science, knowledge and the promotion of sustainable development. It focuses on the role and the use of science 'in supporting the prudent management of the environment and development for the daily survival and future development of humanity'. In order to fulfil this requirement, the UN has set out a programme for action under Agenda 21.

The UN has given a new role to experts and their scientific knowledge, both in identifying and in providing solutions to environmental problems. Some would argue that this has enabled science to play a dominant role in setting the priorities of the international environmental agenda (Blowers, 1997). This has led to several controversies, including for example in relation to the role of science in the development of new, commercial uses for plant and animal genetic resources.

The controversies over biotechnology reveal how knowledge is embedded in wider systems and practices that impact upon sustainable development. The application of science can have social and economic effects, as shown by the way in which distortions in the granting of intellectual property rights can marginalise the knowledge, practices and traditional economies of many Third World communities (Shiva, 1993).

There are other elements to the debate. While scientific knowledge is presented as authoritative, it is more often uncertain and contestable (Yearley, 1997). It can make mistakes, it can bequeath new environmental problems to future generations (for example, nuclear waste), and it can result in the reduction of biodiversity, as has happened with the

---

Box 11.14

Science, biotechnology and indigenous knowledge

The debate over biotechnology provides a good example of disputes over the relationship between science, economy and ecology. On the one hand, the biotechnological revolution is presented as an answer to the loss of biodiversity caused by intensive agriculture and deforestation. By giving a use value to plant and animal genetic resources, biotechnology provides a rationale for halting the destruction of the planet's rich biodiversity. These resources can then be put to use to develop new medicines and products.

In the contrasting view, the application of science is seen as an example of ongoing colonial relations between the North and the South, especially when account is taken of the fact that two thirds of the planet's species are within the Third World.

One particularly contentious issue is that the development of the biotechnology industry has led to the creation of intellectual property rights in relation to natural resources: for example, patents on seeds. The knowledge of pharmaceutical companies is protected by international law through the patent system. However, international law sees the knowledge of indigenous peoples – for example their knowledge of the medicinal properties of plants – as traditional and not 'novel' knowledge and therefore can be obtained without payment. This is seen to serve the interests of corporations at the expense of peasant and farmers rights and fails to guarantee the long-term survival of species. It privileges one form of knowledge over another. This has led to a growing controversy over Trade-Related Aspects of Intellectual Property Rights (TRIPS). Such a system also legitimises private intellectual property rights over life forms, which many find ethically contentious.

(*Source*: Banerjee, 2003)

---

development of scientific agriculture. Scientific agriculture has led to modern practices of mono-crop production in many Third World countries, such as coffee or banana plantations. This form of intensive farming is dependent upon the application of scientific methods of pest control and artificial soil enhancement. Intensive agriculture has created environmental problems that, in turn, require the further application of scientific solutions, leading to a spiral of dependency. There is also concern that Western knowledge is used to devise conservation management regimes, that bypass the indigenous ecology that has been used to 'manage' local environments over an extended time (Redclift and Woodgate, 1997). Scientific ecology, for example, has played a role in the designation of vast tracts of land as 'nature reserves'. There are several documented cases where this had displaced the communities that have depended on these lands for subsistence, while opening the reserves up for the pleasure of Western tourists. This happened for example in the designation of the land of the Chenchu community in southern India as a tiger reserve.

But, despite these tendencies, the decades since the Rio Earth Summit have seen growing recognition of the value of traditional knowledge for the promotion of sustainable development. The CBD, for example, specifically acknowledges the importance of traditional knowledge in preserving biodiversity and in achieving sustainable development.

Traditional knowledge refers to the knowledge, innovations and practices of indigenous and local communities. Developed from experience gained over the centuries and adapted to the local culture and environment, traditional knowledge is transmitted orally from generation to generation. It tends to be collectively owned and takes the form of stories, songs, folklore, proverbs, cultural values, beliefs, rituals, community laws, local language, and agricultural practices, including the development of plant species and animal breeds.

(*Source*: UNEP, n.d.b)

Traditional knowledge can make a significant contribution to sustainable development. Most indigenous and local communities are situated in areas where the vast majority of the world's genetic resources are found. Many of them have cultivated and used biological diversity in a sustainable way for thousands of years. Some of these practices have been proven to enhance and promote biodiversity at the local level and support healthy ecosystems. However, the contribution of indigenous and local communities to the conservation and sustainable use of biological diversity goes far beyond their role as natural resource managers. Their skills and techniques also have the potential to provide valuable information to the global community as it faces threats from climate change. One way in which this knowledge transfer is promoted is through the International Partnership for the Satoyama Initiative. At the CoP meeting of the CBD in Japan in 2010, members endorsed the Satoyama Initiative, *sato* literally meaning arable and liveable land and *yama* meaning hill or mountain. Satoyama has been developed through centuries of small scale agricultural and forestry use in Japan. Similarly, in many parts of the world, people have developed ways to utilise and manage their surrounding natural environment to sustain and improve their daily lives and production activities, such as agriculture, forestry and fisheries. Based on the knowledge and practices that have been accumulated over the generations at the local level, elaborate landscape management and production systems have been developed that support local communities

Box 11.15

UNESCO Local and Indigenous Knowledge Systems

- Secure an active and equitable role for local communities in resource management;
- Strengthen knowledge transmission across and within generations;
- Explore pathways to balance community-based knowledge with global knowledge in formal and non-formal education;
- Support the meaningful inclusion of local and indigenous knowledge in biodiversity conservation and management, and climate change assessment and management.

by providing foods, fuels and other materials, nurturing traditions and culture, and maintaining ecosystems and biodiversity. The Satoyama Initiative explores these ways and means of using and managing natural resources sustainably so that this knowledge can continue to be put to use for the benefit of current and future generations (IPSI, 2014).

Similarly, UNESCO, through its Local and Indigenous Knowledge Systems Programme, is demonstrating the link between cultural and biological diversity. Unique ways of knowing that encompass language, systems of classification, resource use practices, social interactions, ritual and spirituality are increasingly seen as providing a foundation for locally appropriate sustainable development actions (UNESCO, n.d.)

Theme 6: finance

In the run-up to the Rio Earth Summit, the Global Environmental Facility (GEF) was established by the United Nations Environment Programme (UNEP), the UNDP and the World Bank. The GEF receives funding from multiple donor countries, including the United States, and is an important part of the institutional structures established through Rio. It forms the lynchpin in the financial deal reached between the industrial and the Third World at the Summit. It serves as the primary funding mechanism for dealing with global environmental issues and it also acts as the financial mechanism for both the UN FCCC and the CBD. Its mandate is to fund the costs to the Third World of efforts to limit the global impacts in four key environmental areas: ocean pollution, climate change, biodiversity loss and ozone depletion. It provides grants and loans to cover the additional or 'incremental' costs, that is, the difference between a less costly, more polluting project and a costlier, more environmentally sound option. At the Summit itself, developed countries agreed to provide 'new and additional' financial resources to the Third World to enable it to meet the costs involved in implementing the Rio agreements. Chapter 33 of Agenda 21 reiterated the target of industrialised countries transferring 0.7 per cent of GNP as aid to poorest countries.

In practice, however, funding the promotion of sustainable development has proved to be both a difficult and protracted problem. Despite intense discussions at Rio and the expectations of the Third World, there is no satisfactory financial arrangement for funding the measures agreed in Agenda 21. In addition, the US soon fell into arrears in meeting its financial commitments to the GEF. Five years after the Rio Summit, the New York Summit came close to collapse on the issue of North–South finance. As a result, there were fears that the WSSD would be embroiled in acrimonious discussions over finance, which would doom the Johannesburg Summit to failure. Consequently, a strategic decision was made to get heads of state and government to attend an International Conference on Finance for Development in Monterrey, Mexico in 2002. As a result only very limited discussion took place at Johannesburg about financing the promotion of sustainable development.

The Monterrey meeting resulted in the so-called Monterrey Consensus, a commitment to a broad-based development agenda that takes into account poverty reduction and environmental sustainability as well as economic growth. Trade was emphasised as the 'critical engine of growth'. The Monterrey Consensus is distinguished by its recognition of both the need for developing countries to take responsibility for their own poverty reduction and the necessity for rich nations to support this endeavour with more open trade and increased financial aid. Monterrey also saw both the US and the EU pledge additional resources for poverty alleviation. Countries also reached agreements on other issues, including debt relief, fighting corruption, and policy coherence. Funding was subject to the implementation of good governance principles in the beneficiary countries, similar to the conditionalities imposed by the EU on Eastern Europe, as discussed in Chapter 9. Monterrey also provided an impetus for the private sector to take responsibility for promoting sustainable development. Though many of the key commitments of the Monterrey Consensus remain unfulfilled, it remains a key framework for international action on poverty reduction and underpins many of the current and ongoing discussions about ODA and the MDGs (UN, 2002).

The Johannesburg Summit subsequently agreed a replenishment of the GEP by US$3 billion.

However, the GEF continues to be plagued by low levels of funding from donor countries. It has also found it hard to leverage funds from the private sector, despite the increased role that the sector is expected to play in the promotion of sustainable development. Meeting financial commitments is just one problem. There are also problems regarding lack of trust in the GEF. The World Bank houses and manages the GEF secretariat and the GEF Trust Fund. The central role that the World Bank plays keeps the GEF under critical and constant scrutiny (Elliott, 2002). From the perspective of the Third World and many environmental and developmental NGOs, the relationship between the GEF and the World Bank is too close. Despite the restructuring of the GEF in 1994, concerns over the transparency and accountability of its governance structure remain. The GEF has also been criticised for its lack of strategic approach and for long delays in project initiation and implementation. Difficulties in defining what is meant by 'incremental' costs have also resulted in countries arguing over the requirements of additionality. Critics also claim that access to GEP money is used as a leverage to get recipient governments to commit to larger loans with political conditionality. It has thus been argued that 'These and other accusations of institutional torpor, inflexibility, and opaqueness have hindered GEF's objectives as an international site of coordination' (Lattanzio, 2010: 12–13).

The World Bank

Given the historical connection between structural adjustment policies and World Bank lending, the Bank has become subjected to relentless attention from NGOs and grassroots organisations. Critics point to the environmental impact of projects funded by the Bank, especially infrastructure projects, such as large dams, that have resulted in the dislocation of local, often tribal people. While it has become institutionally critical to the promotion of sustainable development, it has, in the words of one critic, 'become materially central to continued environmental decline in those countries in which it funds projects and programmes' (Elliott, 2002: 65).

Having been subject to severe criticism, the World Bank has undertaken several changes, including introducing safeguards and sustainability policies to prevent or mitigate the adverse impacts of its projects on people and the environment (World Bank, 2010). As a result of these changes, many now argue that it is has become a leader among multilateral institutions regarding its efforts to promote sustainable development and in the elaboration of indicators to monitor progress. But the Bank's understanding of sustainable development – as economic growth, environmental stewardship, and social inclusion – is closely aligned with the principles of ecological modernisation and not sustainable development, as discussed in Chapter 2. Its 2013 mission statement, for example, places strong emphasis on the need to support projects that promote green growth, with stress on what it refers to as shared prosperity, inclusive development and sustainable methods of natural resource management (World Bank, 2013). This sees the Bank providing support for low-carbon urban development projects dealing with energy efficiency improvements in buildings and the development of public transport; or supporting what it refers to as resource-efficient, climate-smart agricultural practices (World Bank, 2014).

It is fair to say that these reforms are more ambitious and inclusive than the institutional changes implemented by any other multilateral development bank (Reed, 1997). They have enabled the World Bank to set the pace and standard for other international organisations whose behaviour has environmental consequences. However, there is still a disjuncture between the formal requirements set by the Bank and actual respect for those standards in the design and implementation of projects. This is very often because many Third World countries lack the will or ability to respect the standards in practice (Reed, 1997). More fundamentally, there is the argument that the World Bank helps to accelerate the integration of the Third World into the dominant, growth-orientated development paradigm as discussed in Chapter 1. This condemns the Third World to a subordinate position in the international economic order. It also represents an example of what *Ecologist* has referred to as the 'closure of the commons', that is, bringing more and more areas, countries and resources under the remit of the international economic order, a position examined in Chapter 2.

There is also the criticism that the World Bank imposes a West-centric idea of development that supports a neoliberal, growth-oriented approach. Alongside its counterpart, the IMF, both these Bretton Woods institutions are driven by the neoliberal belief that through deregulation and privatisation, which allow markets to determine resource allocation, rapid economic growth would ensue. This Washington Consensus, as it is known, is now widely judged as having resulted in a 'lost decade' for Africa and other places (Hulme and Turner, 2014: 634). It also fails to recognise that humankind has inherited limited environmental stocks, which imposes irreducible constraints on the economic system. This weak sustainable development position

> allows the human community to cling to unrealistic expectations regarding achievable standards of living for the great majority of humanity and to believe that global inequalities and poverty can be addressed by more growth in both the North and the South.
>
> (Reed, 1997)

From this perspective, the World Bank forms part of the structural causes of unsustainability.

While the Bretton Woods institutions drive an economic growth model, the increased involvement of civil society in sustainable development initiatives has focused attention on new forms of finance. Microfinance, that is small-scale lending, is playing an increasingly important role in supporting local development opportunities, especially for woman, although, as discussed above, there are still some inequalities in access and treatment. Microfinance is also seen to act in support of bottom-up social innovations that can boost local community adaptation to global environmental change (Baker and Mehmood, 2013). It also gives recognition to institutions that go beyond the interests of elites as key agents in sustainable development (Acemoglu and Robinson, 2012).

This review of six key themes – of poverty, food security, gender, trade, knowledge and finance – reveals that Third World countries have a range of sustainable development priorities, which reflect their social, political and economic contexts. Many of these priorities stand in sharp contrast to the priorities of the industrialised world. It has also helped to make visible the interconnections that exist between these issues, such as between water, food and energy security, climate change, urbanisation, poverty, inequality and the empowerment of women. Discussion reveals that at the heart of efforts to promote sustainable development there is both recognition that economic, social and environmental objectives need to be treated in common, but that managing the trade-offs, such as between the maintenance of biodiversity and poverty alleviation strategies, or the growth of low carbon fuels and the pursuit of food security, present very fundamental sustainable development challenges.

Conclusion

In this chapter, the promotion of sustainable development in the Third World was shown to be a difficult task that triggers multiple, complex and highly contested issues. While Third World countries need to be differentiated, they nonetheless experience common constraints imposed by an international economic and political system that is orientated towards the promotion of values, norms and principles, such as the principle of free trade, which, in their present form, are incompatible with the principles of sustainable development. The trade system promoted by the WTO is not easily reorientated towards a sustainable development agenda, and neither are the growth-orientated economic development models pursued by the Bretton Woods institutions.

Exploration of the ways in which structural adjustment policies, the burden of external debt and the liberalisation of trade have contributed to unsustainable development patterns in the South has pointed to a recurring theme in this book: the need to address the inequitable basis of the global economic and political system, including its gendered expression. Promoting sustainable development requires breaking the causal connections between environmental degradation, poverty, and population growth. Not only does this require a reduction in consumption levels in the high consumption societies of the West, but ultimately calls forth a fundamental restructuring of the system of international political economy. Without addressing this fundamental need, the promotion of sustainable development will continue to face difficult challenges. Trade-offs will heighten tensions, as international policy tries to address different priorities, such as maintaining biodiversity, addressing poverty, shifting to low carbon fuel, and ensuring food security. These trade-offs are a product of the unwillingness of the international community to address the problem at source by imposing limits to growth on our finite planet.

Adopting a Third World perspective, some critics are highly sceptical of the discourse on, and engagement with, sustainable development that has emerged since the Rio Earth Summit. The process assumes that the resolution of global environmental problems requires large-scale capacity transfers to the poorer countries. Critics argue that the values that this approach promotes are so overwhelmingly Western in origin and interest orientation that they undermine an equity and justice agenda. They point out that the Western project to modernise post-colonial societies has contributed to poverty, an increase in economic and gender inequality, and to environmental degradation, which, in turn, further diminishes the life chances of the poor. While the sustainable development agendas of international institutions are flawed, the Brundtland formulation still contains powerful arguments for radical change.

Summary points

- The environmental crisis in the South is also a warning signal about the development model in the North.
- The promotion of sustainable development in the Third World requires policies that address poverty and food security, are sensitive to the gender dimension of development, embrace different types of knowledge, and reform global institutions.
- Addressing this array of issues is not straightforward and, under the current system, involves making difficult trade-offs between competing priorities.
- There are opposing views as to how to promote sustainable development in the Third World. One view seeks technological transfer, capacity enhancement and the transfer of funds. The second view calls for a fundamental restructuring of the international political and economic order.
- The Brundtland model contains a radical agenda of political and economic transformation in support of the sustainable development of the Third World.

Further reading

Acemoglu, D. and Robinson, J. (2012), *Why Nations Fail: The Origins of Power, Prosperity, and Poverty* (London: Profile Books).

Adams W. M. (2008), *Green Development: Environment and Sustainability in a Developing World* (London: Routledge, 3rd edn).

Greig, A., Hulme, D. and Turner, M. (2007), *Challenging Global Inequality: Development Theory and Practice in the 21st Century* (Basingstoke: Palgrave Macmillan).

UN Department of Economic and Social Affairs, *World Economic and Social Survey 2013: Sustainable Development Challenges*, Report E/2013/50/Rev. 1, ST/ESA/344 (New York: UN), available online at http://sustainabledevelopment.un.org/content/documents/2843WESS2013.pdf

References

Acemoglu, D. and Robinson, J. (2012), *Why Nations Fail: The Origins of Power, Prosperity, and Poverty* (London: Profile Books).

Agarwal, B. (1997), 'Gender Perspectives on Environmental Action: Issues of Equity, Agency and Participation', in J. Scott, K. Caplan and D. Keates (eds), *Transitions, Environments, Translations: Feminisms in International Policies* (London: Routledge), 189–225.

Baker, S. (1994), 'Structural Adjustment and the Environment: The Gender Dimension', in P. Rajput, and Hem Late Swarup (eds), *Women and Globalisation: Reflections, Options and Strategies* (New Delhi: Ashish, 1994).

Baker, S. (2004), 'The Challenge of Ecofeminism for European Politics', in J. Barry, B. Baxter and R. Dunphy (eds), *Europe, Globalization and Sustainable Development* (London: Routlege), 15–30.

Baker, S. and Mehmood, A. (2013), 'Social Innovation and the Governance of Sustainable Places', *Local Environment: The International Journal of Sustainability*, doi: 10.1080/13549839.2013.842964.

Banerjee, S. B. (2003), 'Who Sustains Whose Development: Sustainable Development and the Reinvention of Nature', *Organization Studies*, 24 (1): 143–180.

Billé, R., Lapeyre, R. and Pirard, R. (2012), 'Biodiversity Conservation and Poverty Alleviation: A Way Out of the Deadlock?' *Sapiens: Surveys and Perspectives Integrating Environment and Society*, 5 (1), available online at http://sapiens.revues.org/1452

Blowers, A. (1997), 'Environmental Policy: Ecological Modernisation or the Risk Society', *Urban Studies*, 34 (5–6): 845–871.

Boserup, E. (1970), *Woman's Role in Economic Development* (London: Allen and Unwin).

Braidotti, R., Charkiewicz, E., Häusler, S. and Wieringa, S. (1994), *Women, the Environment and Sustainable Development: Towards a Theoretical Synthesis* (London: Zed Books).

Buckingham-Hatfield, S. (2000), *Gender and Environment* (London: Routledge).

Cudworth, E. (2013), 'Feminism', in C. Death (ed.), *Critical Environmental Politics* (Abingdon: Routledge), 91–99.

Dresner, S. (2002), *The Principles of Sustainability* (London: Earthscan).

Earth Institute (2012), *World Happiness Report*, edited by J. Helliwell, R. Leyard and J. Sachs, available online at http://issuu.com/earthinstitute/docs/world-happiness-report

Elliott, L. (2002), 'Global Environmental Governance', in R. Wilkinson and S. Hughes (eds), *Global Governance: Critical Perspectives* (London: Routledge), 57–74.

FAO (Food and Agricultural Organisation of the UN) (2003), 'Trade Reform and Food Security: Conceptualising the Linkages' (Rome: FAO), available online at ftp://ftp.fao.org/docrep/fao/005/y4671e/y4671e00.pdf

FoE (Friends of the Earth) (2008), *Undercutting Africa: Economic Partnership Agreements, Forests and the European Union's Quest for Africa's Raw Materials* (London: FoE), available online at www.foe.co.uk/sites/default/files/downloads/undercutting_africa.pdf

Goodland, R. and Daly, H. (1996), 'Environmental Sustainability: Universal and Non-Negotiable', *Ecological Applications*, 6 (4): 1002–1017.

The Green Belt Movement (2014), 'Who We Are; What We Do', available online at www.greenbeltmovement.org/what-we-do

Grubb, M., Koch, M., Munson, A., Sullivan, F. and Thomson, K. (1993), *The Earth Summit Agreements: A Guide and Assessment* (London: Earthscan).

Hoekman, B. (2014), 'Global Trade Governance', in T. G. Weiss and R. Wilkinson (eds), *International Organisations and Global Governance* (Abingdon: Routledge), 552–563.

Hulme, D. and Turner, O. (2014), 'Poverty Reduction', in T. G. Weiss and R. Wilkinson (eds), *International Organisations and Global Governance* (Abingdon: Routledge), 632–643.

ICSU (International Council for Science) (2012), 'Human Well-being for a Planet Under Pressure: Transition to Social Sustainability', Policy brief no. 6, commissioned by the international conference 'Planet Under Pressure: New Knowledge Towards Solutions', available online at www.icsu.org/rio20/policy-briefs/Wellbeing_LR.pdf

IPSI (International Partnership for the Satoyama Initiative) (2014), 'The IPSI', available online at http://satoyama-initiative.org/partnership/

IUCN (International Union for Conservation of Nature (n.d.), 'About the EGI', available online at http://environmentgenderindex.org/about/

IUCN (2010), 'Environment and Gender Equality: The Keys to Achieving Millennium Development Goals', available online at www.iucn.org/about/work/programmes/social_policy/?6065/Environment-and-gender-equality-the-keys-to-achieving-Millennium-Development-Goals

Lattanzio, R. K. (2010), *Global Environment Facility (GEF): An Overview*. CRS Report for Congress, 10 June, Report R41165 (Washington, DC: Congressional Research Service), available online at www.cnie.org/NLE/CRSreports/10Jun/R41165.pdf

Lehtonen, M. (2004), 'The Environmental–Social Interface of Sustainable Development: Capabilities, Social Capital, Institutions', *Ecological Economics*, 49: 199–214.

Maathai, W. (2003), *Green Belt Movement: Sharing the Approach and Experience* (Herndon, VA: Lantern Books).

Mama-86 (n.d.), 'Mama-86: About', avalable online at www.mama-86.org.ua/index.php/en.html

Marshall, K. (2014), 'Global Development Governance', in T. G. Weiss and R. Wilkinson (eds), *International Organisations and Global Governance* (Abingdon: Routledge), 564–579.

Von Moltke, K. (1997), 'Institutional Interactions: The Structure of Regimes for Trade and the Environment', in O. R. Young (ed.), *Global Governance: Drawing Insights from the Environmental Experience* (Cambridge, MA: MIT Press), 247–271.

Momsen, J. H. (2004), *Gender and Development* (London: Routledge).

Oxfam International (2008), 'Another Inconvenient Truth: How Biofuel Policies Are Deepening Poverty and Accelerating Climate Change' (Oxfam International, Briefing paper, June), available online at www.oxfam.org/sites/www.oxfam.org/files/bp114-inconvenient-truth-biofuels-0806.pdf

Pepper, D. (1996), *Modern Environmentalism: An Introduction* (London: Routledge).

Perkins, R. (2003), 'Environmental Leapfrogging in Developing Countries: A Critical Assessment and Reconstruction', *Natural Resources Forum*, 27: 117–188.

Pisupati, B. and Warner, E. (2003), *Biodiversity and the Millennium Development Goals* (Colombo, Sri Lanka: IUCN), available online at https://www.cbd.int/doc/books/2009/B-03186.pdf.

Poku, N. K. and Whitman J. (2011), 'The Millennium Development Goals and Development after 2015', *Third World Quarterly*, 32 (1): 181–198.

Redclift, M. (1997), 'Development and Global Environmental Change', *Journal of International Development*, 9 (3): 391–401.

Redclift, M. and Woodgate, G. (1997), 'Sustainability and Social Construction', in M. Redclift and G. Woodgate (eds), *The International Handbook of Environmental Sociology* (Cheltenham: Edward Elgar), 55–70.

Reed, D. (1997), 'The Environmental Legacy of Bretton Woods: The World Bank', in O. R. Young (ed.), *Global Governance: Drawing Insights from the Environmental Experience* (Cambridge, MA: MIT Press), 227–246.

Royal Government of Bhutan (2012), *The Report of the High-Level Meeting on Wellbeing and Happiness: Defining a New Economic Paradigm* (New York: The Permanent Mission of the Kingdom of Bhutan to the United Nations), available online at www.uncsd2012.org/content/documents/519BhutanReport_WEB_F.pdf

Sachs, J. (2012), 'Introduction', in Earth Institute, *World Happiness Report* (Helliwell, J., Leyard, R. and Sachs, J.), available online at http://issuu.com/earthinstitute/docs/world-happiness-report; 1–59.

Sachs, J. D., Baillie, J. E. M., Sutherland, W. J., Armsworth, P. R., Ash, N., Beddington, J., Blackburn, T. M., Collen, B., Gardiner, B., Gaston, K. J., Godfray, H. C. J., Green, R. E., Harvey, P. H, House, B., Knapp, S., Kümpel, N. F., Macdonald, D. W., Mace, G. M., Mallet, J., Matthews, A., May, R. M., Petchey, O., Purvis, A., Roe, D., Safi, K., Turner, K., Walpole, M., Watson, R. and Jones, K. E. (2009), 'Biodiversity Conservation and the Millennium Development Goals', *Science*, 325 (5947): 1502–1503.

Sachs, W. (1997), 'Sustainable Development', in M. Redclift and G. Woodgate (eds), *The International Handbook of Environmental Sociology* (Cheltenham: Edward Elgar), 71–82.

Said, E. (1979), *Orientalism* (New York: Vintage Books).

Salleh, A. (2009), 'Ecological Debt: Embodied Debt', in A. K. Salleh (ed.), *Eco-Sufficiency and Global Justice; Women Write Political Ecology* (London: Pluto Press), 1–41.

Sampson, G. P. (2005), *The WTO and Sustainable Development* (Tokyo: United Nations University Press).

Sauter, R. and Watson, J. (2008), *Technology Leapfrogging: A Review of the Evidence*. University of Sussex, Sussex Energy Group SPRU, Science and Technology Policy Research Report for DFID, 3 October, available online at www.sussex.ac.uk/.../documents/dfid-leapfrogging-reportweb.pdf

SDSN (The Sustainable Development Solutions Network) (2013), *Second World Happiness Report*, edited by John F. Helliwell, Richard Layard and Jeffrey D. Sachs, available online at http://unsdsn.org/wp-content/uploads/2014/02/WorldHappinessReport2013_online.pdf

Sen, A. K. (1999), *Development as Freedom* (New York: Anchor Books).

Shiva, V. (1989), *Staying Alive: Women, Ecology and Development* (London: Zed Books).

Shiva, V. (1993), *Monocultures of the Mind: Perspectives on Biodiversity and Biotechnology* (London: Zed Books).

Shiva, V. (2000), *Stolen Harvest: The Hijacking of the Global Food Supply* (New York: South End Press).

Spitzner, M. (2009), 'How Global Warming Is Gendered', in A. K. Salleh (ed.), *Eco-Sufficiency and Global Justice; Women Write Political Ecology* (London: Pluto Press), 218–229.

Taylor, I. (2014), 'The Global South', in T. G. Weiss and R. Wilkinson (eds), *International Organisations and Global Governance* (Abingdon: Routledge), 279–291.

UN (United Nations) (n.d.), 'Poverty', available online at www.un.org/cyberschoolbus/briefing/poverty/poverty.pdf

UN (1992), *Agenda 21: Chapter 35: Science for Sustainable Development*, available online at http://habitat.igc.org/agenda21/a21-35.htm

UN (1998), 'Indicators of Poverty and Hunger', Statement of Commitment for Action to Eradicate Poverty Adopted by Administrative Committee on Coordination, Press release ECOSOC/5759 (New York: UN), available online at www.unesco.org/most/acc4pov.htm

UN (2002), *Report of the International Conference on Financing for Development, Monterrey, Mexico, 18–22 March 2002*, Report A/CONF.198/11 (New York; UN), available online at www.ipu.org/splz-e/ffd08/monterrey.pdf

UN (2004), *Human Rights and Poverty Reduction: A Conceptual Framework*, Report 2004 HR/PUB/04/1 (New York/Geneva: United Nations/OHCHR), available online at www.ohchr.org/Documents/Publications/PovertyReductionen.pdf

UN (2006), 'What They Are' (New York: UN), available online at www.unmillenniumproject.org/goals/

UN (2011a), *Economic and Social Council Ten-year Appraisal and Review of the Implementation of the Brussels Programme of Action for the Least Developed Countries for the Decade 2001–2010*, Report A/66/66–E/2011/78 (New York: UN General Assembly), available online at www.un.org/wcm/webdav/site/ldc/shared/SG%20Report%20appraisal%20REPORT.pdf

UN (2011b), *Istanbul Programme of Action for Least Developed Countries for the Decade, 2011–2020*, A/CONF.219/3/Rev.1 (New York: UN), available online at http://documents.wfp.org/stellent/groups/public/documents/eb/wfpdoc061607.pdf

UN (2012), 'Secretary-General Ban Ki-moon at High-level Meeting on *Happiness and Well-being*, UN, April 2, 2012: Secretary-General, in Message to Meeting on 'Happiness and Well-being' Calls for 'Rio+20' Outcome that Measures More than Gross National Income', Press release (New York: UN), available online at www.un.org/News/Press/docs/2012/sgsm14204.doc.htm

UN (2013a), *Human Development Report 2013: The Rise of the South: Human Progress in a Diverse World* (New York: UN), available online at http://hdr. undp.org/sites/default/files/reports/14/hdr2013_en_complete.pdf

UN (2013b), 'LDC Information: The Criteria for Identifying Least Developed Countries' (New York: UN), available online at www.un.org/en/development/ desa/policy/cdp/ldc/ldc_criteria.shtml

UN (2014), 'Millennium Development Goals and post-2015 Development Agenda' (New York: UN), available online at www.un.org/en/ecosoc/about/ mdg.shtml

UNDP (United Nations Development Programme) (1990), *Human Development Report* (New York: Oxford University Press), available online at http://hdr. undp.org/sites/default/files/reports/219/hdr_1990_en_complete_nostats.pdf

UNDP (2013), *Accelerating Progress, Sustaining Results: The MGDs to 2015 and Beyond* (New York: UNDP, September), available online at www.undp. org/content/dam/undp/library/MDG/MDG%20Acceleration%20Framework/ Accelerating%20Progress%20-%20October%2002.pdf

UNEP (United Nations Environment Programme) (n.d.a), 'What Is Gender and Biodiversity?' (Montreal: Secretariat of the Convention on Biological Diversity), available online at https://www.cbd.int/gender/what/default.shtml

UNEP (n.d.b), 'Traditional Knowledge and the Convention on Biological Diversity' (Montreal: Secretariat of the Convention on Biological Diversity), available online at www.cbd.int/traditional/intro.shtml

UNESCO (n.d.), 'Local and Indigenous Knowledge', available online at www. unesco.org/new/en/natural-sciences/priority-areas/links/

UN Women Watch (2009), *Women, Gender Equality and Climate Change* (New York: UN), available online at www.un.org/womenwatch/feature/ climate_change/downloads/Women_and_Climate_Change_Factsheet.pdf

WCED (World Commission on Environment and Development) (1987), *Our Common Future* (Oxford: Oxford University Press).

WEDO (Women's Environment and Development Organisation) (2001), *Primer: Women and Sustainable Development: A Local Agenda*, available online at www.wedo.org/wp-content/uploads/localagenda_primer.htm

WEDO (2004–2008), 'Women's Environment and Development Organisation: About', available online at www.wedo.org/about

WEDO (2014), *Sustainable Development*, available online atwww.wedo.org/ category/themes/sustainable-development-themes

WHO (World Health Organisation) (2014a), *World Health Organisation, Structural Adjustment Programmes* (Geneva: WHO), available online at www.who.int/trade/glossary/story084/en/

WHO (2014b), 'Food Security' (Geneva: WHO), available online at www.who.int/trade/glossary/story028/en/.

World Bank (2010), *Safeguards and Sustainability Policies in a Changing World: An Independent Evaluation of World Bank Group Experience* (Washington, DC: World Bank), available online at http://siteresources.worldbank.org/extsafandsus/Resources/Safeguards_eval.pdf

World Bank (2013), *The World Bank Group Goals End Extreme Poverty and Promote Shared Prosperity* (Washington, DC: World Bank), available online at www.worldbank.org/content/dam/Worldbank/document/WB-goals2013.pdf

World Bank (2014), *Sustainable Development Overview* (Washington, DC: World Bank), available online at www.worldbank.org/en/topic/sustainabledevelopment/overview

WTO (World Trade Organisation) (2014), *The Doha Mandate on Multilateral Environmental Agreements* (New York: WTO), available online at www.wto.org/english/tratop_e/envir_e/envir_neg_mea_e.htm

Yearley, S. (1997), 'Science and the Environment', in M. Redclift and G. Woodgate (eds), *The International Handbook of Environmental Sociology* (Cheltenham: Edward Elgar), 227–236.

Emerging economies
China's model of sustainable development

Key issues

- Emerging economies
- Environmental consequences of rapid growth
- China's model of sustainable development
- National Environmental Model City
- Ecological restoration
- Payment for ecosystem services
- Outsourcing
- Significance of China for global sustainable development.

Introduction

The term 'emerging economies' is used to refer to countries that are restructuring their economies along market lines, entering globalised markets through opening up to trade, technology transfers, and foreign direct investment (OECD, 2009). The biggest emerging economies are Brazil, Russia, India, China and South Africa, collectively known as BRICS (O'Neill, 2011). They are currently making a critical transition from a developing to that of a developed country. From the perspective of sustainable development, each country is important but so too is their combined effect, because as a group their emergence will change the face of global economics and politics. Their development will affect the intensity of natural resource use, pollution emissions, consumption patterns and the generation of waste not only within their own borders but also globally.

Positively, these economies will inherit a highly integrated, technologically driven global society with extraordinary potential to support their burgeoning population. Negatively, they inherit a legacy of over-use of natural resources; continued dependency on, and vested interests in, protecting unsustainable economic models; widespread poverty and inequality; institutions that are at best fit for an earlier time; and fragmented pathways for dealing with the risks and dysfunctions arising from the contemporary model of development (Zadek, 2010).

Emerging nations are understandably suspicious about the approaches of today's leading nations, including those that speak to the sustainable development agenda. They fear that this agenda is built on norms that are not suitable to their own circumstances and values and that it risks bringing competitive disadvantage to their rising economies (Zadek, 2010). Instead, many seek to integrate economic, ecological and social considerations through their own, distinctive sustainable development models. How emerging economies will influence the pursuit of sustainable development forms the subject matter of this chapter. The chapter will focus on China.

Given its sheer size, China is a particularly important emerging economy. This chapter explores the approach taken in China to the pursuit of sustainable development. It begins with a brief overview of China's emergence, and outlines the environmental and social consequences of its rapid development. The Chinese model of sustainable development is then explored in detail, looking at its principles, exploring governance arrangements, and then examining actions across a range of economic sectors and issue areas. Examples and case studies are provided by way of illustration. The chapter concludes by discussing the significance of China's emergence for the pursuit of sustainable development at the global level.

China in the contemporary period

Rapid industrialisation and urbanisation

Over the past three decades, China has experienced unprecedented levels of industrialisation and urbanisation. This growth is mainly due to market-orientated reforms unleashed under Deng Xiaoping (1978–1992) (Vogel, 2011). These allowed the development of non-state-owned enterprises, liberalised foreign trade and investment regulations, relaxed state price controls and gave rural households greater autonomy with respect to land use and crop selection (Zhang, 2012). A raft of government policies also encouraged the wholesale urbanisation of the rural population. The results of these reforms are staggering: China has maintained a double-digit annual growth rate for more than three decades and by 2010 had become the world's second largest economy, after the USA. Furthermore, the economy has the highest volume of production in the world for major industrial products, including crude steel, coal, electricity, cement, fertiliser, and woven cotton fabrics, and by 2014 the country had become the world's largest trading nation (*The Guardian*, 10 January 2014). China is now also heavily involved in foreign direct investment, including on the African continent. In the period 2000–2010, the urban population increased from 460 million to 670 million, with almost half of the population now living in urban areas, resulting in several megacities. Reforms also resulted in the development of tens of thousands of small-scale, rural industries alongside an influx of international investment (Economy, 2006).

Poverty and uneven development

As a result of its staggering growth, hundreds of millions of Chinese have been lifted out of poverty. However, its per capita GDP is very low, about 100 million people live below the national poverty line, and there are widespread income disparities (World Bank, 2014). In addition, apart from the eastern coastal areas, most regions of China remain underdeveloped. The country with the world's largest population (1.354 billion people in 2013) continues to face tremendous development pressures. The need to raise the bulk of its population out of poverty and address the problem of uneven development across its regions poses major threats to the pursuit of sustainable development, not only within China but also at a global level. Further growth will require increased resource input, including both raw materials and energy, and see more land-take for city growth and transport needs. Changing lifestyles will bring rising levels of consumption, especially increased demand for consumer durables, technological devices and meat, all of which require resource- and energy-intensive inputs and create mounting volumes of waste.

The natural environment in China is extremely fragile, which means that growth poses even further threat, particularly to the promotion of the ecological pillar of sustainable development. The country's geographical and geological environment is diverse, with a high proportion of land unsuitable for human habitation (The People's Republic of China, 2012). Furthermore, it regularly suffers from major natural disasters, which in turn pose high risks to people's lives, property, and economic and social development potential. As we will see in this chapter, rapid growth is not only putting pressure on the fragile environment in China, but has global ecological impacts as well.

The environment and development in China

Many have spoken about the extraordinary increase in economic growth as the 'China miracle' (Zhang, 2012). However, this growth has come at a price: the growth 'miracle' has created major social and environmental problems. Rapid industrialisation has relied on ever growing inputs of energy and natural resources, including water and other environmental services, and has led to widespread pollution. It has also caused severe social disruption. Environmental degradation is now so widespread that it is beginning to undermine the very future of the Chinese economy and also threatens the political stability of the state.

Before we discuss this in greater detail, it is important to remember that environmental problems pre-date China's rapid industrialisation. Under the regime of Chairman Mao Zedong (1949–1978) the state implemented development policies which were responsible for causing extreme and widespread environmental damage (Tobin, 2013). The Great Leap Forward (1958–1961), a state policy which aimed to rapidly transform both agriculture and industry, saw environmental considerations almost completely abandoned. Mao Zedong prioritised steel and grain production, resulting in numerous backyard steel furnaces, which in turn led to deforestation, pollution, and waste. Similarly, state targets for grain production led to massive dam construction projects, over-exploitation of groundwater, extinction of wildlife, and destruction of vegetation (Zhang, 2012). Mao's initiative proved a failure, with agriculture devastated and millions of tools destroyed in the production of worthless 'steel' (Tobin, 2013). While these radical policies strengthened the power and reach of the state, they resulted in a devastating famine that left tens of millions of Chinese dead or imprisoned and the economy in shambles (Economy, 2006). In response to such a devastating loss of life, Mao incentivised relentless population growth, believing that a larger population would accelerate production, and as a result the population doubled between 1950 and 1980 to over one billion (Tobin, 2013). Economic growth in the past three decades has come amidst the social and environmental legacies inherited from the turbulent years of Mao Zedong's regime.

Environmental consequences of development

China's pollution makes world headlines. Air pollution is a major problem in cities and acid rain falls on much of the country. Reliance on coal has made the country home to five of the ten most polluted cities in the world. In January 2013, China experienced an unprecedented high pollution episode occurring in hundreds of its major cities. In the same year, smog choked the northern city of Harbin in Heilongjiang Province, forcing schools and highways to shut and disrupting flights. In January 2104, air pollution readings in China's capital Beijing registered more than twenty times the recommended exposure levels of the World Health Organisation (*BBC News*, 2014). The number of lung cancer cases in Beijing has increased by more than 50 per cent over the last decade, much of it attributable to air pollution (*BBC News*, 2013c). Widespread discontent among urban residents and growing international concerns are putting the state under increased pressure to curb pollution. But, in a country where political stability rests on the ability to deliver continued growth, the state has been reluctant to take steps, fearing job losses and closure of factories and electricity generating plants.

Growth has also brought a rapid increase in solid waste generation, especially in its cities, and in 2004 China surpassed the USA as the world's largest waste generator. The quantity of municipal solid waste surged from about 31 million tons in 1980 to 157 million tons in 2009, and is projected to reach 585 million tons by 2030 (World Bank, 2013). Water problems are very serious in China and almost all of the nation's rivers are polluted to some degree, bringing widespread adverse impacts on both human and ecosystem health (Gleick, 2008a). The growing industrial and agricultural sectors now face serious water shortages, often resulting in factory closures (Gleick, 2008a). Water problems have been compounded by the rapid increase in demand for water in cities and towns. State efforts to address this problem, under the rubric of its sustainable development agenda, are discussed in the next section.

While cities continue to experience declining environmental quality, pollution has also intensified in rural areas. Although rapid economic growth has brought unforeseen wealth to many rural communities and helped China reach some of its MDGs, the tens of thousands of small-scale operations, such as paper and pulp, electroplating, dyeing, and chemical factories, have polluted both the water and the air (Economy, 2006). Many of these small firms lack pollution abatement facilities (Zhang, 2012) and are very difficult to monitor. Rural areas also face

Case study 12.1

The Songhua River spill

On 13 November 2005 an explosion occurred at a petrochemical plant of the Jilin Petrochemical Corporation, in Jilin Province, China. Bordering Russia and North Korea, the provinces of Jilin and Heilongjiang are part of a heavy industry belt in northeastern China.

The explosion led to a spill of an estimated 100 tons of toxic substances made up of a mixture of benzene, aniline and nitrobenzene. The pollution entered the Songhua River and a long toxic slick drifted downstream, and at one point the benzene level recorded was 108 times above national safety levels. The slick passed through several counties and cities, converged into the Amur River at the mouth of the Songhua on the border between China and Russia, and eventually made its way into the Pacific Ocean.

Exposure to benzene and nitrobenzene poses cancer risks. The contamination also deprived 9 million people of water in the region around the city of Harbin.

Initially, Jilin Petrochemicals denied that the explosion had leaked pollutants into the Songhua River. The Chinese press was critical of the authorities' response to the disaster. Eventually, the vice-governor of Jilin province and the China National Petroleum Company apologised to the city. On 6 December, the vice-mayor of Jilin was found dead in his home. This followed a threat by the Chinese government to severely punish anyone who had covered up the severity of the accident. The minister of the State Environmental Protection Administration resigned.

The Songhua River spill was a major transboundary and international pollution event.

(*Sources*: Bhattacharya, 2005; UNEP, 2005)

ecological degradation from agricultural pollution and soil erosion from over-grazing, over-ploughing and deforestation. By 2012, the government admitted that 90 per cent of its natural grasslands have varying degrees of degradation (The People's Republic of China, 2012; Qiu, 2011; Xiaoyi and Qian, 2013). In turn, this has further increased problems of water shortage, led to sandstorms and contributed to the migration of millions of Chinese into the cities (Economy 2006; Qiu, 2011). To multiply these problems, 57 per cent of the country's coastal wetlands have disappeared since the 1950s, mostly due to land reclamation. Over the same period, the area covered by mangrove forest and coral reef fell by 73 per cent and 80 per cent, respectively (Qiu, 2011).

Rural problems are compounded by the emergence of new opportunities to reap large financial rewards from investment ventures that have seen local government officials seize farmers' lands for development projects. Opposition to industrial pollution, land seizures, corruption and income differentials have led to a massive rise in protests in the countryside (Jahiel, 2006). Local protests over land rights around Kunming, capital of Yunnan Province in southwestern China, for example, continue to grow as local government officials try to seize land to build a massive tourist development (*BBC News*, 2013d). The traditional *hukou*, or household registration system, is also a rising cause of grievance among rural communities. Under the system, everyone is registered in their home town and only there can they access education, housing and welfare. This prevents rural migrants and their children from accessing social services if they migrate to the city. As a result, a growing underclass is emerging in several major cities, while rural areas see rising numbers of abandoned children. *Hukou* reform would be popular with China's 260 million migrant workers. Some individual cities, such as Chengdu, have already begun reforms, and under the current political leadership (2014) national reform guidelines have been drafted. However, many cities, including Beijing, wish to keep tight controls in place to limit their population growth and avoid the costs associated with providing social services to migrants.

Social instability and political fragility

The regime of Chairman Mao Zedong created a monolithic political structure, a legacy which continues to shape politics today. Dissent is not tolerated, as evidenced by the Tiananmen Massacre in 1989 and the continuing dearth of press, internet freedom and human rights into the twenty-first century. Lacking political democracy, both the legitimacy of the Chinese Communist Party and the stability of its regime have come to rely on the state's ability to deliver economic growth and prosperity to the people. In those parts of China 'left behind' by the industrial boom, there is high political imperative to pursue classic 'developmentalist' strategies, even at environmental cost. However, as environmental conditions deteriorate, the state faces a dilemma: on the one hand, it is under intense pressure to pursue developmentalist policies that prioritise economic growth; on the other hand, these very policies have intensified socio-economic disparities, heightened self-serving actions by local political elites, increased environmental degradation, led to a surge of social discontent and, by undermining the natural resource base for future development, threaten China's economic and political future (Jahiel, 2006).

Box 12.1

The trade-off between poverty eradication and environmental protection

China hopes to reduce its poverty rate by furthering economic growth. Under its current development model, further economic growth will degrade the environment even more.

Poverty and environmental problems are interrelated. For example, deforestation, overgrazing, and overdevelopment of agricultural land leads to resource degradation and increasing natural disasters, which disproportionally occur in the poor regions and reduce their developmental capacities.

The worst-case scenario is a vicious cycle: on the one hand, poverty alleviation requires economic development that puts further pressure on the fragile ecosystem; on the other hand, environment and natural resource protection act as constraints on the ability of low-income regions to emerge from poverty.

(*Source*: Zhang, 2012)

There are also new values emerging in Chinese society that multiply these difficulties, as a culture of consumption develops within the growing class of wealthy Chinese. At the other end of the socio-economic spectrum, the increase in relative deprivation also poses environmental threats. Experiences from other parts of the world suggest that when people's livelihoods are threatened and they perceive uncertainty regarding their future, they often resort to strategies that degrade the environment in order to survive (Jahiel, 2006).

China has long realised that an extensive growth model that relies heavily on natural resource input and creates negative environmental and social externalities is not sustainable.

The Chinese model of sustainable development

China claims to be one of the first developing countries to implement a national sustainable development strategy (The People's Republic of China, 2007). In 1993, following participation in the Rio Earth Conference, the government issued *China's Agenda 21: White Paper on China's Population, Environment and Development in the 21st Century*, and two years later incorporated sustainable development into its national development strategies (The People's Republic of China, 2012).

The idea of harmony between human beings and nature is an important component of the traditional values of Chinese civilisation. As the Chinese foreign minister, Wang Yi, said in 2013:

> Concepts of harmony between man and nature and not to drain the pond to catch the fish have always been part of the fine traditional Chinese culture.
>
> (Xinhua News Agency, 2013a)

Wang went on to stress that sustainable development is a basic state policy and that 'The Chinese government makes it clear that we cannot repeat the old way of "pollution first, treatment afterwards" and that we must be responsible for our future generations and the international community' (Xinhua News Agency, 2013a). However, while official state doctrine declares commitment to sustainable development, there are several key differences between the Chinese approach and the classic Brundtland formulation.

Principle of diversifying development models

Sustainable development in China is based on the 'principle of diversifying development models'. The state believes that, as countries are in different development stages, there is no universal model for sustainable development. Each nation needs to choose a suitable development path that best suits its own context (The People's Republic of China, 2012). China also wishes to see the international community adhere to the principle of common but differentiated responsibility, discussed in Chapter 2. This means that developed countries should honour their commitments to help developing countries achieve sustainable development (The People's Republic of China, 2012). This was reiterated in 2013 by Foreign Minister Wang when he insisted on the need for the international community to take into account the differences between developed and developing countries in level, resources, and means of development (Xinhua News Agency, 2013b).

The scientific outlook on development

China's own development path is based on the so-called Scientific Outlook on Development. This ideology was proposed by President Hu Jintao in 2003 in the context of growing state recognition of the range of problems associated with development, including excessive consumption of resources, serious environmental pollution and a widening gap between the rich and poor, the city and the countryside (Kang and Li, 2013). In this sense, it serves, at least in part, as a rebuke to the growth-at-all-costs economic strategy unleashed by Deng Xiaoping more than three decades earlier and, prior to that, to Chairman Mao Zedong (Roberts, 2012). The Scientific Outlook on Development is now an official guiding socio-economic ideology for the Communist Party of China and was written into the Communist Party and State Constitutions in 2007 and 2008, respectively (Xinhua News Agency, 2012).

> The Chinese Scientific Outlook on Development advocates a people-oriented development focusing on comprehensive, balanced, and sustainable development. It calls for harmonious development between humans and nature.

The Scientific Outlook on Development supports the achievement of more balanced economic and social development by:

1 Promoting a human-oriented approach that combines economic with social development;
2 Ensuring harmonious development between humans and nature;
3 Developing both urban and rural areas;
4 Addressing regional disparities;
5 Opening up to the outside world.

(The People's Republic of China, 2012)

To drive this policy forward, the Twelfth Five-Year Plan (2011–2015), discussed below, outlines a series of targets and actions for resource conservation, environmental protection, energy saving, and climate change mitigation (British Chamber of Commerce, 2011). The Plan also provides for a massive push for low-cost housing for the urban poor, as well as improvements in the lives of rural residents through expanded health care and pension programmes. These developments are to be backed up by the reforms introduced by President Xi Jinping in November 2013, giving renewed attention to rural welfare and land rights, as well as ensuring food security and containing local authority debt (Xinhua News Agency, 2013c).

Focus on growth

Despite pledging sustainable development, the Chinese state retains a strong commitment to economic growth. Achieving growth is central to the career advancement of government officials, giving public policy makers a vested interested in ensuring economic growth (Zhang, 2012). Under the regime of Hu, however, a new focus on developing the green economy emerged, which is being continued under the current leadership of Xi. Xi puts particular stress on the role of local authorities in promoting green development and in shifting industry toward greater resource efficiency (Xinhua News Agency, 2014). The green economy is seen as an important means to achieve sustainable development because it can stimulate economic restructuring, particularly through the application of technology. This, it is believed, will help reduce poverty (The People's Republic of China, 2012).

Strong links to the ideology of ecological modernisation, as discussed in Chapter 2, can be seen here. China's approach to ecological modernisation is based on the pursuit of the so-called 'circular economy', formally approved by central government in 2002 and introduced into law in 2008 (Liu, L., 2008). This emphasises the need to design industrial production to achieve integrated, closed-loop systems, ones that utilise waste streams as inputs into further production (Gallagher and Lewis, 2013). Ecological modernisation is discussed further in the next section through the use of examples.

In summary, China's model of sustainable development has five key features that reflect the particular sustainable development path that the Chinese state wishes to pursue:

Box 12.2

Key principles of China's model of sustainable development

1 It is underpinned by the Scientific Outlook on Development;
2 It aims to promote social harmony and progress;
3 The model has economic growth as its top priority;
4 It is driven by the application of scientific and technological innovations;
5 It requires continued reform and the opening-up of the economy.

(*Source*: The People's Republic of China, 2012)

Having explored the principles that underlie China's model of sustainable development, attention is now turned to examining the governance arrangements in place to enable the state to put these principles into practice.

Governing sustainable development

Central role for the state

Efforts to address environmental problems in China more or less coincided with the start of the economic reforms of the late 1970s (Mol, 2006: 36). A National Environment Protection Office was established in 1974, restructured as the State Environmental Protection Agency and in 2008 upgraded to the Ministry of Environmental Protection. An Environmental Protection Law was enacted in 1979 and revised in 1989. The 1990s brought a period of accelerated development in laws and regulations and, as a result, a national regulatory framework for the environment is now in place (Mol, 2006). This covers an array of environmental concerns, including the marine environment, water, energy conservation and renewable energy, forests, and environmental impact assessment. Furthermore, national standards for air quality and for pollution emissions have been set. There is also a law on the promotion of the 'Circular Economy', an ecological modernisation approach discussed above. In addition, there are a large number of departmental and local laws, regulations and rules (The People's Republic of China, 2012). However, with the state favouring economic growth over environmental protection, the Environment Ministry remains underfunded and there is a lack of capacity to enforce legislation (Tobin, 2013).

As a centrally planned economy, sustainable development actions are framed within five-year national plans, which steer the country's legislation and economic, social, and environment policies. The Twelfth Five-Year Plan for National Economic and Social Development (12th FYP) was adopted in March 2011 and set targets and policies up to 2015. It devotes considerable attention to energy and climate change (British Chamber of Commerce, 2011). There are targets for pollution reduction and improvements in the living environment, including in water, air and soil quality, and the control of pollution from agriculture (Feng et al., 2013). The Plan also includes targets on re-forestation (Lewis, 2011), the further development of nature reserves, improving nuclear safety and enhancing environmental monitoring capabilities.

The promotion of sustainable development is also the duty of the Administration Centre of China Agenda 21, a body responsible for co-ordinating capacity enhancement efforts and information sharing to implement China's Agenda 21 commitments (Pan and Xianli, 2005/2006).

Box 12.3

Limiting plastic shopping bags

In 2007 the Chinese government issued an order limiting the production, sale and use of plastic shopping bags, known as the Plastic Limit Order.

Since the Order, the annual use of plastic bags at major retailers has been reduced by more than 24 billion pieces. Consumption of plastics has been cut by 600,000 tons, the equivalent of 3.6 million tons of oil, over 5 million tons of standard coal and 10 million tons of carbon dioxide emissions.

A small step in a large country can have big impact.

(*Source*: The People's Republic of China, 2012)

As a centralised, planned economy, China favours a command and control approach to both environmental protection and the promotion of sustainable development.

While command and control remains central, by the 1990s the state was increasingly confronted with failures in environmental policy implementation. As a result, it began to make increased use of prices and other market incentives to promote sustainable development, a step made possible by China's transition to a market economy.

Market tools and payments for ecosystem services

Aside from the command and control approach, including regulation and centralised planning, the use of market instruments has also grown. As mentioned above, policy makers in China have become increasingly interested in developing new approaches for environmental governance as an aid to support more effective implementation. China has, for example, launched the first carbon trading scheme as a pilot project in Shenzhen. The test scheme was rolled out to seven areas in 2014, and if successful could be spread across the country after 2015 (*BBC News*, 2013a), as discussed in Chapter 6. There is also an Environment Protection Fund that facilitates the implementation of environmental policy by building closer ties with businesses and other non-state actors (Tang and Zhan, 2008).

There is also a wide range of initiatives under the broad heading of 'eco-compensation' that incorporate, to varying degrees, aspects of market-based approaches (Sterner, 2003). They include a growing number of programmes that utilise Payments for Ecosystem Services (PES) schemes, mentioned in Chapters 3 and 7. PES schemes consist of negotiated contractual arrangements involving direct payments between those who can provide, and those who benefit from ecosystem or environmental services. Indeed the Chinese government is driving some of the largest public payment schemes for ecosystem services in the world, and has more than US$90 billion invested in existing or planned schemes. The ecological restoration initiatives, the Natural Forest Conservation Program and the Grain to Green Program are two of the biggest programs offering PES in both China and worldwide in terms of scale, payment, and duration. Both of these are discussed later in the chapter.

However, a lack of institutional capacity, poor fit with existing institutional arrangements and the continuation of a centralised, planned economy are seriously hampering further development of market-based approaches (Economy, 2006). While the leadership of Xi sees greater acknowledgment of the role of markets, the state remains dominant in shaping the governance of sustainable development in China.

Towards new forms of participatory governance

In China, the state imposes its own 'China Model' of governance, which it sees as an alternative to the West's 'Democracy Model'. This model of a 'Socialist Harmonious Society' combines economic growth achieved through combining free market, mostly state, input with heavy political and media control and restrictions on political freedoms. At the same time, the state also recognises that to promote sustainable development it is also necessary to have the broad participation of civil society, the private sector, business communities and other major groups (The People's Republic of China, 2012). Increased public participation offers one way to overcome weak enforcement of environmental legislation. Promoting participation, while maintaining strong, centralised state control results in a distinctive approach to the governance of sustainable development.

To improve participation, China relies on Government-Organised Non-Governmental Organisations (GONGOs), bodies created by the state as a tool to implement government policy. They differ from typical NGOs in that they are not independent of the state. Despite this, they receive up to 70 per cent of their funding from abroad because of their perceived abilities to facilitate democratisation (Wu, 2002: 42). But, as state-run bodies,

---

Box 12.4

PES in China

Use

China uses PES across a broad range of programmes for the delivery of ecosystem services, including carbon sequestration, biodiversity conservation and anti-desertification services. An increasing number of initiatives aim to protect watershed services and resolve conflicts over the rights and access to water resources. In China, PES are known as 'eco-compensation mechanisms' (*shengtaibuchangjizhi*).

PES-type policies involve direct payments from the government to individual and community-level suppliers of ecosystem/environmental services. They also aim at developing co-operation between various levels of government for the financing and sharing of costs of environmental protection and restoration.

Features

PES operations in China have several distinctive features, including that they are geographically concentrated in the richer coastal regions. In addition, they tend to have a sectoral focus and are often directed at water-related issues. They are marked by limited involvement of the private sector, while exhibiting a high degree of local variation in their design.

Significance

PES use in China is marked by both the breadth and depth of programmes. They have also encouraged the state to pay attention to the need to address issues of property rights. As state-controlled schemes, many PES fall short of being fully market-based instruments. Implemented in non-democratic political systems, their use can involve an element of coercion.

(Source: Bennett, 2009)

---

they have been criticised for displacing genuine NGOs as the voice heard by government (Tobin, 2013). However, despite the state maintaining a strong hand over civil society, reforms under the current leadership of Xi have brought some relaxation of the ban on independent NGOs. Some, such as the Green-web and Greener Beijing, operate only through the web and are unregistered (Yang, 2009). They publicise environmental information, set up discussion groups, mobilise volunteers, organise activities and campaigns and catalyse offline campaigns (Carter and Mol, 2006). However, the state still sets the parameters for these new freedoms and NGOs remain subject to acute supervision. Similarly, the small but growing numbers of international NGOs operating in China do so under significant political constraint. Despite this, they play an important role: international NGOs, such as Greenpeace, highlight pollution problems afflicting Chinese cities and spread green values, such as sustainable consumption (Greenpeace, https://www.facebook.com/gpchina).

There are also signs of independent public engagement at the local level, around specific environmental concerns and land rights issues, as discussed above. However, lack of access to information, and political constraints stemming from the country's authoritarian political system, remain serious barriers. Other obstacles include a culture that favours top-down solutions to public problems, imposed by a centralised state (Johnson, 2010).

The state has nonetheless increased access to environmental information, for example by making monitoring data more accessible and by starting an eco-labelling project, alongside a corporate disclosure programme. These developments are also supported by foreign assistance (Moore and Warren, 2006). There is also a much publicised complaints system, including complaint hotlines, which allows citizens to whistle-blow by informing the authorities of local level infringements of environmental laws. The complaints system, however, has been criticised for offering a limited, reactive role to civil society (Johnson, 2010). Much of civil society participation takes the form of what has been called 'rule based activism' (Johnson, 2010). There are legal requirements for public participation, for example, in local development decisions; if local authorities fail to adhere to these regulations, groups campaign for the rules of participation to be upheld. Such protest is often motivated by single environmental issues, where citizens come in conflict with local, party-state officials, often related to planning decisions. A well known example is provided by the strong media and NGO protests over a plan to build thirteen dams on the Nu River in Yunnan province. Following these protests, the plan was put on hold in 2004 by premier Wen Jiabao. However, fear of the personal consequences of appearing to oppose the regime remains. There is also limited political will to facilitate further participation, lest this open a Pandora's box of grievances (Johnson, 2010).

Nonetheless, China is undoubtedly becoming a freer state, one that recognises that sustainable development cannot be imposed solely from above. While modifications are modest, the changes to its system of environmental governance are remarkable when compared with the stagnation and even de-institutionalisation of environmental governance in Russia, another major emerging economy. Emergence in Russia has not seen the development of new environmental institutions, nor the engagement of more economic and market actors on the environmental stage, nor has it brought more liberties for civil society (Carter and Mol, 2006).

Having discussed the governance arrangements for supporting the pursuit of sustainable development, several key policy areas are explored in the next section, including in relation to population control. More attention is paid to water management than other areas, as water scarcity is one of the most serious barriers to sustainable development, while state efforts to address water problems reflect very clearly the distinctive Chinese approach. Attention is also paid to the pursuit of sustainable development in urban environments and to China's substantial investments in ecological restoration, both of which provide exemplars of best practice.

Promoting sustainable development

Population control

Population control is an important but highly controversial element of the sustainable development strategy of China. China introduced a 'one child-per-couple' policy in 1978 and has successfully kept the population growth rate at a relatively low level. Although it was not initially designed to address sustainable development concerns, the policy has had important consequences in that it has reduced population pressures on resources and on the environment.

However, China's population policy has had several unintended consequences, including skewed sex ratios, an aging society, a non-viable social security system and human rights violations (Zhang, 2012). From this perspective, it can be argued that mandatory population control policy does not align with the social goals of sustainable development (Zhang, 2012). November 2013 saw President Xi bring some relaxation to this policy.

Addressing poverty and uneven regional development

Sustainable development policy in China places a great deal of attention on eradicating poverty. A key way this is done is through policies designed to deal with uneven regional development. As mentioned above, China's economic activity clusters on its eastern seaboard and, in comparison, the western provinces have high poverty rates. However, since 2000 priority has been given to the western regions that are mainly inhabited by ethnic groups. As a result, China has put enormous efforts into revitalising old industrial bases in the northeast and promoting a rise of the central regions.

Most projects in the Grand Western Development Programme have focused on infrastructure construction, such as highways, railroads, airports, dams and gas pipelines (Grumbine, 2007). Landmark projects include the Qinghai–Tibet railway, the West–East natural gas pipeline and the West–East power grid. A regional economy centered round the Chengdu-Chongqing, Guanzhong-Tianshui and Beibu Gulf economic zones has also emerged (The People's Republic of China, 2012).

However, while China's 'go west' strategy has boosted economic growth in the poverty-stricken regions, it has also transferred industrial pollution to the western areas, bringing with it the same environmental degradation that has been experienced in the eastern regions (Zhang, 2012). Local governments are also seriously in debt, amounting to a total of $1.8 trillion in the most recent audit. From this perspective, poverty alleviation through regional development programmes has not supported the pursuit of sustainable development, not least because they have undermined the ecological pillar. It also promotes what is referred to as 'internal outsourcing', which sees rich regions in China consuming high-value goods that depend upon production of low-cost and pollution-intensive goods in poorer regions of the same country (Feng et al., 2013).

Reducing pollution through ecological modernisation

In the face of rising public concerns and international attention, China has recently enhanced its efforts to address pollution, in particular air pollution in cities. One way this has been achieved is through the ecological modernisation of its industrial sector, as discussed above. A good example is provided by the restructuring of the Shougang steel company.

Efforts to reduce pollution have also brought renewed attention to enforcement of environmental legislation and standards. Environmental enforcement is sometimes implemented in a campaign style by local governments, resulting in very public factory closures and decisions to

---

Case study 12.2

Ecological modernisation of the Shijingshan steel plant

The state-owned Shougang Group is one of the world's top 500 companies and operates within the global market. It had a large steel production plant in the Shijingshan district of Beijing that was a major employer, but also a major source of pollution.

As one of the efforts made by the Chinese government to improve Beijing's air quality under the Beijing Urban Master Plan (2004–2020), the Shougang Group began in 2005 to relocate its facilities to Hebei Province, some 200 km east of Beijing. The new plant was completed by 2010. In the lead-up to the 2008 Olympic Games the Beijing factory was finally shut down with substantial job losses. The original factory site in Beijing is being transformed into an industrial theme park.

Modernisation resulted in large reductions in China's emissions of both sulphur dioxide and nitrogen oxide, and in levels of coal consumption. The shutdown and relocation is part of a more general state strategy for the modernisation of the steel sector. This is aimed at rationalisation of production, industry consolidation and achieving environmental improvements through the application of advanced technology.

(*Sources*: The People's Republic of China, 2012; OECD, 2011)

---

ban driving to reduce pollution rapidly within a particular time frame. Often such actions take place in the lead-up to major international events, such as the 2008 Beijing Olympic Games, the 2010 Shanghai Expo, and the 2010 Guangzhou Asian Games (Streets et al., 2008). Although these campaigns help cities maintain a better image during high profile, public events, their contribution to long-run sustainable development goals is questionable. For example, such actions cannot be replicated in other cities and sometimes have unintended consequences: the driving ban based on license plate numbers may have encouraged households to buy a second car (Zhang, 2012).

Attention has also been paid to addressing air pollution by dealing with pollution arising from coal-fired power stations. China is heavily reliant upon coal as a primary source of energy.

A growing amount of China's burgeoning waste is also being sent to new, 'waste to energy' plants. There are also plans to increase the share of renewable energy in its energy supply mix, including through wind power, as discussed in Chapter 6. The use of renewable energy is stressed in the 12th FYP so as to help meet binding targets for emission reductions.

Box 12.5

Curbing coal

The threat from coal

Coal is a major cause of pollution in China. Coal fires more than two thirds of the country's power plants. Smoggy skies and high pollution alerts in the largest cities are visible to a rising and more assertive middle class, and to the outside world.

Actions

In 2013, China announced new plans to control the use of coal to combat the dangerous levels of air pollution in its major cities. The state now accepts that air pollution is harming people's health and affecting social harmony.

New targets were set to reduce coal-fired energy capacity by some 5 per cent between 2013 and 2017 and replace it with cleaner fuels, including natural gas and nuclear power. The plan calls for a reduction of about a quarter in the dangerous fine particles that blanket Beijing. Beijing is also aiming to reduce the amount of carbon dioxide emitted per unit of gross domestic product by 40–45 per cent from 2005 levels by 2020. The state has also banned the construction of new coal-fired power plants in areas around Beijing, Shanghai and Guangzhou. There are also further plans to close down some of the worst-offending factories. Further agreement to cap emissions by 2030 was reached with the USA in 2014, as discussed in Chapter 6.

Concerns about implementation

The 5 per cent target is considered ambitious because China still has a growing demand for energy. Lack of advanced technology in energy production is also a hindrance.

Environmental groups have applauded the ambition of the targets, but some have expressed doubts about implementation, given growing energy needs and state concerns about the potential job losses that may result from actions to reduce pollution.

(*Sources*: *BBC News*, 2013b; Mou et al., 2013)

While renewable energy has a key role to play in the promotion of sustainable development, the use of renewable energy in China should be seen as a plank of industrial policy, serving to support future economic growth. However, it can also be seen as a key component of environmental policy, as it supports efforts to replace polluting coal-fired power plants with cleaner energy. Furthermore, it also forms a key part of the country's climate policy, helping China comply with its pledge to reduce carbon intensity by 40–45 per cent by 2020 compared to 2005, as was announced during the Copenhagen climate talks in 2009 (Zhang, 2012) and further developed following a joint agreement with the USA in November 2014.

Even though China has substantially increased the supply of energy obtained through renewable sources, renewable energy retains only a negligible share of the country's total power generation. In addition, economic growth has dramatically increased electricity consumption, such that the rebound effect is clearly at play. In addition, China's economic growth still relies heavily on the use of coal (World Bank, 2007).

Water management

China has limited water supply and, as mentioned above, the demand for water has increased substantially over the last decades, both as a result of growing domestic consumption and as a consequence of industrial development. While China can rightly celebrate achieving the MDGs of 'halving the number of people without sustainable access to safe drinking water' (The People's Republic of China, 2012) most people have no awareness of the need to save water and casual wasting of water is common (Tobin, 2013). Chinese water is also so polluted that 75 per cent of supply is unfit for human consumption (Economy, 2006: 18). Such acute pollution exacerbates an already significant water shortage problem. Factories in some provinces have been forced to close due to insufficient water supplies, and the availability of water for domestic consumption is also threatened. Water scarcity is now one of the foremost concerns of the Chinese state. There are three main causes of water shortages in China: insufficient supplies for a population of over 1.3 billion, excessive pollution and inefficient consumption.

The Chinese state has made some effort to clean up some of its most polluted waterways. A typical example is found in the clean-up of the waters in the Taihu Lake basin. The complex nature of pollution clean-up can be seen from the case of Lake Taihu. The need not just to engage in end-of-pipe measures but to deal with pollution at source is clearly evident. Such complexity points to the importance of integrating environmental considerations into planning at an early stage and to ensure

Case study 12.3

The clean-up of Lake Taihu

Pollution problems

Lake Taihu is China's third largest freshwater lake. The lake is heavily contaminated, and blooms from the cyanobacterium *Microcystis aeruginosa* produce toxins that can damage the liver, intestines and nervous system. In May 2007 a massive *Microcystis* bloom overwhelmed the waterworks that supplies Wuxi city on Taihu's northern shore, leaving more than two million people without drinking water for a week. At the time, Wuxi was drawing 80 per cent of its drinking water from the lake.

Clean-up

In 2008 a clean-up of the lake began. Local authorities set about controlling pollution and reducing emissions into the waterway.

To cut down on pollution, authorities forced hundreds of small chemical and manufacturing plants near Taihu to close or relocate to northern Jiangsu. They also instituted stricter monitoring of effluents from those factories permitted to stay. Ecological restoration within the basin was also undertaken. In addition, the province erected sewage treatment facilities on Taihu's tributaries and dredged tributary mouths to remove nutrient-rich sediments. As a result, nutrient concentrations in the lake's water column are beginning to taper off. The water quality has also substantially improved.

Future efforts

Local authorities are eager to do more, but this requires addressing sources of nutrients from the agricultural sector. Nitrogen runoff into the lake spikes in the spring, after tea farmers apply chemical fertilisers. The use of chemical fertilisers has soared in China in recent decades. Agriculture is not the only source of nitrogen. About a quarter of Taihu's nitrogen inputs come from the atmosphere, most probably from vehicle emissions.

Other proposed elements of the clean-up plan include further improving sanitation in the countryside; and offering tax incentives to try to shift the region's industrial profile from one dominated by chemical companies to one centred on renewable energy and internet technologies.

(*Sources*: The People's Republic of China, 2012; Stone, 2011)

sectoral policy integration, including in this case, in the transport and agricultural sectors. We can also see how efforts to address environmental concerns in one region can sometimes lead to short-term thinking, such as, in this case, the demand that companies relocate to northern Jiangsu. This does not address environmental harm; it merely results in its spatial displacement.

Water management is also inhibited by political considerations. The Chinese economy is polarised by huge corporations and small town and village enterprises. Regulations of larger employers remain low so as to maximise employment and avoid potential threats to the regime by thousands of unemployed workers (Ma and Ortolano, 2000). Moreover, many international businesses relocate factories to China so as to avoid pollution regulations, and China fears that tighter pollution controls could result in the loss of potentially lucrative foreign investment (Jahiel, 2006). Similarly, small-scale town and village enterprises provide jobs to over 20 million people and are important in maintaining stability in poorer regions. However, these enterprises are highly polluting, generating almost half of all water pollution in China. China also has difficulties confronting consumer demands. It could be highly destabilising for a country with widespread poverty to start charging high prices for water in an effort to conserve stocks. As such, domestic water consumption remains unchecked.

Struggling to persuade the citizenry to reduce its water consumption and trying to avoid more stringent industrial regulation for fear of harming employment, the state has been forced to find other ways to meet demand.

The urban environment and the creation of eco-cities

China began to build what it refers to as 'experimental zones' for sustainable communities in 1986, each exploring different pathways to sustainable development (The People's Republic of China, 2005). One of the flagship sustainable development efforts can be found in the National Model City (NMC) for Environmental Protection programme. The programme builds on an earlier Urban Environmental Quality Examination System, which the State Environmental Protection Administration introduced in 1989. The NMC, initiated in 1997, develops this system further by implementing an urban sustainable development strategy through the

Box 12.6

Water management through mega engineering

One way China has dealt with its water problem is through mega infrastructure projects involving large scale water transfer. 'Prestige' infrastructure works often garner the support of the Party, and it is prudent for engineers and local politicians alike to support such projects.

One of the most infamous is the Three Gorges Dam on the Yangtze River. This is the largest water-supply development in the history of humankind. Another project is the South–North Water Transfer Project. This involves moving vast quantities of water from the southern Yangtze River to the northern Yellow River through a series of grand aqueducts carved through mountainsides and stretched across deserts. Another project aims to provide Beijing with a water supply by way of a major water transfer project on the eastern and central waterways.

Critique

Such projects come at high social and environmental cost. Over 330,000 people have already been forced from their homes as a result of the central waterway project.

Large-scale river diversion projects have led to both domestic and international protests. The environmental impacts associated with the Three Gorges Dam project have generated controversy among environmentalists inside and outside China. The dam resulted in the forced relocation of millions of Chinese, caused further ecological degradation of the Yangtze ecosystem and fisheries, and presents growing risks of landslides.

Mega projects only provide short-term benefit as they do not improve conservation or address pollution issues. Water shortages in other countries have often led to public policy efforts to conserve water and to reduce water pollution at source. However, in China, we see a totalitarian solution, where the authoritarian nature of the state has facilitated enormous water projects regardless of human and environmental consequences, without fear of electoral punishment.

(*Sources*: Tobin, 2013; Gleick, 2008b)

establishment of a number of demonstration cities. In addition to the model city programme, there are also similar programmes operating at different scales, for example, for provinces and townships, and there is also an eco-agricultural village programme. The UN regards the eco-agricultural villages as exemplars of best practice that could provide sustainable development models for developing countries. As of 2010, China had built 104 experimental sustainable communities.

Box 12.7

National Model Cities for Environmental Protection

Origins

Introduced in 1997, the NMC programme was initially part of an agreement between China and Japan, the 'Japan–China Environmental Development Model City Scheme'. The Chinese cities of Dalian, Chongqing and Guiyang were the test cases and financial support was provided by Japan.

Achieving the accolade

To attain status as a National Model City, a city must meet specific environmental and development targets. These include metrics of how the city handles energy supply and waste, preserves green space, funds environmental protection, and rates among surveys of its citizens. Cities are re-evaluated on a regular basis.

Becoming an NMC typically takes several years. For example, it took Guangzhou eight years to attain the status.

Desirable status

The NMC programme is used to demonstrate the ability of Chinese local officials to develop their economies while protecting their cities' environments. Attainment of NMC status has become highly desirable for political leaders, who equate the title with the ability to attract foreign investment, as well as enabling their municipalities to host large-scale international gatherings.

Programme expansion

Since the initial three test cases, the programme has blossomed. By 2012, eighty-four Chinese cities and three urban districts had been awarded NMC status.

Shenyang provides an example. In 1984, Shenyang was ranked one of the ten most polluted cities in China. In 2001, in an effort to achieve model environmental status, officials began upgrading or relocating downtown factories and building wastewater treatment plants. In the following three years, Shenyang closed down over 600 factories, upgraded 300 more, removed industries from the downtown area, replacing them with residential and business districts, and increased the number of wastewater treatment plants. The change was dramatic: in 2001, Shenyang experienced 162 good air quality days; by 2003 that number had increased to 298. By 2004 it had achieved NMC status.

With the prestige of the award, the mayor was then able to attract the International Horticultural Exposition in 2006 and entice further foreign investment.

Another example is provided by the Tianjin Eco-City, started in 2008 and involving collaboration between the Chinese government and the government of Singapore. The eco-city covers about thirty square kilometres. The city has built sewage treatment plants and completed several photovoltaic power generation projects. A renovation was undertaken on the Ji canal, a landscape construction project was implemented, and surrounding wetlands restored. The eco-city has nearly 100 enterprises registered under the category of environmental technology. In September 2010, the first International Eco-City Forum and Expo was held in the city.

International assistance

International assistance is important for the Programme. For example, co-operation with Singapore assisted Zhongshan in its bid, and the EU helped Nanjing attain NMC status by supporting the development of low-carbon industries. Similarly, Singapore's Surbana Urban Planning Group is involved in the Tianjin Eco-City and the UK's Arup Group is designing the proposed Dongtan Eco-City outside of Shanghai.

Displacement activity?

Despite their evident local environmental value, NMCs are subject to several criticisms. One main problem is that NMC status may be achieved at the expense of outlying areas. Zhongshan, for example, was able to win its status, in part, because officials moved environmentally-polluting industries beyond the city's evaluation area. Poorer regions then become the repository for the polluting industries moved from the eco-cities, displacement activities that breach sustainable development equity principles. The Programme has also been criticised for its lack of transparency and limited opportunities for citizen participation.

It is also difficult to obtain accurate data so as to fully assess the contribution of the programme to the promotion of sustainable development.

(*Sources*: Bremer, 2011; Economy, 2006; Liu, 2008; Zhao, 2011)

Ecological safety: two shelters and three belts

In China, nature conservation policy is governed by a mixture of command and control approaches and the use of new environmental policy instruments, as discussed in Chapter 3. Command and control is epitomised by the use of regulations, as is evident by the array of laws aimed at nature conservation, including, *inter alia*, the Law on the Protection of Wildlife (revised 2004), the Grasslands Law (1985) and the Law on Marine

Environmental Protection (1983). In turn, these laws support an array of policy developments, including more recently the National Biodiversity Conservation Strategy and Action Plan (2011–2030); the Plan for Development of Nature Reserves (1996–2010); the National Ecological Environment Protection Programme; and the National Plan for Biological Species Resources Protection and Use (2006–2020) (The People's Republic of China, 2008). Such plans set out the overall targets and strategic tasks and identify priority areas for actions and projects. In 2000, China formulated an Action Plan for Wetland Conservation and in 2003 it introduced a National Wetland Conservation Programme (2002–2030). From 2006 onwards, the government has earmarked funds to implement demonstration projects for wetland protection, restoration and sustainable utilisation.

China's approach to nature conservation is framed in terms of ecological safety and is referred to as a 'two shelters and three belts' approach. This approach has been integrated into land use planning, which divides the land into four ecological function categories: areas for conserving water sources; for maintaining water and soil; for preventing sandstorms; and for protecting biodiversity. These 'Ecological Function Conservation Areas' aim to protect the delivery of ecosystem services (Yukuan et al., 2010). The zones cover a total area of about 3.86 million square kilometres and account for around 40 per cent of the nation's total land area. Development that disrupts the priority ecological functions in these areas is restricted.

Land use planning also supports the earmarking of land as nature reserves. China's first nature reserve was established in 1956 and by 2010 a total of 2,588 reserves had been established, covering about 15 per cent of the country's land area (Liu et al., 2008). There are also more than 550 wetland protection areas, 41 wetlands of international significance and 213 national wetland parks on a trial basis (The People's Republic of China, 2012). In addition, by 2010, over 70,000 hectares of wetlands of various types had been restored. Indeed, China can now be said to be a world leader in ecological restoration practices.

Several ecological restoration initiatives in China are driven by the realisation that both droughts and floods are, at least in part, caused by farming on steep slopes and by deforestation. China now spends millions of dollars on the restoration of its key ecosystems. The 'Three-North' Shelterbelt is the largest and most distinctive artificial ecological engineering project in China, underway since 1978 in the three northern regions – Northeast China, North China and Northwest China (Li, M-M. et al.,

2012). Forest shelter belts in the Yangtze River basin have also been established. The state has also introduced new projects for controlling sources of dust storms affecting Beijing and Tianjin, through afforestation. A number of projects to restore grasslands, such as in the southwest karst areas, have also been launched. These aim at conserving water and preventing soil erosion and sandstorms, with additional monies earmarked up to 2015 to pay farmers and nomadic people to conserve grasslands. Some of these projects have focused on the production of a fast-growing and high-yielding timber forest base and are thus not primarily aimed at maintenance of biodiversity. However, other ecological restoration projects focus on building national parks as well as urban wetlands.

Among the best known and most important of the country's ecological restoration initiatives are the Natural Forest Conservation Programme (NFCP, also known as the Natural Forest Protection Programme) and the Grain to Green Programme (GTGP, also known as the Sloping Land Conversion Programme and the Farm to Forest Programme). The NFCP conserves natural forests through logging bans and reforestation, whereas the GTGP converts cropland on steep slopes to forest and grassland by providing farmers with grain and cash subsidies.

While they are major components of China's forest conservation programmes, the NFCP and GTGP also have important global implications. These generate many benefits by protecting biodiversity, while also helping to advance both the science and practice of ecological restoration.

The significance of China's emergence for global sustainable development

China is arguably the most geopolitically important developing nation in the world and has a major influence on the prospects for global sustainable development. Attention is now turned to the significance of China's emergence for global efforts to promote sustainable development. This is done in the context of the increasing integration of China into international governance regimes and global trade networks.

Integration into international environmental regimes

China is increasingly integrated into international and regional organisations, such as the UN, and is involved in dozens of international

Case study 12.4

Ecological restoration in China: the Sloping Land Conversion Programme

Background

Accidents and disasters have strongly influenced the introduction of ecological restoration policies and projects. A key event was the 1998 Yangtze River floods, which caused thousands of deaths and left ten million homeless. The flooding had its roots in the late 1960s, when farmers in the mountainous western provinces began clearing vast stretches of land to make way for more crops. The increased agricultural production helped feed a growing nation. However, during the monsoon rains in 1998, soil from the agricultural fields washed down the mountain slopes, killing thousands of people in the villages below. The unprecedented damage caused by the floods prompted China to reconsider the wisdom of replacing forests with farms, especially in steeply sloping terrain. This forced the country to start a project of massive ecosystem recovery, including returning farmland to forests.

The Sloping Land Programme

The Sloping Land Conversion Programme began in 2000 and aimed to return more than 37 million acres of cropland on steep slopes back to forest or grassland. The government pays villagers in cash and rice to give up farming and find new sources of employment. Several programmes help farmers find new work in surrounding cities.

Ecologically speaking the Programme has been positive and has helped to decrease soil erosion by as much as 68 per cent in some areas. But economically, the benefits have been less pronounced. On average, families that participated in the programme reported doing better financially than those who did not, but some farm workers have trouble finding new work. The programme has therefore had some negative socio-economic effects. Prohibition on logging in China has also had negative impacts in other countries, as discussed below.

(*Sources*: The People's Republic of China, 2012; Li et al., 2011; Qiu, 2011; Xu et al., 2006)

co-operation programmes, including Local Agenda 21, the EU–China Environmental Management Co-operation Project and the UN-Habitat Sustainable Cities Programme. Far from being a passive member of such regimes, China's influence on the making of global environmental politics is growing (Carter and Mol, 2006). In recent years China has signed up to more than fifty multilateral environmental agreements.

Significantly, it ratified the Biosafety Protocol in 2005, which was essential for the success of the CBD treaty. Similarly, China's compliance with the ozone treaty has proven critical in ensuring that it remains one of the few genuine success stories of environmental diplomacy (Carter and Mol, 2006).

China believes that developed countries should take the lead in changing unsustainable patterns of production and consumption so as to set an example for developing countries. Moreover, they should also help developing countries advance a green economy, including funding technology transfer and capacity building. In line with its Principle of Diversifying Development Models, however, China also believes that countries should pursue economic development strategies in keeping with their own national conditions. Such a position could be construed to give states licence to pursue developmentalist strategies, irrespective of their consequences for the pursuit of sustainable development. However, international pressure has been brought on China to fall in line with international standards, especially those that relate to environmental protection. This is particularly the case since China entered the WTO in 2001. WTO membership is directly responsible for the introduction of tougher environmental standards, as these are essential if China is to export goods to Western markets (Jahiel, 2006). The opening of China has also seen the country increasingly exposed to global environmental discourse and to new standards set by international partnerships with foreign companies.

Growing concern for planetary limits

There is also growing concern that growth is leading to breaches of planetary boundaries, threatening the future of life on Earth, as discussed in Chapter 2. China is now adding to such breaches, making it more difficult to find a safe operating space for humanity. It has, for example, overtaken the USA as the world's biggest producer of carbon dioxide, the chief greenhouse gas (World Bank, 2014), as discussed in Chapter 6. As a result, China's problems have become the planet's problem. China has also now surpassed the USA as the world's largest consumer, and acts as a major player in many commodity markets and global product chains. This growth has seen China's ecological footprint rise above its biological capacity (Clifton, 2013: 141). By 2012, China's per capita footprint was twice as big as its available biocapacity, and its ecological overshoot is continuing to increase year by year (WWF

Box 12.8

China: trade openness and sustainable development

Openness

Openness, including to international investment and trade, is a core element of China's sustainable development model.

Openness has seen China embrace globalisation, create special economic zones and offer favourable tax and policy treatment to foreign capital. As a result, China has attracted massive foreign direct investment (FDI), contributing to economic prosperity. However, it is also blamed for environmental degradation and resource depletion.

Openness and sustainable development are interrelated, but exactly how they interact is the subject of considerable dispute.

Two conflicting hypotheses

The 'race to the bottom' hypothesis posits that international trade and investment create downward pressure on environmental regulations in host countries. As different jurisdictions compete to chase investment and raise competitiveness, they tend to lower their environmental standards to reduce costs of production, thereby appearing more attractive. The consequence is that international trade and investment bring environmental deterioration and weaken efforts to promote sustainable development.

In the case of China, local politicians are incentivised to attract FDI, not least because their careers may depend on such performance. Furthermore, although environmental standards are set at the national level, local government can play a significant role by relaxing or tightening environmental enforcement, in accordance with need. The least developed regions in western China are particularly vulnerable to developmentalist pressures, heightening the potential for a race to the bottom.

On the contrary, the 'gains from trade' hypothesis states that openness has a positive effect. International trade enables countries to attain cleaner technologies in a cost-effective manner. Accordingly, openness contributes to a better environment. Thus openness can allow China to assimilate the advanced scientific, technological, and managerial innovations created by developed economies.

Example: membership of the WTO

It was predicted that accession to the WTO in 2001 would bring China significant environmental rewards. Increased international competition would see the updating of technology in the metallurgy, construction, chemicals and energy sectors. To stay competitive, the steel and automobile industries would also have to adopt newer technologies.

Ecological modernisation, to take the example discussed above, has taken hold in the steel sector. However, WTO accession has increased the vulnerability of China's biodiversity. As trade has blossomed, so has the number of exotic and potentially invasive species unintentionally introduced into China. In 2001, flowers imported from the Netherlands inadvertently carried a non-native insect which spread throughout Yunnan province, affecting the unique ecosystem and devastating its horticulture industry. The lowering of trade barriers is spreading environmental effects beyond the country's borders, as seen for example in its impacts on forests in Africa.

(*Sources*: Jahiel, 2006; Zhang, 2012)

China, 2010). China can thus no longer supply the resource inputs needed to support its consumption demands and therefore has to draw resources from outside its own borders. However, from a planetary perspective there is no spare biocapacity for China to draw upon, as humanity is currently heavily in ecological debt.

China is also causing regional pollution problems. The sulphur dioxide and nitrogen oxides from China's coal-fired power plants fall as acid rain on Seoul in South Korea, and on Tokyo. Similarly, research has also shown that on some days 25 per cent of airborne particulates in southern California originate from China (Grumbine, 2007). The escalation of transboundary environmental problems has already led to international conflicts. The most recent case is the fisheries disputes between China and its East and Southeast Asian neighbours. These conflicts are not just the result of unsettled territorial and maritime disputes. They are also attributable to China's ever-increasing demand for seafood amidst its dwindling domestic fishery resources (Zhang, 2012). The 'go west' strategy is also driving numerous major regional infrastructural projects, funded through outward FDI by Chinese companies. Several new railway lines and large scale highways connecting China with Central Asian countries are under construction. China and Russia are planning at least one large oil pipeline and two natural gas pipelines. China is also heavily involved with Pakistan and Bangladesh in the construction of two huge port complexes on the Indian Ocean to service the shipment of oil and other goods (Grumbine, 2007).

The growing influence of China in the world can also be seen in natural resource exploitation. It is increasingly relying on neighbouring states in Indonesia, but is also travelling further afield, especially to Africa, to source raw materials, including timber.

Case study 12.5

External outsourcing: the forest sector

More stringent domestic policies on deforestation following the 1998 flooding have forced many Chinese logging companies abroad, both to the East Asia region and beyond. The reduction of tariffs and elimination of quotas under the WTO has allowed a rapid growth in timber imports into China.

As a result, China has been able to meet its rising domestic demand for wood products, while at the same time becoming a major international base for furniture production and wood processing.

While imports have eased pressures on Chinese forests, much of this wood comes from areas of the world practising poor forest stewardship, such as Indonesia, Malaysia, Burma, Cambodia, Russia, and various South American and African countries, and is often illegally purchased.

Economic globalisation and trade liberalisation have promoted globalised resource allocation, making it possible for one country to transfer stress on its forests to other countries. A 2005 World Wildlife Fund report highlights the severity of this transfer, warning that 'surging demand [in China] … threatens to have a devastating impact on forests around the world'.

(Sources: Sun, 2004, Watts, 2005; WWF, 2005)

Timber is but one example of how China seeks cheap raw materials overseas for processing at home. Increasingly, Chinese firms are tapping into a wide range of natural resources in Africa, the Middle East, South America, Southeast Asia and Russia. By doing so, certain environmental conditions in China may improve, but it serves to deepen environmental problems elsewhere. Unfortunately, this pattern mirrors that of developments in the industrialised world, which brought ecological destruction to countries in the Global South and to the global ecosystem. However, we also need to be mindful that wealthy, industrial countries are fully implicated in China's ascendency (Grumble, 2007). China's export economy is fuelled by developed-world investment and consumer demand. China finances US debt, while the growing economy in China is seen as a source of lucrative investment opportunities by foreign companies from the industrialised world.

Conclusion

China introduced strategies to promote sustainable development at a very early stage of its modernisation and industrialisation. It continues to pursue its commitment to sustainable development through a unique blend of strong state steering realised through central planning and market control, technologically driven ecological modernisation and highly constrained forms of civil society engagement. It also provides examples of best practice, especially with those projects pursuing sustainable development in urban contexts, and through highly ambitious ecological restoration initiatives.

While economic growth has lifted a substantial proportion of the population out of poverty, it has also brought unprecedented ecological decline. The deterioration of the overall state of China's environment is of global concern. While formally China has embraced a sustainable development framework, actions are hindered by the prioritisation of economic growth needed to ensure regime legitimacy. Retaining an 'unflappable sense of technological optimism' (Tobin, 2013), the state continues to hold to a belief that it can pursue sustainable development without having to reform its authoritarian political system.

China's rapid economic expansion is having major impacts on the global environment and on the potential for global sustainable development. Its path of development bodes ill as a model for other emerging economies. India is projected to pass China as the world's fastest growing economy by 2020. The pursuit of sustainable development at a global level would be seriously undermined should India and other emerging economies adopt development trajectories similar to that of China.

Summary points

- China has its own, distinctive model of sustainable development, which relies on strong state control, belief in technology and market oriented growth.
- In pursuit of sustainable development China has reduced poverty; maintained population control; integrated environmental protection into national economic and social development planning; and implemented legislation and regulation, including through the use of market mechanisms and controlled forms of civil society engagement.
- Central planning has enabled China to experiment. The National Model City for Environmental Protection programme stands out as an exemplar of good practice, as do its ecological restoration practices.
- China's rapid economic development has contributed to ecological deterioration and severe pollution. Air and water pollution, land

degradation, desertification and the declining availability of resources, such as water, are beginning to affect the country's development, public health and social stability.

- As a result of its growth, the income gap in China is widening, intensifying consumerism on the one hand, and relative deprivation on the other, both of which threaten the pursuit of sustainable development.
- The country's integration into global governance regimes and trade networks has helped the pursuit of sustainable development, not least through the diffusion of sustainable development norms and practices. It has also widened opportunities to engage in resource exploitation in other countries and regions across the globe.

Further reading

Carter, N. and Mol. A. P. J. (2008), *Environmental Governance in China* (London: Routledge).

Gallagher, K. S. and Lewis, J. I. (2013), 'China's Quest for a Green Economy', in Norman J. Vig and Michael K. Kraft (eds), *Environmental Policy: New Directions for the 21st Century* (London: Sage), 321–343.

Detailed information on state plans, legislation and projects can be found at the Ministry of Environmental Protection, official website: http://english.mep.gov.cn/

References

BBC News (2013a), 'China in Carbon Trading Experiment', 18 June, available online at www.bbc.co.uk/news/business-22931899

BBC News (2013b), 'China to Curb Coal Use to Combat Air Pollution', 12 September, available online at www.bbc.co.uk/news/world-asia-china-24068519

BBC News (2013c), 'Smoggy Beijing Sees Lung Cancer Cases Soar', 9 November, available online at www.bbc.co.uk/news/magazine-24880737

BBC News (2013d), 'Tensions Flare Over Government "Land Grabs" in China', 9 November, available online at www.bbc.co.uk/news/blogs-china-blog-24865658

BBC News (2014), 'Caution Urged as Beijing Smog Levels Soar', 16 January, available online at www.bbc.co.uk/news/world-asia-china-25744682

Bennett, M. T. (2009), *Markets for Ecosystem Services in China: An Exploration of China's "Eco-Compenstion" and other Market-Based Environmental Policies*, Forest Trends, available online at www.forest-trends.org/documents/files/doc_2317.pdf

Bhattacharya, S. (2005), 'Chemical Spill in Chinese River May Pose Cancer Risk', *New Scientist*, 25 November, available online at www.newscientist.

com/article/dn8379-chemical-spill-in-chinese-river-may-pose-cancer-risk.
html#.Ux9ko_l_vTo

Bremer, M. (2011), 'China National Model Cities for Environmental
Protection', *Green Explored*, 20 February, available online at www.
greenexplored.com/2011/02/china-national-model-cities-for.html

British Chamber of Commerce (2011), *China's Twelfth Five Year Plan
(2011–2015)*, available online at www.britishchamber.cn/content/chinas-
twelfth-five-year-plan-2011-2015-full-english-version

Carter, N. and Mol, A. P. J. (2006), 'China and the Environment: Domestic and
Transnational Dynamics of a Future Hegemony', *Environmental Politics*, 15
(2): 330–344.

Clifton, D. (2013), 'Sustainable Development: A Way Forward or an Illusion?'
in C. Anderssen, M. Rahamathulla and W. Xiaoyi (eds), *Sustainable
Development in China* (Abingdon: Routledge), 133–151.

Economy, E. (2006), 'Environmental Governance: The Emerging Economic
Dimension', *Environmental Politics*, 15 (2): 171–189.

Feng, K., Davis, S. J., Sun, L., Li, X., Guan, D., Liu, W., Liu, Z. and Hubacek, K.
(2013), 'Outsourcing CO_2 within China', *PNAS, Proceedings of the National
Academy of Science of the United States of America*, 110 (28): 11654–11659,
available online at www.pnas.org/cgi/doi/10.1073/pnas.1219918110

Gallagher, K. S. and Lewis, J. I. (2013), 'China's Quest for a Green Economy',
in Norman J. Vig and Michael K. Kraft (eds), *Environmental Policy: New
Directions for the 21st Century* (London: Sage), 321–343.

Gleick, P. H. (2008a), 'China and Water', in P. H. Gleick, with Heather
Cooley, Michael J. Cohen, Mari Morikawa, Jason Morrison and Meena
Palaniappan, *The World's Water 2008–2009* (Washington, DC: Island
Press), 79–100.

Gleick, P. H. (2008b), 'Three Gorges Dam Project, Yangtze River, China', in
P. H. Gleick, with Heather Cooley, Michael J. Cohen, Mari Morikawa, Jason
Morrison and Meena Palaniappan, *The World's Water 2008–2009*
(Washington, DC: Island Press), 139–150.

Grumbine, R. E. (2007), 'China's Emergence and the Prospects for Global
Sustainability', *BioScience*, 57 (3): 249–255.

The Guardian (2014), 'China Surpasses US as World's Largest Trading
Nation', 10 January.

Jahiel, A. R. (2006), 'China, the WTO, and Implications for the Environment',
Environmental Politics, 15 (2): 310–329.

Johnson, T. (2010), 'Environmentalism and NIMBYism in China: Promoting
Rule-Based Approaches to Public Participation', *Environmental Politics*, 13
(3): 430–448.

Kang, L. and Li, C. (2013), 'Discussion on Internal Scientificity of Scientific
Outlook on Development', *Asian Culture and History*, 5 (2): 142–146.

Lewis, J. (2011), 'Energy and Climate Goals of China's Twelfth Five-Year
Plan', Pew Centre on Global Climate Change, March, available online

at www.c2es.org/international/key-country-policies/china/energy-climate-goals-twelfth-five-year-plan

Li, J., Feldman, M. W., Li, S. and Daily, G. C. (2011), 'Rural Household Income and Inequality under the Sloping Land Conversion Program in Western China', *PNAS*, April 25, doi: 10.1073/pnas.1101018108.

Li, M-M., Liu, A-t., Zou, C-j., Xu, W-d., Shimizu, H. and Wang, K-y. (2012), 'An Overview of the "Three-North" Shelterbelt Project in China', *Forestry Studies in China*, 1 (14 March): 70–79.

Liu, J., Li, S., Ouyang, Z., Tam, C. and Chen, X. (2008), 'Ecological and Socioeconomic Effects of China's Policies for Ecosystem Services', *PNAS*, 105 (28): 9477–9482.

Liu, L. (2008), 'Sustainability Efforts in China: Reflections on the Environmental Kuznets Curve through a Locational Evaluation of "Eco-Communities"', *Annals of the Association of American Geographers*, 98 (3): 604–629.

Ma, X. and Ortolano, L. (2000), *Environmental Regulation in China: Institutions, Enforcement, and Compliance* (Oxford: Rowman and Littlefield).

Mol, A. J. P. (2006), 'Environment and Modernity in Transitional China: Frontiers of Ecological Modernisation', *Development and Change*, 37 (1): 29–56.

Moore, A. and Warren, A. (2006), 'Legal Advocacy in Environmental Public Participation in China: Raising the Stakes and Strengthening Stakeholders' (Washington, DC: Woodrow Wilson International Center for Scholars, China Environment Series, issue 8), available online at www.wilsoncenter.org/sites/default/files/CEF_SpecialReport.8.pdf#page=10; 3–26.

Mou, W., Jiahau, P. and Ruiying, Z. (2013), 'Addressing Climate Change in China: Challeges and Opportunities', in C. Anderssen, M. Rahamathulla and W. Xiaoyi (eds), *Sustainable Development in China* (Abingdon: Routledge), 164–174.

OECD (2009), 'Globalisation and Emerging Economies: Brazil, Russia, India, Indonesia, China and South Africa', Policy brief, March, available online at www.oecd.org/regional/searf2009/42576801.pdf

OECD (2011), *Developments in Steelmaking Capacity of Non-OECD Economies 2010* (Paris: OECD).

O'Neill, J. (2011), 'Building Better Global Economic BRICs', Goldman Sachs Global Economics Papers, no. 6630, November, available online at www.content.gs.com/japan/ideas/brics/building-better-pdf.pdf

Pan, J. and Zhu, Xianli (2005/06), 'Energy and Sustainable Development in China', Helio International: Sustainable Energy Watch, 2005/2006, available online at www.rcsd.org.cn/NewsCenter/NewsFile/Attach-20061107125719.pdf

The People's Republic of China (2005), 'Creating a Human Centred Sustainable Community', The Administrative Centre for China's Agenda 21, 2005 Local Agenda 21 Division of ACCA21, available online at www.acca21.org.cn/local/encase/case/jiangan.htm

The People's Republic of China (2007), *Program of Action for Sustainable Development in China in the Early 21st Century* (China: National

Development and Reform Commission (NDRC)), available online at http://en.ndrc.gov.cn/newsrelease/200702/t20070205_115702.html

The People's Republic of China (2008), *China's Fourth National Report on Implementation of the Convention on Biological Diversity*, Ministry of Environmental Protection, 2 November, available online at https://www.cbd.int/doc/world/cn/cn-nr-04-en.pdf

The People's Republic of China (2012), *National Report on Sustainable Development*, available online at www.china-un.org/eng/zt/sdreng/

Qiu, J. (2011), 'China Faces Up to "Terrible" State of Its Ecosystems', *Nature*, 471: 19.

Roberts, D. (2012) 'After Soothing Bromides, China Will Unveil Leadership', *Newsweek*, 14 November, available online at www.businessweek.com/.../after-soothing-bromides-china-will-unveil-lea...

Sterner, T. (2003), *Policy Instruments for Environmental and Natural Resource Management* (Washington, DC: Resources for the Future Press).

Stone, R. (2011), 'On Lake Taihu, China Moves To Battle Massive Algae Blooms', *Yale Environment 360*, 21 July, available online at http://e360.yale.edu/feature/on_lake_taihu_china_moves_to_battle_massive_algae_blooms/2429/

Streets, D. G., Fu, J. S., Jang, C. J., Hao, J., He, K., Tang, X., Zhang, Y., Wang, Z., Li, Z., Zhang, Q., Wang, L., Wang, B. and Yu, C. (2008), 'Air Quality during the 2008 Beijing Olympic Games', *Atmospheric Environment*, 41: 480–492.

Sun, Y. (2004), *Corruption and Market in Contemporary China* (Ithaca, NY: Cornell University Press).

Tang, Shui-Yan and Zhan, Xueyong (2008), 'Civic Environmental NGOs, Civil Society, and Democratisation in China', *Journal of Development Studies*, 44 (3): 425–448, doi: 10.1080/00220380701848541.

Tobin, P. (2013), 'It's Not Easy Being Green: Sustainable Development in China', *White Rose Politics Review*, 1 (1): 1–41.

UNEP (United Nations Environment Programme) (2005), *The Songhua River Spill, China*, Field Mission Report, December (New York: UNEP), available online at www.unep.org/PDF/China_Songhua_River_Spill_draft_7_301205.pdf

Vogel, E. F. (2011), *Deng Xiaoping and the Transformation of China* (Cambridge, MA: Harvard University Press).

Watts, J. (2005), 'China Consumes Forests of Smuggled Timber', *The Guardian*, 22 April, available online at www.theguardian.com/world/2005/apr/22/china.jonathanwatts

World Bank (2007), *Cost of Pollution in China: Economic Estimates of Physical Damages* (Washington, DC: World Bank), available online at http://documents.worldbank.org/curated/en/2007/02/7503894/cost-pollution-china-economic-estimates-physical-damages

World Bank (2013), 'China's Millions to Benefit from Improved Waste Management in Cities', available online at www.worldbank.org/en/news/

press-release/2013/05/31/china-3-million-to-benefit-from-improved-solid-waste-management-in-city-of-ningbo

World Bank (2014), *China Overview* (Washington DC: World Bank), 1 March, available online at www.worldbank.org/en/country/china/overview

Wu, F. (2002), 'New Partners of Old Brothers? GONGOs in Transitional Environmental Advocacy in China', *China Environment Series*, 5: 45–58.

WWF (World Wildlife Fund) (2005), 'China's Rising Wood Imports a Threat to the World's Forests', available online at http://wwf.panda.org/about_our_earth/about_forests/forest_news_resources/?unewsid=19031

WWF China (2010), *China Ecological Footprint Biocapacity, Cities and Development*, Report by Lin, L., Gaodi, X., Shuyan, C., Zhihai, L., Humphrey, S., Shengkui, C., Liqiang, G., Haiying, L. and Ewing, B. (Beijing: WWF China).

Xiaoyi, W. and Qian, Z. (2013), 'How Climate Change Affected the Herders' Livelihood in a Semi-Arid Pastoral Community: The Case of Gonger in Inner Mongolia', in C. Anderssen, M. Rahamathulla and W. Xiaoyi (eds), *Sustainable Development in China* (Abingdon: Routledge), 175–186.

Xinhua News Agency (2012), 'Scientific Outlook on Development Becomes CPC Theoretical Guidance', 8 November, available online at www.china.org.cn/china/18th_cpc_congress/2012-11/08/content_27041783.htm

Xinhua News Agency (2013a) 'China Actively Backs, Pursues Sustainable Development: Chinese FM', available online at http://news.xinhuanet.com/english/china/2013-09/25/c_132749530.htm

Xinhua News Agency (2013b), 'Sustainable Development Is the Only Road to Take to Realize the Chinese Dream, Despite the Many Difficulties China Will Face, Chinese Foreign Minister Wang Yi Said Tuesday', 24 September, available online at http://news.xinhuanet.com/english/china/2013-09/25/c_132749594.htm

Xinhua News Agency (2013c), 'Top Chinese Leaders Attend Third Plenary Session of Eighteenth CPC Central Committee in Beijing', available online at www.xinhuanet.com/english/special/cpcplenum2013

Xinhua News Agency (2014), 'Chinese Leaders Join Lawmakers in Panel Discussions', available online at http://news.xinhuanet.com/english/special/2014-03/09/c_133172970.htm

Xu, J., Yin, R., Li, Z. and Liu, C. (2006), 'China's Ecological Rehabilitation: Unprecedented Efforts, Dramatic Impacts, and Requisite Policies', *Ecological Economics*, 57: 595–607.

Yang, G. (2009), *The Power of the Internet in China: Citizen Activism Online* (New York: Columbia University Press).

Yukuan, W., Bin, F., Colvin, C., Ennaanay, D., McKenzie, E. and Min, C. (2010), *Mapping Conservation Areas for Ecosystem Services in Land-Use Planning, China*, TEEB Case Report, available at www.teebweb.org/wp-content/uploads/2013/01/Mapping-conservation-areas-for-ecosystem-services-in-land-use-planning-China.pdf

Zadek, S. (2010), 'Emerging Nations and Sustainability: Chimera or Leadership?' *Notizie di Politeia*, XXVI (98): 153–167.

Zhang, J. (2012), *Delivering Environmentally Sustainable Economic Growth: The Case of China*, Asia Society Policy Report, September, available online at http://asiasociety.org/files/pdf/Delivering_Environmentally_Sustainable_ Economic_Growth_Case_China.pdf

Zhao, J. (2011), *Towards Sustainable Cities in China: Analysis and Assessment of Some Chinese Cities in 2008* (London: Springer).

Conclusion

The promotion of sustainable development: what has been achieved?

Constructing a new development paradigm

Our exploration of the prospects for, and barriers to, the promotion of sustainable development began with an examination of environmental critiques of the conventional model of development. This served to differentiate the new model of sustainable development from the conventional approach, while also clarifying the problems that the new model is designed to address. The discussion showed that the rise of environmentalism brought with it a fundamental questioning of the basic tenets of the Western development model, in particular as pursued through the rapid economic growth of the post-war period.

Environmental critiques point to the irrationality of an approach to societal development that threatens the bases upon which future development depends (Dryzek, 1983). Based on these critiques, it is no longer possible to see development in isolation from its ecological and social consequences. From an ecological point of view, this threat has become evident in biodiversity loss, climate change, deforestation and desertification, and growing problems of water shortages. It also leads to a rejection of the idea of equating human progress with the domination of nature. When judged from a social perspective, deteriorating environmental quality is causing hardships that can weaken the capacity of communities to respond to risks and vulnerabilities. This undermines the assumption of a continuous, more or less harmonious development for society. Repudiating the equation of development with growth, environmentalism also discards the idea that consumption is the most important contributor to human welfare. The conventional model is also criticised for relying too heavily upon markets to distribute goods and services. Market access depends upon the ability to pay, while reaping the profits of production and service provision depends upon the ability to commodify more and more areas of life, from food supply to leisure activities and from seeds to biological resources. Most significantly, by showing that the model of development pursued by the Western industrial societies cannot be carried into the future, either in its present forms or at its present pace, environmentalism makes it imperative for society to construct a new development model.

The term 'sustainable development' forms the core, organising theme that integrates environmental, economic and social considerations into a new development model. The model is built upon normative principles that promote equity in access to the planet's limited resources in order to promote human needs, be they physical, cultural, spiritual or social. Equity extends across space, for example, between different geographical locations, as well as across time, for example, between generations. In order to promote sustainable development, a halt has to be put to the practice, typical within the conventional model of development, which allows the present generation to adopt a policy of *temporal displacement*, that is, passing the risks and problems of modernity down to future generations. The *spatial displacement* of the negative environmental consequences of traditional development has also to be stopped. Spatial displacement is a process whereby a more powerful state or actor imposes environmental harm upon another, less politically or economically powerful, state or actor (Blowers, 1997). This can include displacing industrial pollution to other areas, or depleting the environmental assets, such as biodiversity, of another region or country for one's own benefit. In addition, the stronger the form of sustainable development, the more weight is given to the further commitment of ensuring equity between species, that is, between human and other life forms.

These normative principles drive a model of development that protects the planetary resources, be they physical, in the form for example of fresh water, or systemic, in the form, for example, of the ecosystems and of the climate system, while also promoting their use. Thus, while nature is seen to have instrumental value, recognition of the intrinsic value of nature forms the bedrock upon which stronger models of sustainable development are built. At the same time, because of the recognition of ultimate limits to growth, all models of sustainable development accept a hierarchical interdependency between economy, society and nature: society is possible without a market economy, but neither society nor the market economy is possible without the natural environment. This rationality sees nature ultimately trump the use value society may attribute to her goods and services.

While there are many, and often competing, versions of the model of sustainable development, they share a common belief that there are ultimate, biophysical limits to growth. New theorising about planetary limits, including efforts to define a safe operating space for humanity, and new revelations about the extent of our impact, including through footprint analysis, provide fresh impetus to the argument that we need a politics and economics of limits. As well as ensuring the basic survival of us and other species, such limits also serve a justice agenda: they create the conditions necessary for ecologically legitimate development, particularly in the Third World. An agenda of limits to growth calls upon industrial societies to reduce the resource intensity of production (sustainable production) and undertake new patterns of consumption, that not only reduce the levels of consumption but change what is consumed and by whom (sustainable consumption). The sustainable development model thus challenges conceptions of development that prioritise individual self-advancement and replaces it with a focus on the common good. This means that the promotion of the common good takes precedence over the wants of the favoured few.

Box 13.1

Key characteristics of the model of sustainable development

- Recognising the ultimate value of the planet's biophysical and ecological resource system;

- Imposing limits to growth;

- Prioritising the common good over and above individual interest;

- Understanding development in terms of quality of life;

- Promoting socially and ecologically legitimate development, especially in the Third World;

- Reducing consumption in the industrialised world;

- Accepting shared responsibility across multi-levels of governance;

- Participating in open-ended dialogue to identify and agree priorities;

- Respecting diversity as development trajectories are implemented across different social, cultural and ecological contexts.

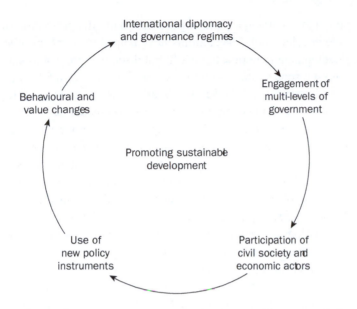

Figure 13.1 *Promoting sustainable development involves multi-dimensional actions.*

Promoting strategic and political engagement

Understanding how the model of sustainable development works in practice involves studying the strategic and political context within which actions to promote sustainable development take place. In this book, awareness of the *outer* limits of the Earth's environment has gone hand-in-hand with a new awareness of the ways in which the *internal* organisation of society, be it at the local or the international level, shapes the prospects for a sustainable future. Attention has thus been given to the interlinked spheres of authority and influence that shape the way society is constructed and policies are made. These operate from the international down to the local levels, from the transnational corporation to the individual consumer, and from application of technology to the pursuit of a more spiritual engagement with the natural world.

What has been achieved?

Beginning with the international level, the involvement of the UN in the promotion of sustainable development is subject to two conflicting interpretations. On the one hand, the UN, particularly through holding environmental summits, can be seen as having played a key, positive role in shaping understanding of, and engagement with, the promotion of sustainable development. The summits have helped the concept of sustainable development to permeate the official discourse not just of states but also of civil society and economic actors. The Rio Earth Summit was particularly important and the Rio Declaration has provided an authoritative set of normative and governance principles to guide a new era of development. Since Rio, there has been progress in environmental institution building and in the development of new governance patterns, including at the national and sub-national levels. Today, nearly all countries have government ministries and/or agencies in charge of the environment and several have established participatory sustainable development forums. This has led to the development of both hard and soft laws aimed at ensuring that the ecological conditions necessary for the promotion of sustainable development are maintained, including as they relate to limiting climate change and ensuring the maintenance of biodiversity (Ross, 2009). There has also been a proliferation of engagement from within civil society and major social groups, including local authorities, business and industry, women and indigenous communities. Especially at the local, community level, there has been an explosion of activities, including, but not limited to, those taking place under the banner of LA21.

Of late, there has been a shift in emphasis within the UN from the elaboration of principles to promote reconciliation between the economy and the environment to more practical considerations of implementation strategies. The WSSD epitomised this shift, with its focus on implementation partnerships. Monitoring has also taken on a new urgency, as witnessed by the use of sustainable development goals and indicators. In addition, there have been new policy issues added to the sustainable development agenda. Globalisation has led to recognition of the importance of trade in the promotion of global sustainable development, while simultaneously enhancing the need for international organisations, such as the WTO, to bring their actions in line with sustainable development principles and practices. New tools have also been made available, including the use of ecological restoration for both climate change mitigation and adaptation and for addressing biodiversity and related ecosystem services loss.

Judged from this perspective, the UN can be seen as having made a positive contribution to the shaping of our collective future. It has helped structure the legal, institutional, political, economic and social engagement with sustainable development from the international down to the local levels. Nevertheless, it is fair to say that more progress has been made in environmental institution building that in actually protecting the environment or in implementing effective policies, as shown across many of the chapters in this book. There are also shortcomings in the institutions and the resources available for global environmental governance. The political will among UN member states to translate declaratory commitments into practical policies is also lacking. This has meant that the UN process is marred by the failure of its members to commit the necessary funds, as member states continue to enter into negotiations driven by narrow, short-term, and national interests. Increased prioritisation of and reliance upon markets for the promotion of sustainable development, despite the strong risk of market failure, is also noted.

Thus, the more positive view of the role of the UN runs alongside an opposite view, one that criticises the organisation as a management agent, helping to promote a system of global environmental governance that is preoccupied with means and not ends. This preoccupation displaces a more fundamental critique of the flaws of conventional development policies, the structure of international politics, and of Western-orientated, environmental management practices. This questions whether the institutions of environmental governance that fall under the remit of UN can address the underlying causes of unsustainable forms of development. Some would go further, arguing the UN has become a conduit for the spread of practices of unsustainability by facilitating marketisation and the spread of global capital downwards to the local level and outwards to more and more areas of life.

Viewed in this more fundamental way, promoting sustainable development is not just about finding more effective and efficient institutions of environmental governance. It is also about genuine commitment to a common interest, developing new ecologically and socially based values and focusing on community rather than state security. It is ultimately about the distribution of power between the global and the local and between the privileged and the marginalised, and about the priority given to the economic over and above the social and the environmental, at the present and into the future. For many, the system of environmental governance promoted by the UN is incapable of addressing these more fundamental requirements. Viewed as such, the UN governance practices are seen as having opened up structures of governance, without making changes to either the processes of governance itself or its ultimate aims.

At the root of these conflicting interpretations lies deep conflict over whether sustainable development is a tool for the construction of radically different environmental futures or whether it should be rejected out of hand, representing little more than an anthropocentric management tool, useful to help capitalism to find a way out of its environmental crisis. For radical environmentalism, ensuring a sustainable future mandates the empowerment of the local and those most directly affected by environmental degradation as a way to hold public and private power and authority accountable to present and future generations. This requires new patterns of politics, including at the global level. For the less radical, market-led solutions, operating alongside technological innovations, can promote a new era of development.

Ecological modernisation: promoting weak sustainable development

The study of the engagement with sustainable development within different social, political and economic contexts has been particularly helpful in exposing the limitations of current practices. The commitment by the EU, for example, has allowed it to move beyond a policy approach that was dominated by the imposition of an ever-tighter regulatory framework governing economic activity, especially production, to a new approach that sees environmental protection as a positive goal of economic activity. Over time, the EU has shifted its understanding of the nature of the environmental problem: it is no longer merely an issue of pollution control and, hence, environmental regulation. Instead, attention has shifted from earlier concerns with resource management and ensuring that economic development resulted in an 'improvement in the quality of life' (decoupling) to elaborating new forms of environmental governance (participatory, consensus-driven). Despite these advances, however, new understandings of public, environmental citizenship that could result in de-materialisation, especially with respect to consumption, are slower to develop. In addition, when the focus is moved beyond the declaratory commitments and treaty obligations of the EU, a less positive picture emerges. Both in relation to its recent eastern enlargement and in terms of its sectoral policies, the EU has promoted a weaker form of sustainable development, premised upon a belief in the advantages of ecological modernisation.

A model of development based on resource efficiency, pollution control and waste reduction cannot be generalised to the planet as a whole. Despite the advances of ecological modernisation, growth still needs to be limited to within the biophysical carrying capacity of the planet.

Furthermore, consideration of equity principles lies outside the scope of the ecological modernisation agenda. Corporate 'greening' is particularly suspect when considered from a Third World perspective, because it does not address issues of social justice and the equitable access to, and distribution of, finite resources (Blowers, 1997).

Ecological modernisation gives a key role to the relationships between government and industry, one that affords the latter an important say in influencing the emerging agenda of sustainable development. The World Business Council for Sustainable Development was formed for precisely this reason and it is evident in the Type II partnership deals struck at the WSSD, and more recently in the development of the UN Global Compact. However, the danger of this approach is that it can allow multinational corporations, globally responsible for extensive environmental degradation and resource depletion, to be recast as corporate environmentalists upon whom society can rely to promote sustainable development. At the same time, in trade relations, the structures of international finance and in the generation of debt, the poor are cast as the perpetrators of environmental decline and as a barrier to a more progressive future.

Embedding the local in the global

Viewed from the local level, sustainable development is about promoting social change within the community, to take account of locally agreed upon ecological, cultural, political and social preferences. LA21 practices have helped to put flesh, as it were, on what this means in practice. Among the key ways to identifying sustainable development priorities at the local level is to open up the policy making processes to wider groups from within society. However, this is not a simple task, as it requires a public that has learned a civic spirit and that no longer sees the public sphere as a forum for the expression of narrow, vested self-interest.

Engagement at the local level has expanded the agenda of sustainable development to encompass consideration of the ways in which development exploits physical space and shapes place making through land-use planning – an increasingly important component of the sustainable development agenda. This, in turn, is premised upon a willingness to dismantle inter-departmental rivalries within local authorities and to change existing institutional practices and weaken entrenched policy coalitions that have a vested interest in maintaining the status quo (Bulkeley and Betsill, 2005). Where local authorities successfully address these challenges, we find new moves to create sustainable development at the urban level that not only create the conditions for the integration of sustainable development considerations into the planning process, but also enhance the contribution that urban forms make to the construction of our sustainable future.

The promotion of sustainable development places a great deal of emphasis on the role of governments at the local level as well as upon actors from within civil society. The advantage of this twin emphasis is that it enhances the chances of generating examples of sound practice. As these successes become a tangible aspect of everyday life, commitment to sustainable development will acquire increased legitimacy and public acceptance (Bridger and Luloff, 1999). However, the discussions raised in this book, particularly as they relate to LA21, point to the necessity to pursue local development needs in ways that take account of the wider governance and ecological systems in which the local level is embedded.

Promoting sustainable development cannot rest on the weight or the input of traditional political authority alone, particularly that which is vested in national governments, including as actors at the international level. In its place is required a governance process that engages state and non-state actors, the public and the private sectors, as they wrestle to agree priorities and devise action plans to put the commitment into practice through concrete development projects. Only through governance structures that are invigorated through the sense of partnerships and shared responsibility, through the expression of empathy for the needs of others over and above the wants of the few, and through the acceptance of humans as part of, not dominant over, nature, can the conditions be created to bring this development model to fruition.

Returning to the Brundtland formulation

The Brundtland formulation of sustainable development represents a radical agenda for social change. This book began by pointing out that the Brundtland conceptualisation now commands authoritative status, acting as a guiding principle of economic and social development. It ends by arguing that precisely because of the radical nature of its agenda, those that have engaged with the promotion of sustainable development have not adhered to all of its principles nor its recommended practices.

However, let us not throw the baby out with the bath water! Because international political and economic processes have restricted the agenda of sustainable development, this does not mean that the promotion of sustainable development is itself a limited agenda for change. Rather, the model recognises that every human interaction made with the world brings change, but challenges society to find ways to ensure that these changes are for the betterment of all. This necessitates the adoption of a spirit of compassion, not only for other human beings but also for all life forms. With such steps, a model of development can be constructed that opens up a future for the coming generations who will inherit the Earth.

The conditions for the promotion of sustainable development have become far more challenging in the decades since the publication of the Brundtland Report. Climate change and biodiversity loss reveal the devastating consequences of the growth oriented, Western model of development. Emerging economies put increasing pressure on finite resources, while replicating the irrationality of Western development. These changes threaten life on Earth. While necessary, attempts to address the manifestations of global environmental change, such as through ecological restoration and the development of low-carbon fuels, may deflect attention away from the need to address their underlying causes. Policy responses may also exacerbate existing inequality. At the same time, global environmental change can provide a wake-up call, a time to reflect on the construction of a new way of being and doing that enhances, rather than diminishes, the conditions necessary for our collective future. The model of sustainable development, based on recognition of the finite nature of planetary resources and founded on strong, ethical principles, provides a sound framework within which to construct a much needed, new politics and economy of limits, one that can sustain our collective futures in an equitable manner and over the long term.

References

Blowers, A. (1997), 'Environmental Policy: Ecological Modernisation or the Risk Society', *Urban Studies*, 34 (5–6): 845–871.

Bridger, J. C. and Luloff, A. E. (1999), 'Towards an Interactive Approach to Sustainable Community Development', *Journal of Rural Studies*, 15: 377–387.

Bulkeley, H. and Betsill, M. M. (2005), 'Rethinking Sustainable Cities: Multilevel Governance and the "Urban" Politics of Climate Change', *Environmental Politics*, 14 (1): 42–63.

Dryzek, J. S. (1983), 'Ecological Rationality', *International Journal of Environmental Studies*, 21 (1): 5–10, doi: 10.1080/00207238308710058.

Ross. A. (2009), 'Modern Interpretations of Sustainable Development', *Journal of Law and Society*, 36 (1): 32–54.

Index